Y0-CBO-218

Design Guidelines for Surface Mount Technology

Design Guidelines for Surface Mount Technology

John E. Traister
Bentonville, Virginia

ACADEMIC PRESS, INC.
Harcourt Brace Jovanovich, Publishers
San Diego New York Berkeley Boston
London Sydney Tokyo Toronto

This book is printed on acid-free paper. ∞

Academic Press, Inc.
San Diego, California 92101

United Kingdom Edition published by
Academic Press Limited
24–28 Oval Road, London NW1 7DX

Library of Congress Cataloging-in-Publication Data

Traister, John E.
Design guidelines for surface mount technology / John E. Traister.
p. cm.
Includes index.
ISBN 0-12-697400-4 (alk. paper)
1. Printed circuits--Design and construction. 2. Surface mount technology. I. Title.
TK7868.P7T73 1989
621.381'531--dc20 89-6788
CIP

Printed in the United States of America
89 90 91 92 9 8 7 6 5 4 3 2 1

CONTENTS

PREFACE

There is a growing trend in the electronics industry away from mounting components on printed circuit boards by inserting the component leads through holes in the boards and then soldering. The trend is to replace this method with a process called surface mount technology, where the component leads, or terminals, are soldered to the top surface of the boards. This trend has accelerated recently due to the rise in the use of high density packages that have greatly reduced terminal spacings. These reduced spacings make through-hole insertion undesirable; hence the move to surface mounting, which makes the boards easier to build, increases their reliability, and cuts labor and manufacturing costs at the same time.

Design Guidelines for Surface Mount Technology was developed and written to address the needs of those using this new technology. Starting with the basics—component selection, space planning, materials and processes—before the mechanics of surface mounted design will give the PC designer/engineer the total concept needed to ensure a manufacturable design. The author's aim is to provide data of the greatest importance from the vast amount of available material and to arrange this material in a way to be helpful in solving problems and difficulties likely to be encountered in the daily work of those involved in surface mounted technology.

The treatment of some subjects has necessarily been brief, but subjects of greater importance have been dealt with more fully. The various tables and charts have been selected with great care, and only those that are most likly to be consulted have been included. The numerous rules and equations are stated as simply and concisely as possible, and their applications are clearly illustrated by the full solution of many examples.

Care has been taken to arrange the chapters in a convenient and logical manner, and a very full index further increases the facility with which any given subject may be located.

I am indeed grateful to the many manufacturers who supplied reference material for use in this book. Names and addresses of these manufacturers appear in Appendix I. I am especially indebted to NuGrafix Group, Inc. of Los Gatos, California, Signetics of Sunnyvale, California, and Universal Instruments Corporation of Binghamton, New York, for use of illustrations, charts, and tables of their products.

A special word of thanks is due John Karns of Martinsburg, West Virginia, who provided many of the drawings for this book. Ruby Updike was responsible for much of the research, typing, and much-needed encouragement.

JOHN E. TRAISTER

chapter 1

SMD ESSENTIALS

Surface mount technology embodies a totally new automated circuit assembly process, using a new generation of electronic components: surface mounted devices (SMDs). Smaller than conventional components, SMDs are placed onto the surface of the substrate, not through it like leaded components. And from this, the fundamental difference between SMD assembly and conventional through-hole component assembly arises; SMD component positioning is relative, not absolute.

When a through-hole (leaded) component is inserted into a printed circuit board (PCB), either the leads go through the holes, or they don't. An SMD, however, is placed onto the substrate surface; its position is only relative to the solder lands. Placement accuracy is therefore influenced by variations in the substrate track pattern, component size, and placement machine accuracy.

Other factors influence the layout of SMD substrates. For example, will the board be a mixed-print (a combination of through-hole components and SMDs) or an all-SMD design? Will SMDs be on one side of the substrate or both? And there are process considerations like what type of machine will place the components and how will they be soldered?

Designing with SMD

SMD Technology is penetrating rapidly into all areas of modern electronic equipment manufacture; in professional, industrial, and consumer applications. Boards are made with conventional print-and-etch PCBs, multilayer boards with thick film ceramic substrates, and with a host of new materials specially developed for SMD assembly.

However, before substrate layout can be attempted, footprints for all components must be defined. Such a footprint will include the combination of patterns for the copper solder lands, the solder resist, and possibly, the solder paste. So the design of a substrate breaks down into two distinct areas: the SMD footprint definition, and the layout and track routing for SMDs on the substrate.

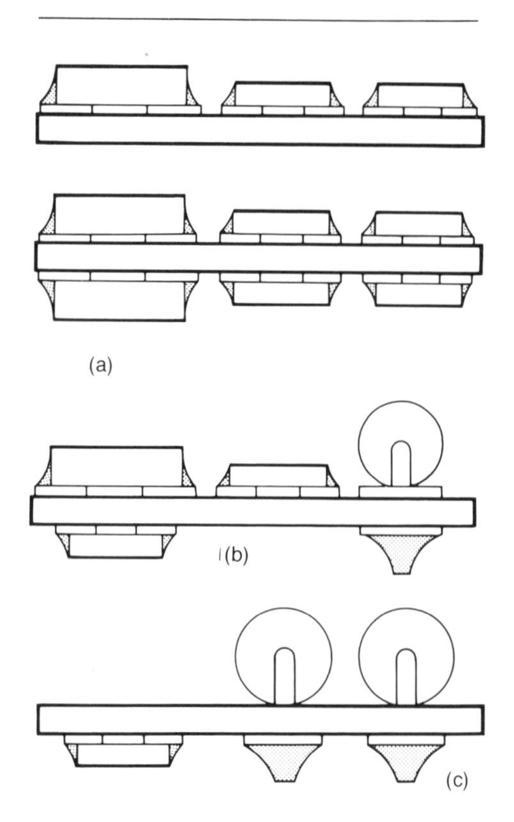

Fig. 1-1: (a) Type 1—total surface mount substrates; (b) Type IIA—mixed print (double-sided) substrate; (c) Type IIB—mixed print (underside attachment) substrate.

Substrate Configurations

SMD substrate assembly configurations are classified as:

Type I — Total surface mount (all-SMD); substrates with no through-hole components at all. SMDs of all types (SM integrated circuits, discrete semiconductors and passive devices) can be mounted either on one side, or both sides of the substrate as shown in Fig. 1-1(a).

Type IIA — Double-sided mixed-print; substrates with both through-hole components and SMDs of all types on the top, and smaller SMDs (transistors and passives) on the bottom. See Fig. 1-1(b).

Type IIB — Underside attachment mixed-print; the top of the substrate is dedicated exclusively to through-hole components, with smaller SMDs (transistor and passives) on the bottom as shown in Fig. 1-1(c).

Although the all-SMD substrate will ultimately be the cheapest and smallest variation as there are no through-hole components, it's the mixed-print substrate that many manufacturers will be looking to in the immediate future, for this technique enjoys most of the advantages of SMD assembly, and overcomes the problem of non-availability of some components in surface mounted form.

The underside attachment variation of the mixed-print (type IIB — which can be thought of as a conventional through-hole assembly with SMDs on the solder side) has the added advantages of only requiring a single-sided print-and-etch PCB, and using the established wave soldering technique. The all-SMD and mixed-print assembly with SMDs on both sides require reflow or combination wave/reflow soldering, and in most cases, a double-sided or multilayer substrate.

The relatively small size of most SMD assemblies compared with equivalent through-hole designs means that circuits can often be repeated several times on a single substrate. This multiple-circuit substrate technique, shown in Fig. 1-2, further increases production efficiency.

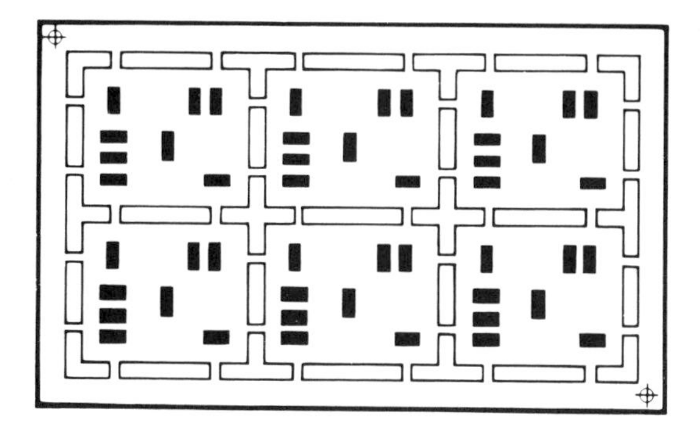

Fig. 1-2: Multiple-circuit substrate.

Mixed Prints

The possibility of using a partitioned design should be investigated when considering the mixed-print substrate option. For this, part of the circuit would be an all-SMD substrate, and the remainder a conventional through-hole PCB or mixed-print substrate. This allows the circuit to be broken down into, for example, high and low power sections, or high and low frequency sections.

Automated SMD Placement Machines

The selection of automated SMD placement machines for manufacturing requirements is an issue reaching far beyond the scope of this book. However, as a guide, the four main placement techniques are outlined as follows:

In-line placement — a system with a series of dedicated pick-and-place units, each placing a single SMD in a pre-set position on the substrate. Generally used for small circuits with few components. See Fig. 1-3(a).

Sequential placement — a single pick-and-place unit sequentially places SMDs onto the substrate. The substrate is positioned below the pick-and-place unit using a computer controlled X-Y moving table (a "software programmable" machine). See Fig. 1-3(b).

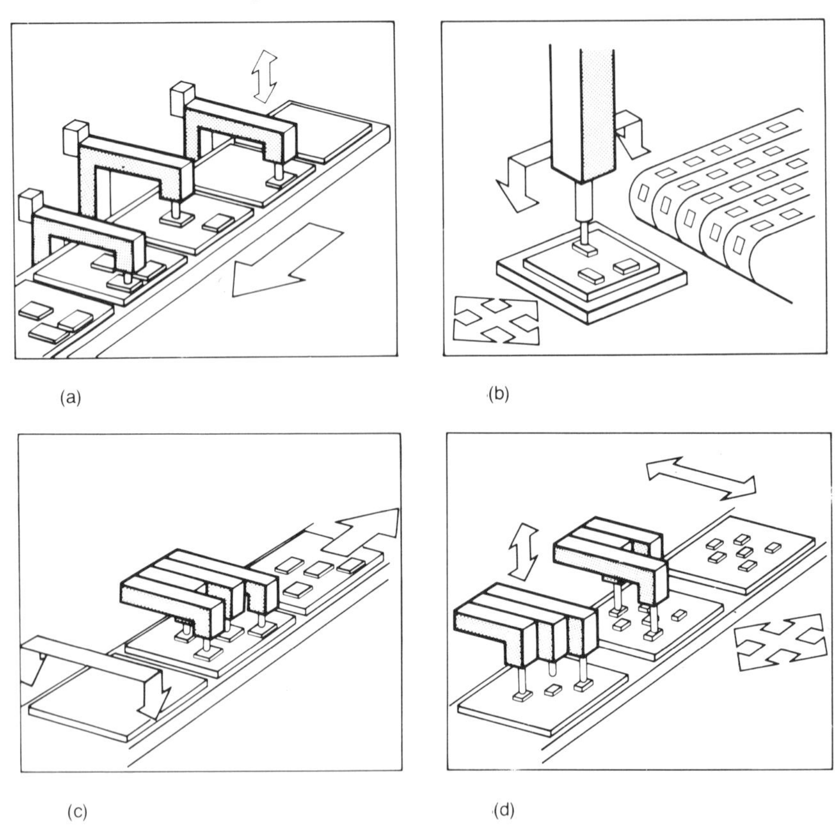

Fig. 1-3: SMD Placement machines: (a) in-line placement; (b) sequential placement; (c) simultaneous placement; (d) sequential/simultaneous placement.

Simultaneous placement — places all SMDs in a single operation. A placement module (or station), with a number of pick-and-place units, takes an array of SMDs from the packaging medium and simultaneously places them on the substrate. The pick-and-place units are guided to their substrate location by a program plate (a "hardware programmable" machine), or by software controlled X-Y movement of substrate and/or pick-and-place units. See Fig. 1-3(c).

Sequential/simultaneous placement — a complete array of SMDs is transferred in a single operation, but the pick-and-place units within each placement module can place all devices simultaneously, or individually (sequentially). Positioning of the SMDs is software controlled by moving the substrate on an X-Y moving table, by X-Y movement of the pick-and-place units, or by a combination of both. See Fig. 1-3(d).

All four techniques, although differing in detail, use the same two basic steps; picking the SMD from the packaging medium (tape, magazine or hopper), and placing it on the substrate. In all cases, the exact location of each SMD must be programmed into the automated placement machine.

Soldering Techniques

The SMD populated substrate is soldered by conventional wave soldering, reflow soldering, or a combination of both wave and reflow soldering. These techniques are covered at length in Chapter 4, but briefly, they can be described as follows:

Wave soldering — the conventional method of soldering through-hole component assemblies where the substrate passes over a wave (or more often, two waves) of molten solder. This technique is favored for mixed print assemblies with through-hole components on the top of the substrate, and SMDs on the bottom.

Reflow soldering — a technique originally developed for thick-film hybrid circuits using a solder paste or cream (a suspension of fine solder particles in a sticky resin-flux base) applied to the substrate which, after component placement, is heated causing the solder to melt and coalesce. This method is predominantly used for Type I (all-SMD) assemblies.

Combination wave/reflow soldering — a sequential process using both the foregoing techniques to overcome the problems of soldering a double-sided mixed-print substrate, with SMDs and through-hole components on the top, and SMDs only on the bottom (Type IIB).

Footprint Definition

An SMD footprint, as shown in Fig. 1-4, consists of:

- a pattern for the (copper) solder lands,

- a pattern for the solder resist,
- if applicable, a pattern for the solder cream.

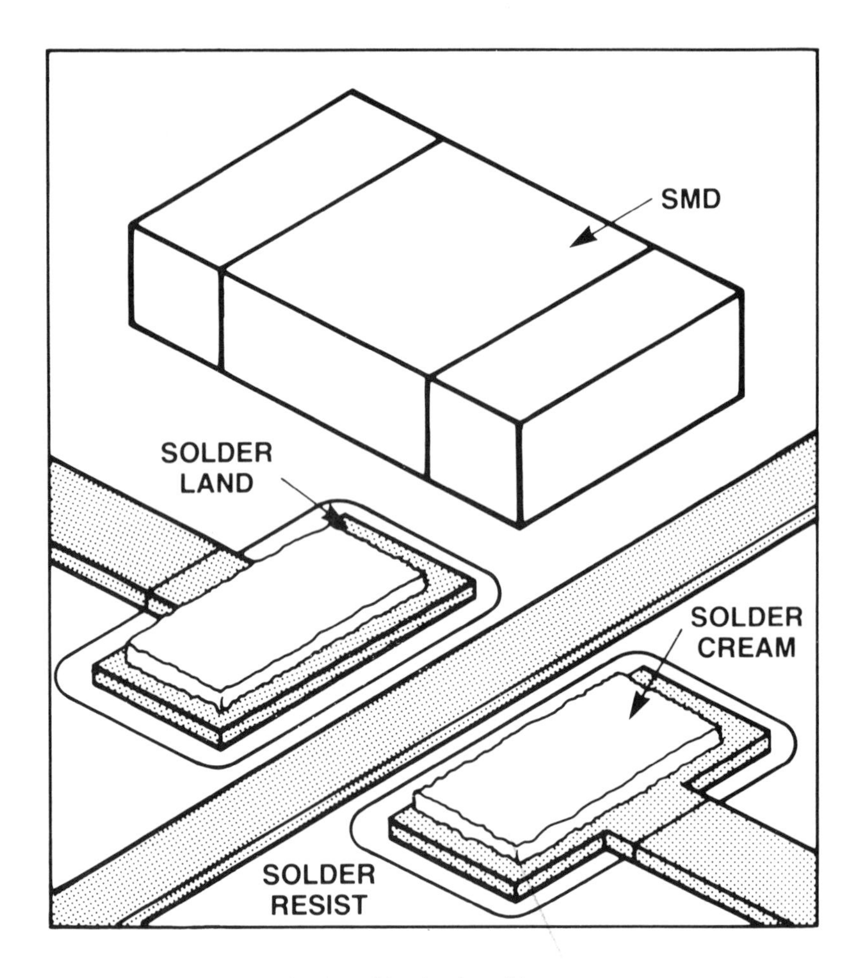

Fig. 1-4: Component lead, solder lead, solder resist and solder cream "footprint."

The Design for the footprint can be represented as a set of nominal coordinates and dimensions. In practice, the actual coordinates of each pattern will be distributed around these nominal values due to positioning and processing tolerances. Therefore the coordinates are stochastic; the actual values from a probability distribution, with a mean value (the nominal value) and a standard deviation.

The coordinates of the SMD are also stochastic. This is due to the tolerances of the actual component dimensions, and the positional errors of the automated placement machine.

The relative positions of solder land, solder resist pattern, and SMD are not arbitrary. A number of requirements may be formulated concerning clearances and overlaps. These include:

- limiting factors in the production of the patterns (for example, the spacing between solder lands or tracks has a minimum value);
- requirements concerning the soldering process (for example, the solder lands must be free of solder resist); and
- requirements concerning the quality of the solder joint (for example, the solder land must protrude from the SMD metallization to allow an appropriate solder meniscus).

Mathematical elaboration of these requirements, and substitution of values for all tolerances and other parameters leads to a set of inequalities that have to be solved simultaneously. To do this manually using worst case design is not considered realistic. A better approach is to use a statistical analysis, and although this requires a complex computer program, it can be done.

Such an approach may deliver more than one solution, and if this is so, then the optimal solution must be determined. Optimization is achieved by setting the following objective: find the solution that:

- minimizes the area occupied by the footprint, and
- maximizes the number of tracks between adjacent solder lands.

The final SMD footprint design also depends on the soldering process to be used. The requirements for a wave soldered substrate differ from those for a reflow soldered substrate, so each is discussed individually.

Footprints for Wave Soldering

To determine the footprint of an SMD for a wave soldered substrate, there are four main interactive factors to consider:

- the component dimensions plus tolerances—determined by the component manufacturer;
- the substrate metallization—positional tolerance of the solder land with respect to a reference point on the substrate;
- the solder resist—positional tolerance of the solder resist pattern with respect to the same reference point; and
- the placement tolerance—the ability of an automated placement machine to accurately position the SMD on the substrate.

The coordinates of patterns and SMDs have to meet a number of requirements. Some of these have a general validity, for example, the minimum overlap of SMD metallization and solderland, and available space for solder maniscus. Others are specifically required to allow successful wave soldering. One has, for example, to take account of factors like the "shadow effect" (missing of joints due to high component bodies), the risk of solder bridging, and the available space for a dot of adhesive.

The "Shadow Effect"

In wave soldering, the way in which the substrate addresses the wave is important. Unlike wave soldering of conventional printed boards, where there are no component bodies to restrict the wave's freedom to traverse across the whole surface, wave soldering of SMD substrates is inhibited by the presence of SMDs on the solder-side of the board. The solder is forced around and over the SMDs as shown in Fig. 1-5, and the surface tension of the molten solder prevents it reaching the far end of the component, resulting in a dry-joint down-stream of the solder flow. This is known as the "shadow effect."

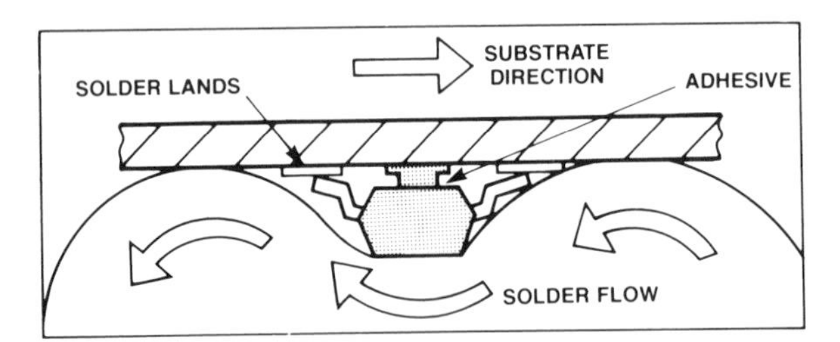

Fig. 1-5: Surface tension can prevent the molten solder reaching the downstream end of the SMD, known as the "shadow effect."

The shadow effect becomes critical with high component bodies. However, wetting of the solder lands during wave soldering can be improved by enlarging each land as shown in Fig. 1-6. The extended substrate metallization makes contact with the solder, allowing it to flow back and around the component metallization to form the joint.

The use of the dual-wave soldering technique also partially alleviates this problem because the first, turbulent wave has sufficient upward pressure to force solder onto the component metallization, and the second, smooth wave "washes" the substrate to form good fillets of solder. Similarly, oil on the sur-

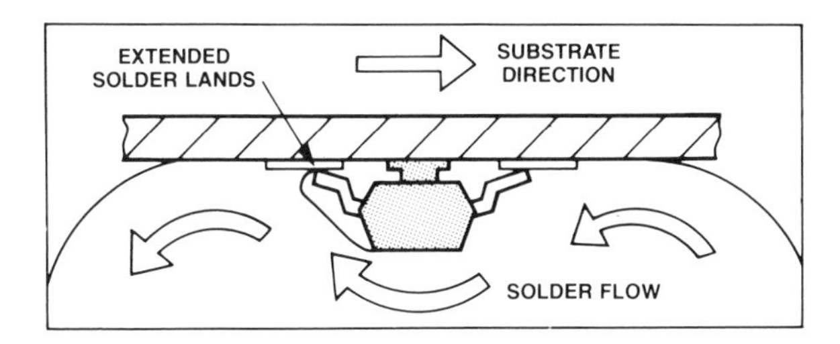

Fig. 1-6: Extending the solder lands to overcome the shadow effect.

face of the solder wave lowers the surface tension, which lessens the shadow effect, but this technique introduces problems of contaminants in the solder when the oil decomposes.

Footprint Orientation

The orientation of SO (small outline) and VSO (very small outline) ICs is critical on wave-soldered substrates for the prevention of solder bridge formation. Optimum solder penetration is achieved when the central axis of the IC is parallel to the flow of solder as shown in Fig. 1-7. The SO package may also be transversely oriented, as shown in Fig. 1-8; but, this is totally unacceptable for the VSO package.

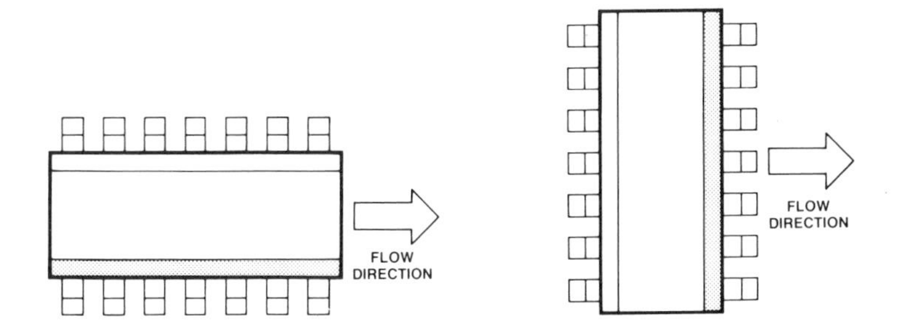

Fig. 1-7: Parallel orientation for for SO and VSO packages.

Fig. 1-8: Transverse orientation for SO packages only.

Solder Thieves

Even with parallel mounted SO and VSO packages, solder-bridges have a tendency to form on the leads down-stream of the solder flow. The use of solder thieves (small squares of substrate metallization), shown in Fig. 1-9 for a 40-pin VSO, further reduces the likelihood of solder-bridge formation.

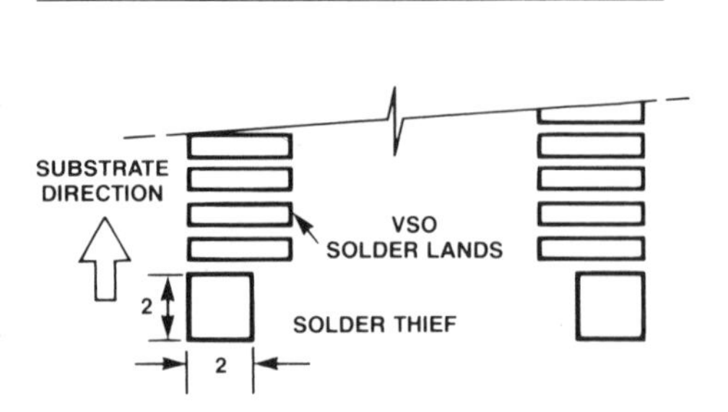

Fig. 1-9: Example of solder thieves for VSO-40 footprints.

Placement Inaccuracy

Another major cause of solder bridges on SO ICs and plastic leaded chip carriers (PLCCs) is a slight misalignment as shown in Fig. 1-10. The close spacing of the leads on these devices means that any inaccuracy in placement drastically reduces the space between adjacent pins and solder lands thus increasing the chance of solder bridges forming.

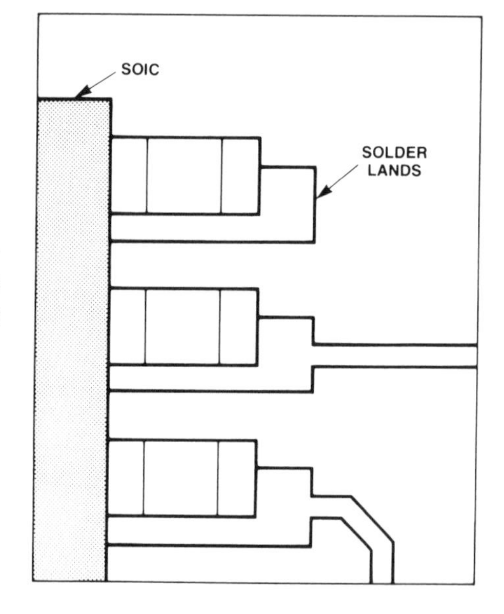

Fig. 1-10: Misaligned placement of SO package increases the possibility of solder bridging.

Dummy Tracks For Adhesive Application

For wave soldering, an adhesive to affix components to the substrate is required. This is necessary to hold the SMDs in place between the placement operation and the soldering process (this technique is covered at length in Chapter 4).

The amount of adhesive applied is critical for two reasons: first, the adhesive dot must be high enough to reach the SMD, and second, there must not be too much adhesive which could foul the solder land and prevent the formation of a solder joint. The three parameters governing the height of the adhesive dot are shown in Fig. 1-11. Although this diagram illustrates that the minimum requirement is $C > A + B$, in practice, $C > 2(A + B)$ is more realistic for the formation of a good strong bond.

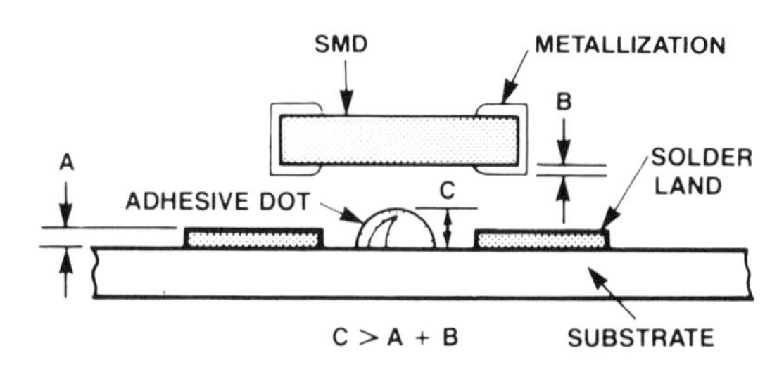

Fig. 1-11: Adhesive dot height criteria.

Taking these parameters in turn, the substrate metallization height (A), can range from about 35 μm for a normal print-and-etch PCB, to 135 μm for a plated-thrugh-hole board. And the component metallization height (B), for example on 1206 size passive devices, may differ by several tens of microns. Therefore, A + B can vary considerably, but it is desirable to keep the dot height (C) constant for any one substrate.

The solution to this apparent problem is to route a track under the device, as shown in Fig. 1-12. This will eliminate the substrate metallization height (A), from the adhesive dot-height criteria. Quite often, the high component density of SMD substrates necessitates the routing of tracks between solder lands, and where it does not, a short dummy track should be introduced.

For bonding small outline (SO) ICs to the substrate, two dots of adhesive are sufficient for SO-8, 14, and 16 pin packages, but the SOL-20, 24, 28 and VSO-40 pin packages need three dots. The through-tracks (or dummy tracks) must be positioned beneath the IC accordingly to support the adhevise dots.

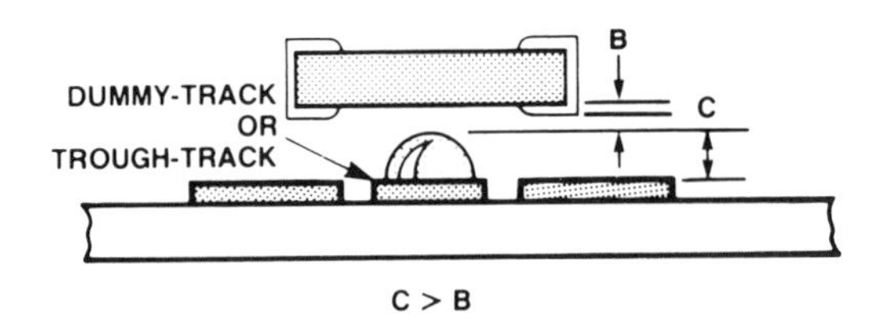

A = Substrate metallization height
B = SMD metallization height
C = Height of adhesive dot

Fig. 1-12: Through-tract or dummy track to modify dot height criteria.

Footprints for Reflow Soldering

To determine the footprint of an SMD for a reflow soldered substrate, there are now five interactive factors to consider. The four that affect the wave solder footprints (although the solder resist may be omitted), plus an additional factor relating to the solder cream application. That is, the positional tolerance of the screen printed solder cream with respect to the solder lands.

Solder Cream Application

In reflow soldering, the solder cream (or paste) is applied by pressure syringe dispensing, or by screen printing. For industrial purposes, screen printing is the favored technique, as it is much faster than dispensing.

Screen Printing

A stainless steel mesh, coated with emulsion except for the solder land pattern where cream is required, is placed over the substrate. A squeegee passes across the screen and forces solder cream through the uncoated areas of the mesh and onto the solder land. As a result, dots of solder cream of a given height and density (in mg/mm^2) are produced.

There is an optimum amount of solder cream for each joint. For example, the solder cream requirements for the C1206 SM capacitor is around 1.5 mg per end, and for the SO IC between 0.5 and 0.75 mg per lead is required.

The solder cream density, combined with the required amount of solder, makes a demand upon the area of the solder land (in mm^2). The footprint dimensions for the solder cream pattern are typically identical to those for the solder lands.

Floating

One phenomenon sometimes observed on reflow soldered substrates is that known as "flowing" (or "swimming"). This occurs when the solder paste reflows, and the force exerted by the surface tension of the now molten solder "pulls" the SMD to the center of the solder land.

When the solder reflows at both ends simultaneously, the swimming phenomenon results in the SMD self-centering on the footprint as the forces of surface tension fight for equilibrium. Although this effect can remove minor positional errors, it is not a dependable feature and cannot be relied upon. Components must always be positioned as accurately as possible.

Footprint Dimensions

The following diagrams (Figs. 1-13 to 1-21) show footprint dimensions for SO ICs, the VSO-40 package, PLCC packages, and the range of surface mounted transistors, diodes, resistors and capacitors. All dimensions given are based on criteria discussed in this chapter.

Please note, however, that these footprints are based on both experimental and actual production substrates, and are reproduced for guidance only. Research is constantly going on to cover all SMDs currently available, and those planned for the future, and the designer must keep abreast of these new developments as they become available.

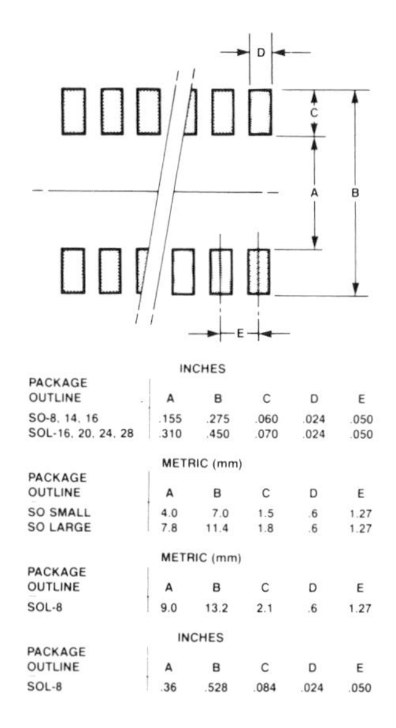

INCHES

PACKAGE OUTLINE	A	B	C	D	E
SO-8, 14, 16	.155	.275	.060	.024	.050
SOL-16, 20, 24, 28	.310	.450	.070	.024	.050

METRIC (mm)

PACKAGE OUTLINE	A	B	C	D	E
SO SMALL	4.0	7.0	1.5	.6	1.27
SO LARGE	7.8	11.4	1.8	.6	1.27

METRIC (mm)

PACKAGE OUTLINE	A	B	C	D	E
SOL-8	9.0	13.2	2.1	.6	1.27

INCHES

PACKAGE OUTLINE	A	B	C	D	E
SOL-8	.36	.528	.084	.024	.050

Fig. 1-13: Footprints for SO ICs.

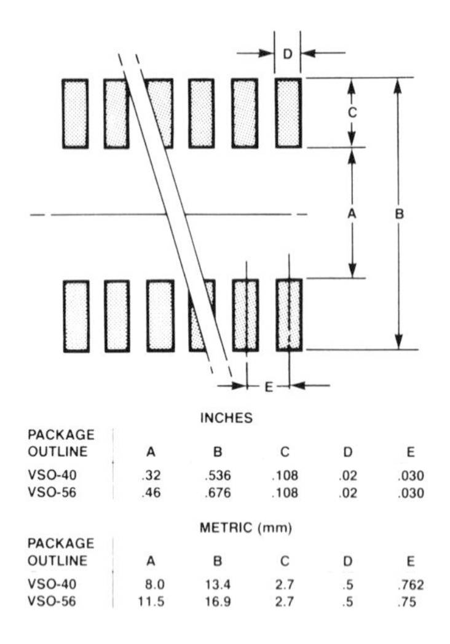

PACKAGE OUTLINE	INCHES A	B	C	D	E
VSO-40	.32	.536	.108	.02	.030
VSO-56	.46	.676	.108	.02	.030

PACKAGE OUTLINE	METRIC (mm) A	B	C	D	E
VSO-40	8.0	13.4	2.7	.5	.762
VSO-56	11.5	16.9	2.7	.5	.75

Fig. 1-14: Footprints for VSO ICs.

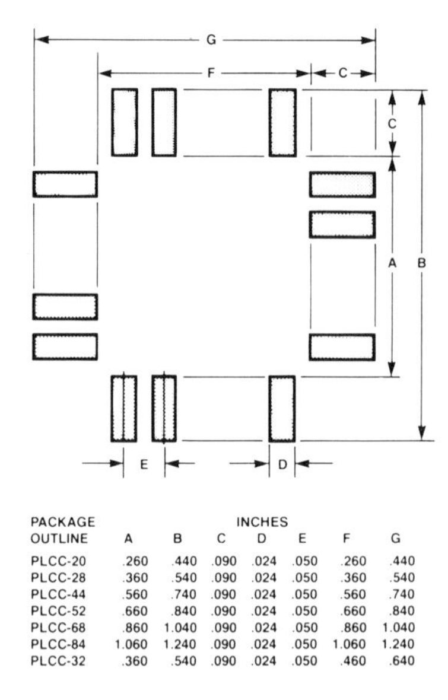

PACKAGE OUTLINE	INCHES A	B	C	D	E	F	G
PLCC-20	.260	.440	.090	.024	.050	.260	.440
PLCC-28	.360	.540	.090	.024	.050	.360	.540
PLCC-44	.560	.740	.090	.024	.050	.560	.740
PLCC-52	.660	.840	.090	.024	.050	.660	.840
PLCC-68	.860	1.040	.090	.024	.050	.860	1.040
PLCC-84	1.060	1.240	.090	.024	.050	1.060	1.240
PLCC-32	.360	.540	.090	.024	.050	.460	.640

Fig. 1-15: Footprints for PLCCs.

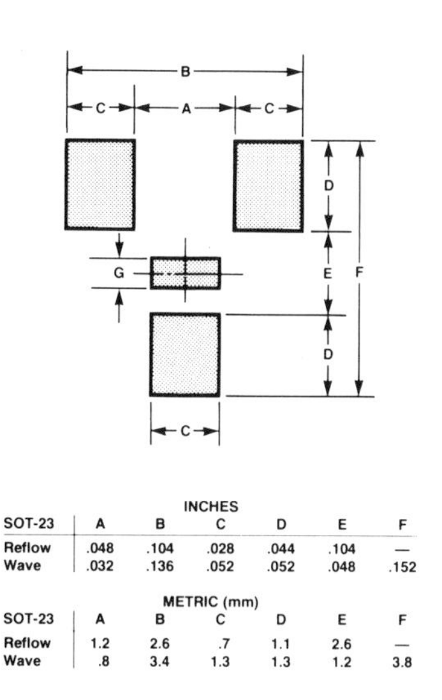

SOT-23	INCHES					
	A	B	C	D	E	F
Reflow	.048	.104	.028	.044	.104	—
Wave	.032	.136	.052	.052	.048	.152

SOT-23	METRIC (mm)					
	A	B	C	D	E	F
Reflow	1.2	2.6	.7	1.1	2.6	—
Wave	.8	3.4	1.3	1.3	1.2	3.8

Fig. 1-16: Footprints for SOT-23 transistors.

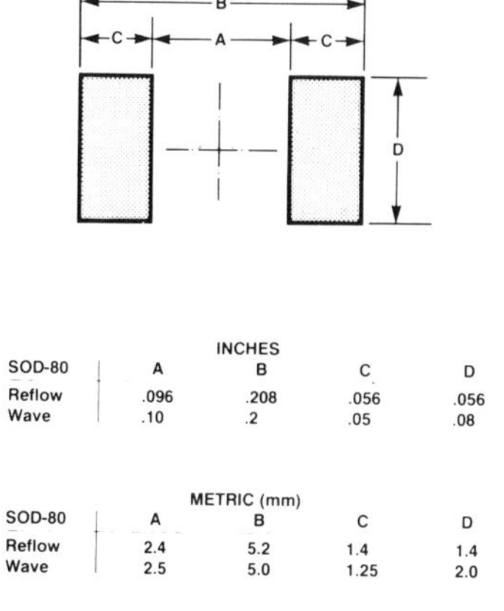

SOD-80	INCHES			
	A	B	C	D
Reflow	.096	.208	.056	.056
Wave	.10	.2	.05	.08

SOD-80	METRIC (mm)			
	A	B	C	D
Reflow	2.4	5.2	1.4	1.4
Wave	2.5	5.0	1.25	2.0

Fig. 1-17: Footprints for SOD-80 diodes.

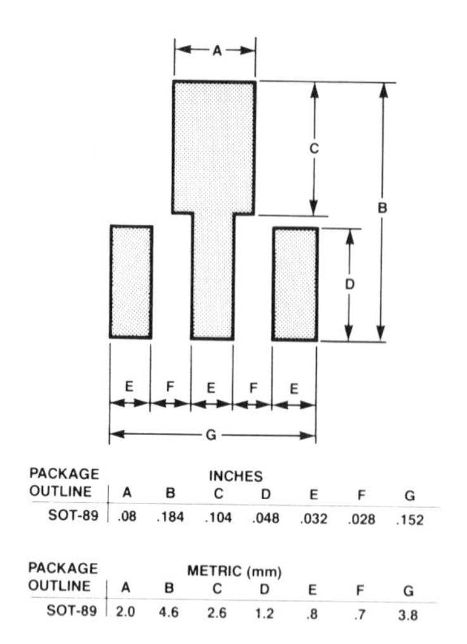

PACKAGE OUTLINE	INCHES						
	A	B	C	D	E	F	G
SOT-89	.08	.184	.104	.048	.032	.028	.152

PACKAGE OUTLINE	METRIC (mm)						
	A	B	C	D	E	F	G
SOT-89	2.0	4.6	2.6	1.2	.8	.7	3.8

Fig. 1-18: Footprints for reflow soldered SOT-89 transistors.

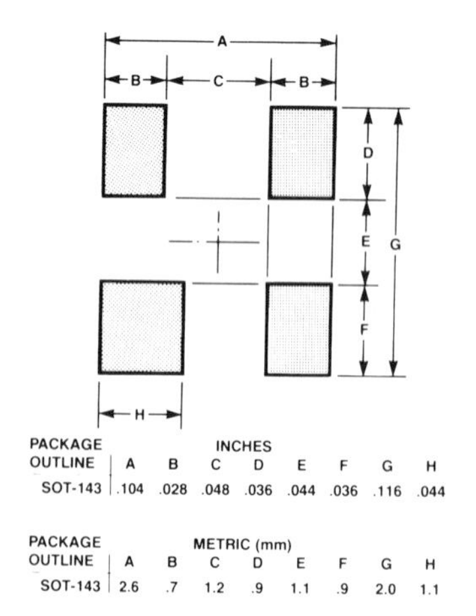

PACKAGE OUTLINE	INCHES							
	A	B	C	D	E	F	G	H
SOT-143	.104	.028	.048	.036	.044	.036	.116	.044

PACKAGE OUTLINE	METRIC (mm)							
	A	B	C	D	E	F	G	H
SOT-143	2.6	.7	1.2	.9	1.1	.9	2.0	1.1

Fig. 1-19: Footprints for reflow soldered SOT-143 transistors.

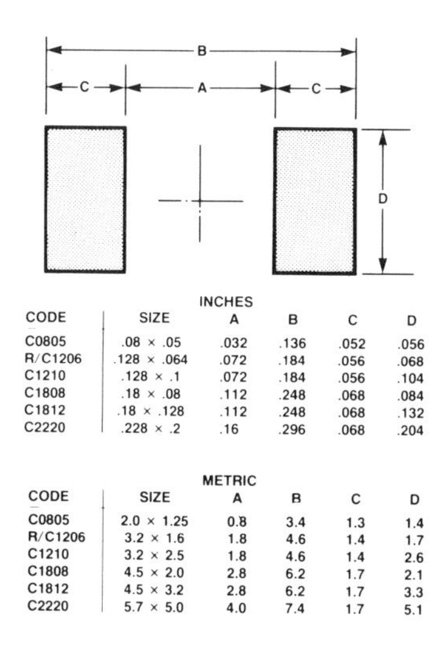

INCHES

CODE	SIZE	A	B	C	D
C0805	.08 × .05	.032	.136	.052	.056
R/C1206	.128 × .064	.072	.184	.056	.068
C1210	.128 × .1	.072	.184	.056	.104
C1808	.18 × .08	.112	.248	.068	.084
C1812	.18 × .128	.112	.248	.068	.132
C2220	.228 × .2	.16	.296	.068	.204

METRIC

CODE	SIZE	A	B	C	D
C0805	2.0 × 1.25	0.8	3.4	1.3	1.4
R/C1206	3.2 × 1.6	1.8	4.6	1.4	1.7
C1210	3.2 × 2.5	1.8	4.6	1.4	2.6
C1808	4.5 × 2.0	2.8	6.2	1.7	2.1
C1812	4.5 × 3.2	2.8	6.2	1.7	3.3
C2220	5.7 × 5.0	4.0	7.4	1.7	5.1

Fig. 1-20: Footprints for reflow soldered surface mounted resistors and ceramic multilayer capacitors.

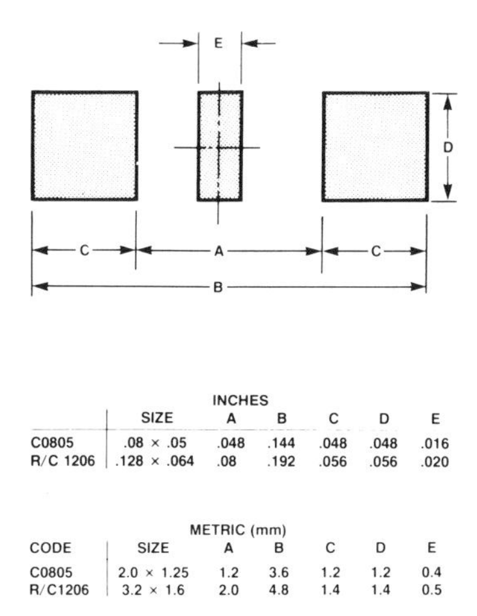

INCHES

	SIZE	A	B	C	D	E
C0805	.08 × .05	.048	.144	.048	.048	.016
R/C 1206	.128 × .064	.08	.192	.056	.056	.020

METRIC (mm)

CODE	SIZE	A	B	C	D	E
C0805	2.0 × 1.25	1.2	3.6	1.2	1.2	0.4
R/C1206	3.2 × 1.6	2.0	4.8	1.4	1.4	0.5

Fig. 1-21: Footprints for wave soldered surface mounted resistors and ceramic multilayer capacitors.

Layout Considerations

Component orientation plays an important role in obtaining consistent solder-joint quality. The substrate layout shown in Fig. 1-22 will result in significantly better solder joints than a substrate with SMD resistors and capacitors positioned parallel to the solder flow.

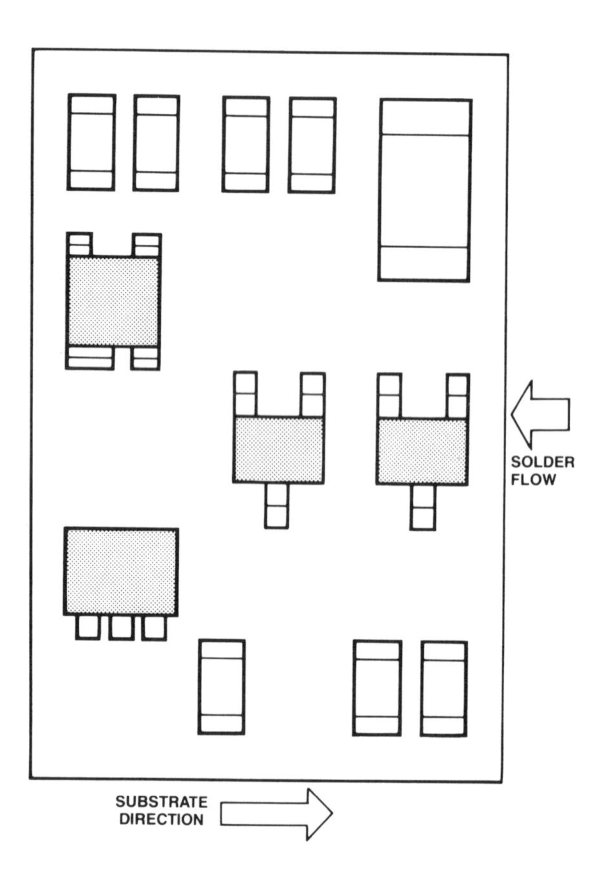

Fig. 1-22: Recommended component orientation for wave soldered substrates.

Component Pitch

The minimum component pitch is governed by the maximum width of the component and the minimum distance between adjacent components. When defining the maximum component width, the rotational accuracy of the placement machine must also be considered. Fig. 1-23 shows how the effective width of the SMD is increased when the component is rotated with respect to the footprint by angle $\phi°$ (for clarity, the rotation is exaggerated in the illustration).

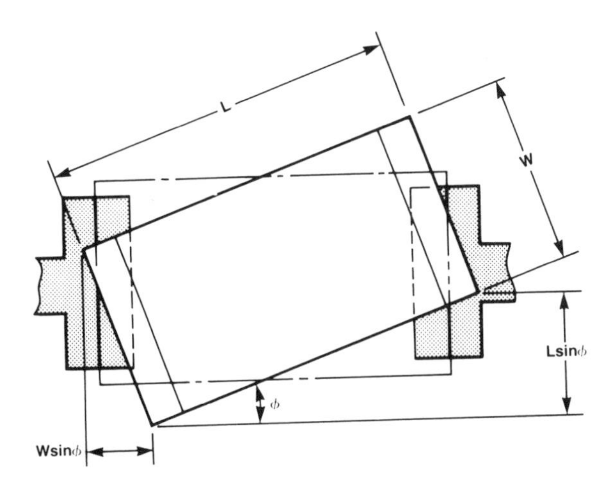

ϕ = Component rotation with respect to footprint
L sin ϕ = Effective increase in width
W sin ϕ = Effective increase in length

Fig. 1-23: The influence of rotation of the SMD with respect to the footprint.

The minimum permissible distance between adjacent SMDs is a figure based upon the gap required to avoid solder-bridging during the wave soldering process. Fig. 1-24 shows how this distance, plus the maximum component width, are combined to derive the basic expression for calculating the minimum pitch (F_{min}).

As a guide, the recommended minimum pitches for various combinations of two sizes of SMDs, the R/C1206 and C0805 (R or C designating resistor or capacitor respectively, the number referring to the component size), are given in Table 1-1. These figures are statistically derived under certain assumed boundary conditions, as follows:

- positioning error (Δp) ± 0.3 mm; ($\pm$ .012″)
- pattern accuracy (Δq) ± 0.3 mm; ($\pm$.012″)
- rotational accuracy (θ) $\pm 3°$;
- component metallization/solder land overlap (M_{min}) 0.1 mm (.004″) (note this figure is only valid for wave soldering);
- the figure for the minimum permissible gap between adjacent components (G_{min}) is taken to be 0.5 mm (.020″).

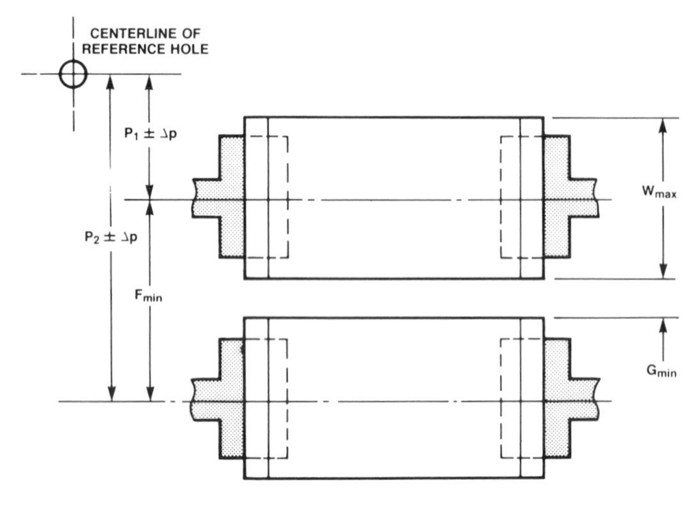

W_{max} = Maximum width of component
G_{min} = Minimum permissible gap
F_{min} = Minimum pitch
P_1 = Nominal position of component 1 (tolerance Δp)
P_2 = Nominal position of component 2 (tolerance Δp)
F_{min} = $W_{max} + 2\Delta p + G_{min}$

Fig. 1-24: Criteria for determining the minimum pitch of SMDs.

As these calculations are not based on worst-case conditions, but on a statistical analysis of all boundary conditions, there is a certain flexibility in the given area.

For example, it is possible to position R/C1206 SMDs on a 2.5 mm pitch, but the probability of component placements occurring with G_{min} smaller than 0.5 mm with increase, hence the likelihood of solder-bridging also increases. Each application must be assessed on individual merit, with regard to acceptable levels of re-work and so on.

Solder Land/Via Hole Relationship

With reflow soldered multilayer and double-sided plated-through-hole substrates, there must be sufficient separation between the via holes and the solder lands to prevent a solder-well forming. If too close to a solder joint, the via hole may suck the molten solder away from the component by capillaric action, resulting in insufficient wetting of the joint.

Table 1-1
Recommended Pitch For R/C1206 and C0805 SMDs

Combination	Component A	Component B: R/C1206	Component B: C0805
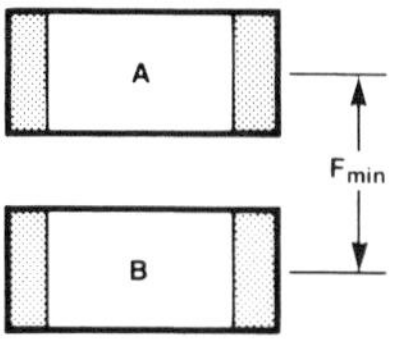 A above B, F_{min} vertical	R/C1206	3.0 (.12″)	2.8 (.112″)
	C0805	2.8 (.112″)	2.6 (.104″)
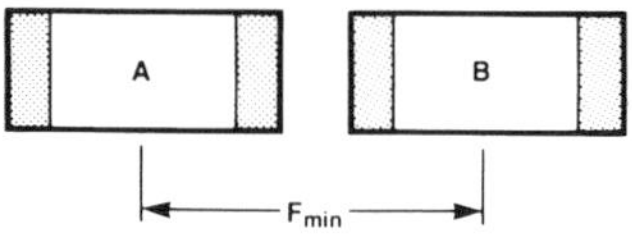 A beside B, F_{min} horizontal	R/C1206	5.8 (.232″)	5.3 (.212″)
	C0805	5.3 (.212″)	4.8 (.192″)
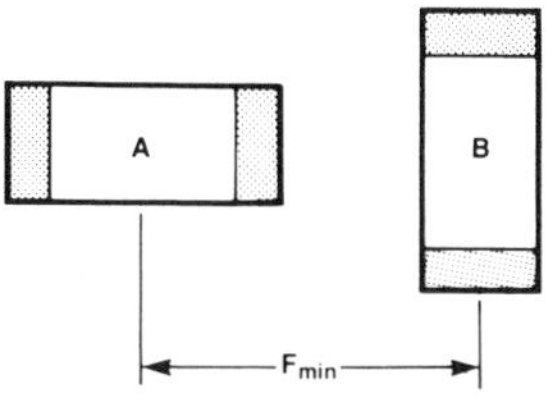 A beside B rotated, F_{min} horizontal	R/C1206	4.1 (.164″)	3.7 (.148″)
	C0805	3.6 (.144″)	3.0 (.12″)

(inches in parentheses)

Solder Land/Component Lead Relationship

A special consideration for mixed print substrate layout is the location of leaded components with respect to the SMD footprints, and the minimum distance between a protruding clinched lead and a conductor or SMD. Figure 1-25 shows typical configurations for R/C1206 SMDs mounted on the underside of a substrate with respect to the clinched leads of a leaded component. Minimum distances between the clinched lead ends and the SMDs or substrate conductors are 1 mm (.04″) and 0.5 (.02″) respectively.

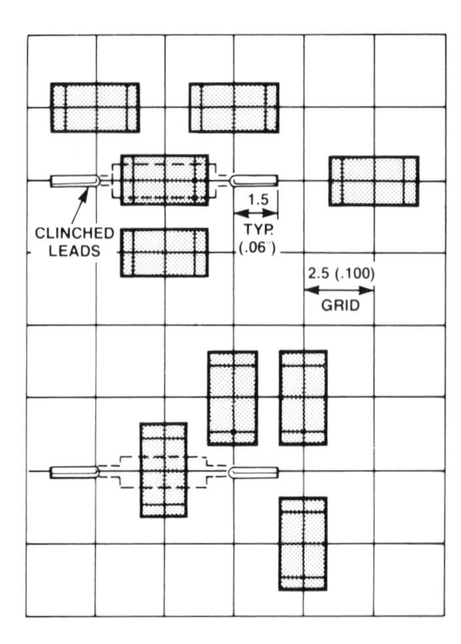

Fig. 1-25: Location of R/C1206 SMDs on the underside of a mixed print substrate with respect to the clinched leads of through-hole components (dimnensions in mm).

Placement Machine Restrictions

There are two ways of looking at the distribution of SMDs on the substrate: uniform SMD placement, and non-uniform SMD placement. With non-uniform placement, center-to-center dimensions of SMDs are not exact multiples of a pre-determined dimension as shown in Fig. 1-26, so the location of each is difficult to program into the machine.

Uniform placement uses a modular grid system with devices placed on a uniform center-to-center spacing, for example, 2.5 (.1" or 5 mm (.2") as shown in Fig. 1-27. This has the distinct advantage of establishing a standard, enabling the use of other automated placement machines for future production requirements without having to redesign boards.

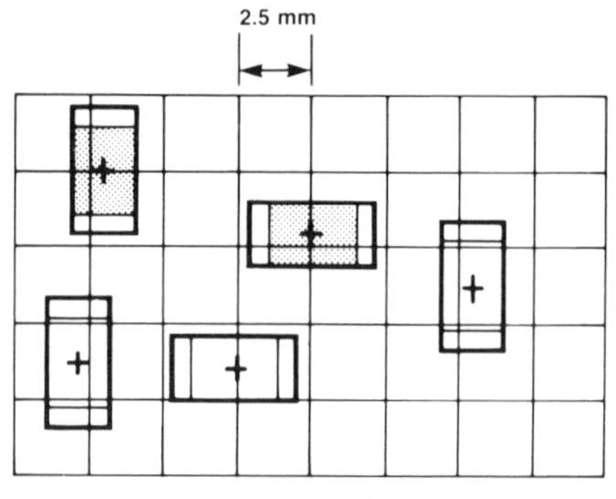

Fig. 1-26: Non-uniform component placement.

Substrate Population

Population density of SMDs over the total area of the substrate must also be carefully considered, as placement machine limitations can create a "lane" or

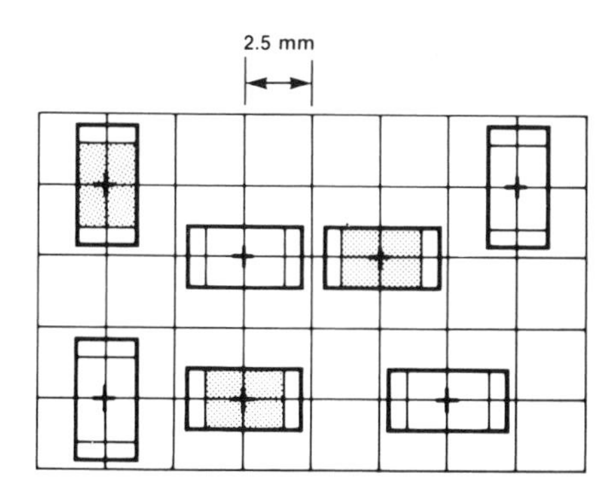

Fig. 1-27: Uniform component placement.

"zone" that restricts the total number of components which can be placed within that area on the substrate.

For example, on a hardware-programmable simultaneous placement machine (see Fig. 1-3(c)), each pick-and-place unit within the placement module can only place a component on the substrate in a restricted lane (owning to adjacent pick-and-place units), typically 10 to 12 mm (.4″ to .48″) wide, as shown in Fig. 1-28.

Placement of the 10 components in the lane on the right of the substrate shown will require a machine with 10 placement modules (or ten passes beneath a single placement module), an inefficient process considering that there are no more than three SMDs in any lane.

Test Points

Sitting of test-points for in-circuit testing of SMD substrates presents problems owning to the fewer via-holes, higher component densities, and components on both sides of SMD substrates. On conventional double-sided PCBs, the via-holes and plated-through component lead-holes means that most test-points are accessible from one side of the board. However, on SMD substrates, extra provision for test-points may have to be made on both sides of the substrate.

Figure 1-29(a) shows the recommended approach for positioning test-points in tracks close to components, and Fig. 1-29(b) shows an acceptable, though not recommended alternative, where the solder land is extended to accommodate the test-pin. This latter method avoids sacrificing too much board space thus maintaining a high-density layout, but can introduce the problem of components moving ("floating") when reflow soldered. The approach in Fig. 1-29(c) is to-

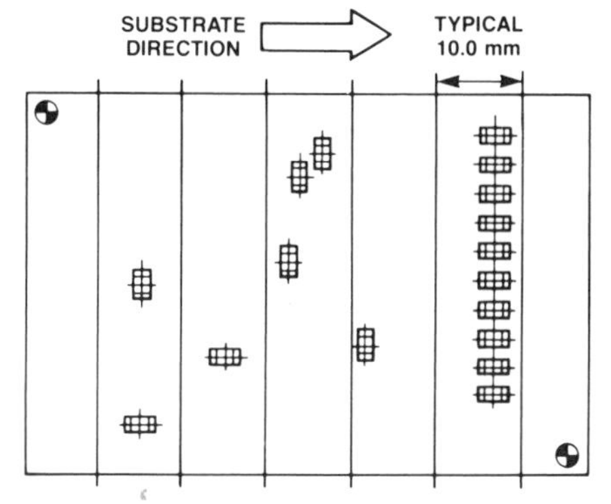

Fig. 1-28: Substrate "lanes" from use of a simultaneous placement machine.

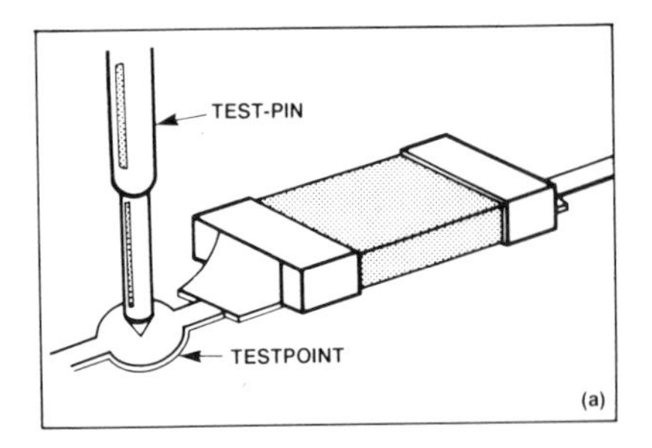

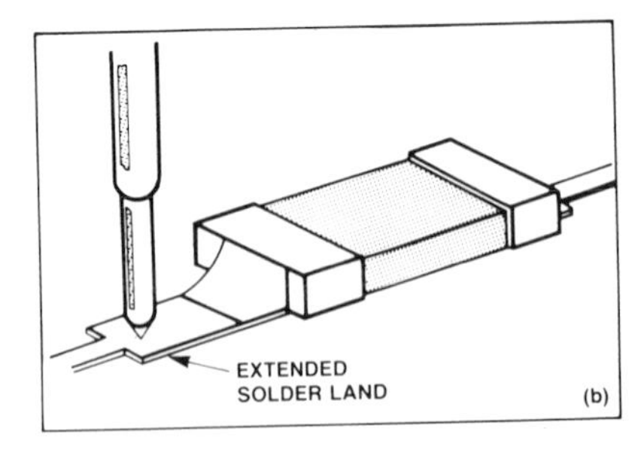

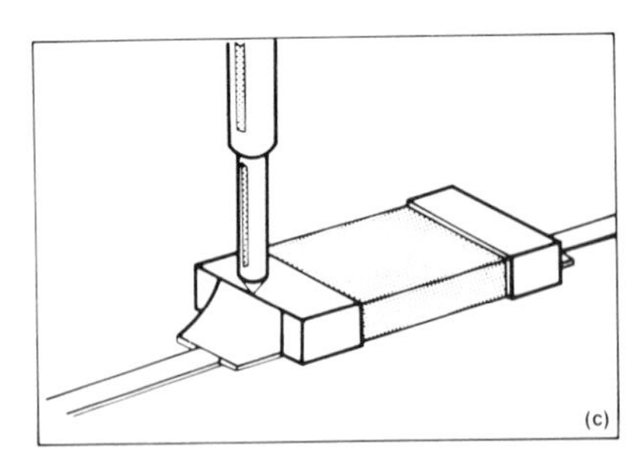

Fig. 1-29: Test points: (a) recommended test point location close to an SMD; (b) acceptable test point location; (c) unacceptable test point location.

tally unacceptable, as the pressure applied by the test-pin can make an open-circuit soldered joint appear to be good, and more importantly, the test-pin can damage the metallization on the component, particularly with small (SO) SMDs; use the test point in (a) whenever possible.

CAD Systems for SMD Layout

At present, about half of all PCBs are laid out using computer-aided design (CAD) techniques, and this proportion is expected to rise to over 95 per cent by 1990. Of the many current CAD systems available for designing PCB layouts for conventional through-hole components and ICs in DIL packages, few are SMD-compatible, and systems dedicated exclusively to SMD substrate layout are still comparatively rare. There are two main reasons for this: some CAD suppliers are waiting for SMD technology to fully mature before updating their systems to cater to SMD-loaded substrates, and others are holding back until standard package outlines are fully defined.

However, updating CAD systems used for through-hole printed boards is not simply a case of substituting SMD footprints for conventional component footprints, since SMD-populated substrates impose far tougher restraints on PCB layout and require a total rethink of the layout programs. For example, systems must deal with higher component densities, finer track widths, devices on both sides of the substrate (possibly occupying corresponding positions on opposite sides), and even SMDs under conventional DILs on the same side of the substrate.

The amount of re-working that a program requires depends on whether it's an interactive (manual) system, or one with fully automatic routing and placement capabilities. For interactive systems, where the user positions the components and routes the tracks manually on-screen, program modifications will be minimal. Automatic systems, however, must contend with the stricter design rules for SMD substrate layout. For example, many autorouting programs assume that every solder land is a plated-through-hole, and therefore can be used as a via-hole. This is not applicable for SMD populated substrates.

CAD programs base the substrate layout on a regular grid. This method, analogous to drawing the layout on graph paper, must have the grid lines on a pitch that is no larger than the smallest component or feature (track width, pitch, and so on). For conventional DIL boards, this is typically 0.635 mm (0.025″), but with the much smaller SMDs, a grid spacing of 0.0254 mm (0.001″) is required. Consequently, for the same area of substrate, a CAD system based on this finer grid requires a resolution more than 600 times greater than that for conventional layout CAD systems.

To handle this, extra memory capacity can be added, or the allowable substrate area can be limited. In fact, the small size of SMDs, and the high-density layouts possible, generally results in a smaller substrate. However, high density layout gives rise to additional complications not directly related to the SMD substrate design guidelines. Most CAD systems, for instance, cannot always completely route all interconnects, and some traces have to be routed manually. This can be particularly difficult with the fewer via-holes and smaller component spacing of SMD boards.

Ideally, the CAD program should have a "tear-up and start again" algorithm that allows it to re-start autorouting if a previous attempt reaches a position where no further traces can be routed before an acceptable percentage of interconnects (and this percentage must first be determined) have been made. This minimizes the manual re-working required.

CAE/CAD/CAM Interaction

Computer-aided production of printed boards has evolved from what was initially only a computer-aided manufacturing process (CAM—digesting a manually generated layout and using a photoplotter to produce the artwork) to fully interactive computer-aided engineering, design and manufacture using a common data base. Fig. 1-31 illustrates how this multi-dimensional interaction is particularly suited to SMD-populated substrate manufacture in its highly automated environment of pick-and-place assembly machines and test equipment.

Using a fully integrated system, linked by local area network to a central data base, will make it possible to use the initial computer-aided engineering (CAE—schematic design, logic verification and fault simulation) in the generation of the final test-patterns a the end of the development process. These test-patterns can then be used with the automatic test equipment (ATE) for functional testing of the finished substrates.

Such a system is particularly useful for testing SMD-populated substrates, as their high component density and fewer via-holes make in-circuit testing ("bed of nails" approach) difficult. Consequently, manufacturers are turning to functional testing as an alternative. (These aspects are covered in another chapter in this book).

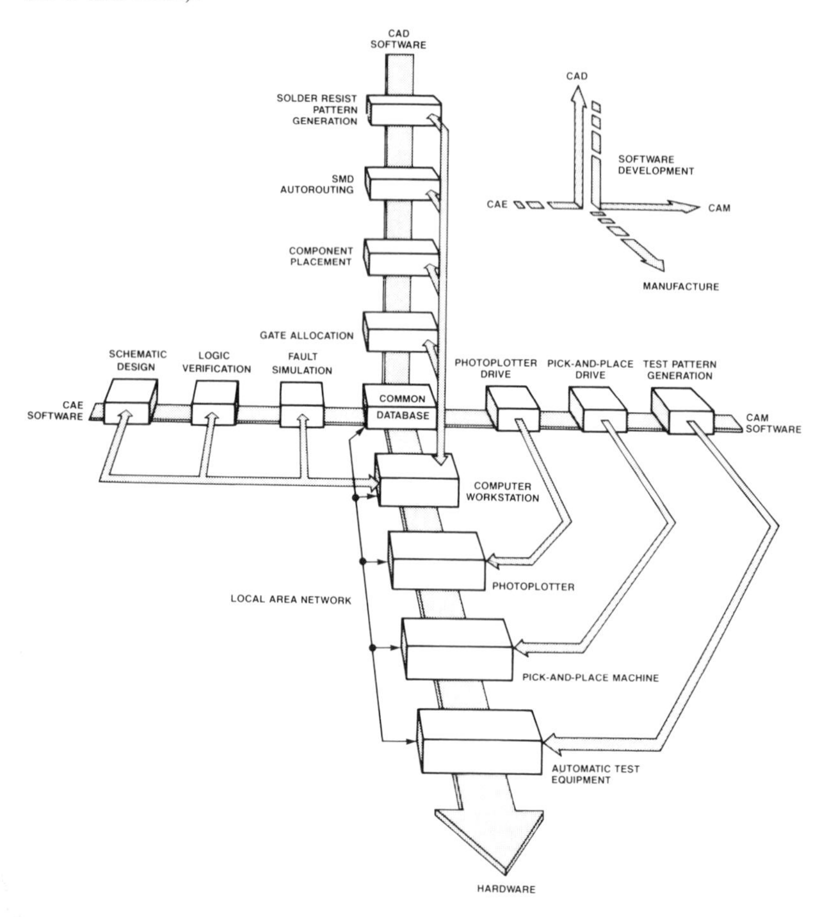

Fig. 1-31: The software-hardware interaction for the computer-aided engineering, design and manufacture of SMD substrates.

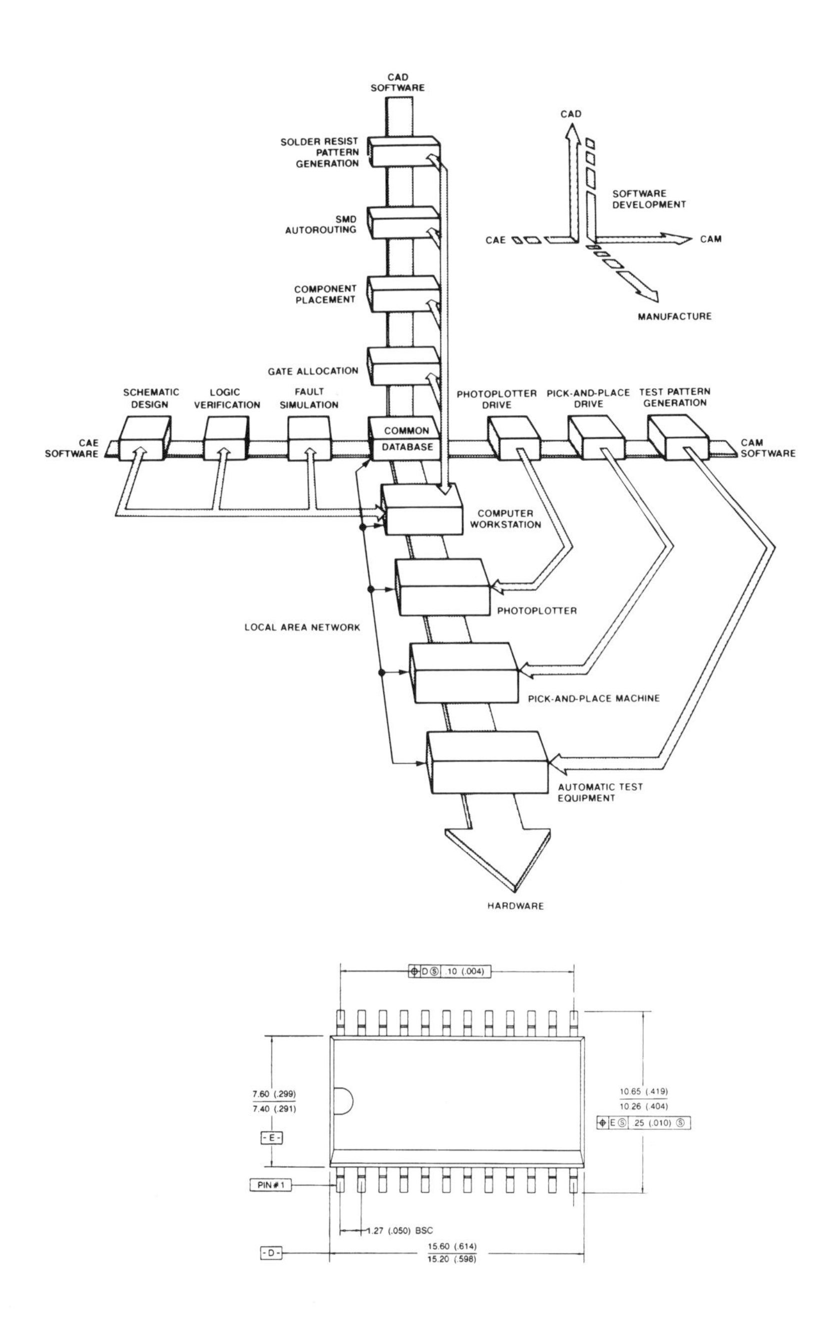

Fig. 1-31: The software-hardware interaction for the computer-aided engineering, design and manufacture of SMD substrates.

chapter 2

SPACE PLANNING AND INTERFACE

When switching from standard insertion mount components to the newer high-density surface mount components there are significant changes to be made in the printed circuit board. Some of the principle points to keep in mind are:

1. An average of three times the component density will be the goal.
2. Conductor lithography will typically have to be improved from 12 mil lines/12 mil spaces to 8 mil lines/8 mil spaces as shown in Fig. 2-1.
3. The elimination of insertion leads reduces the number of through-holes thus providing more space for routing.
4. Higher density boards can be achieved without adding more conductive layers.
5. Both insertion mounted and surface mounted components can be mounted on the same board.
6. Components can be mounted on both sides of the board.
7. Mounting pad designer should consider the following:
   a. Terminal width and spacing of the components.
   b. Minimizing opens and bridging.
   c. Creation of proper fillets.
8. Standard board materials are acceptable for leaded components.

9. Special board materials may be required for leadless components to prevent solder joint failure induced by various mechanical stresses.

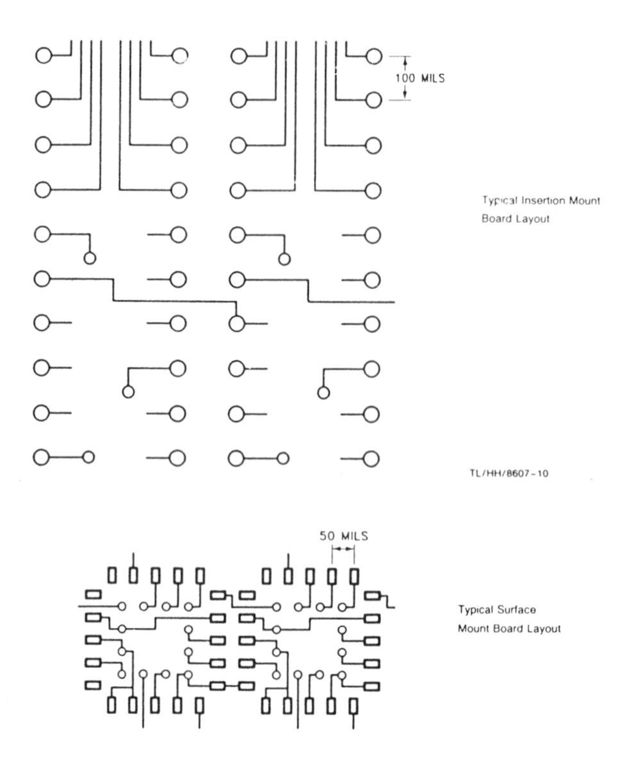

Fig. 2-1: When utilizing SMT conductor lithography will typically have to be improved from 12 mil lines/12 mil spaces to 8 mil lines/8 mil spaces.

Standards for SMT Components

The selection of a suitable surface mounted equivalent to a comparable leaded device is a critical step which requires more than a casual look through the various component directories and data sheets.

Errors occur easily during this phase due to the options and styles of surface mounted component packages which may be unfamiliar to the component engineer and designer. Examples are shown in Fig. 2-2.

While one surface mounted device may look the same as the next, there are subtle differences in package types. For example, the same IC function may be available in two or more package styles from the same manufacturer.

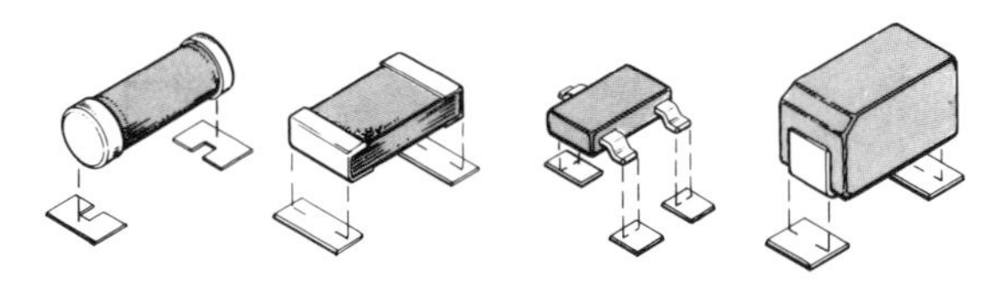

Fig. 2-2: Several options and styles of SMDs are available which sometimes confuses the designer. A careful study of the various types available is necessary to avoid errors.

In the United States, the JEDEC, EIA, IPC and SMTA committees play a valuable role in communicating the importance of standardization to the component manufacturer.

In Japan, the EIAJ is the counterpart of the U.S. JEDEC committee. While the Japanese use the metric system of measurement, they do maintain the .050 inch spacing on many IC components in order to be compatible with the American market.

On the products available with .050 inch lead spacing, significant differences remain. For example, the distance between rows on the contacts of the EIAJ-SOP is not the same as on the JEDEC SO-IC. If it is necessary to use IC devices from multiple sources, it may not be possible to mount both sizes on the footprint or pad geometry recommended by the component manufacturer. We will discuss alternative ways to handle this problem in the Design Section.

Even with the increase in component standardization, a designer may find it necessary to make a few compromises within the standard. For example, some components will not meet the environmental specifications of others and/or the plating may not be suitable for the assembly process used. We will expand on the subtle differences during each individual component section.

Meanwhile, until standards are established for all surface mounted devices and everyone becomes comfortable working within them, component engineers, designers and purchasing personnel must work closely together to insure that the correct component is designed into and purchased for the product. It should be emphasized that the success of your SMT project is often directly linked to the quality of communication between members of this group.

Component Packaging Options

SMT components are supplied in one of three configurations—bulk, tube magazine and tape and reel. Many companies prefer the tape and reel packaging for medium to high volume production. With low volume or prototype production, the tube magazine is recommended. Both containers are clearly marked for material control and should adapt easily to automated placement equipment. Bulk packaging is less desirable because the part must be individually handled or repackaged for the assembly equipment.

Chip resistors and capacitors are supplied from many sources on 8mm tape and will hold up to approximately 4000 chips per reel. Material control is vital. The manufacturer's code number on the package may be the only way to verify contents since not all chip component values are identified on the body.

SOT, diode and transistor packages can also be furnished in tape and reel, as shown in Fig. 2-3. Because these are polarized components, the direction of the feed must be specified. Close communication between the design engineering department and the manufacturing operation is essential to insure planning early in the development of SMT products.

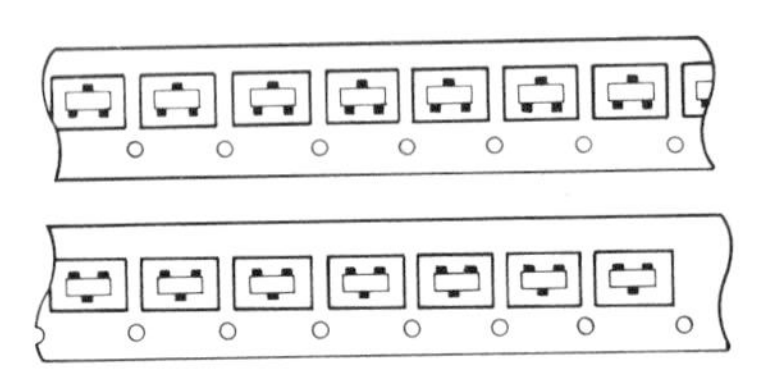

Fig. 2-3: SOT-23 Diodes and transistors are available from several manufacturers in tape-and-reel form; two options are shown here.

Component Selection Guidelines

Surface mounted technology will offer more assembly options than through-hole. However, all the elements must be compatible to insure the reliability of your product. For instance, uniformity of component size, termination area and plating type will affect the producibility and reliability of your product.

Components are usually identified by one of two categories: passive and active. In the following pages passive and active components will be discussed, as well as transistors, potentiometer, connectors and interfacing techniques.

Passive Devices
Monolithic Capacitors

The selection process of capacitors is also more than choosing the value and the dielectric. Choices will include body size, end-cap termination and material or plating. The value range will span more than a dozen sizes from one manufacturer to another. To reduce unnecessary inventory of seldom used sizes, the EIA membership has agreed on five popular sizes which they recommend for use in the United States, as detailed in Fig. 2-4.

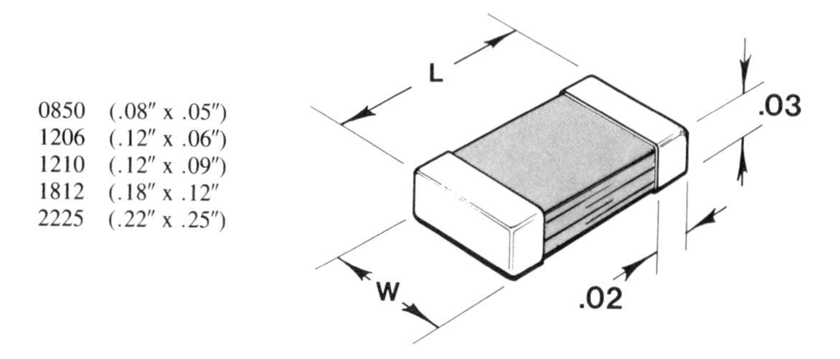

Fig. 2-4: Five popular sizes of monolithic capacitors are currently recommended by EIA membership in the United States.

The 0805 is used in applications requiring maximum miniaturization or very low capacitance applications. The most common size for capacitor values, up to .18 μF, will be the 1206—considered a worldwide standard. To furnish the purchasing department multiple sources for capacitor values of .047 μF through .15 μF, it may be prudent to allow for the 1210 size component. The 1206 will mount on the 1210 footprint pattern. Higher voltage and capacitance values are available in the larger size with several dielectric options.

When specifying capacitors, it is important to select the proper end-cap termination material. The most compatible end-cap terminations plating option for surface mounting to circuit boards is referred to as nickel barrier/solder type.

Tantalum Capacitors

Manufacturers around the world are converging on an accepted set of standards for surface mounted tantalum capacitors. The basis for these standards has been influenceds by increased demand for these devices and the need for multiple sources by the end user.

There are many styles and sizes presently on the market, each with excellent mechanical and electrical characteristics. See Fig. 2-5.

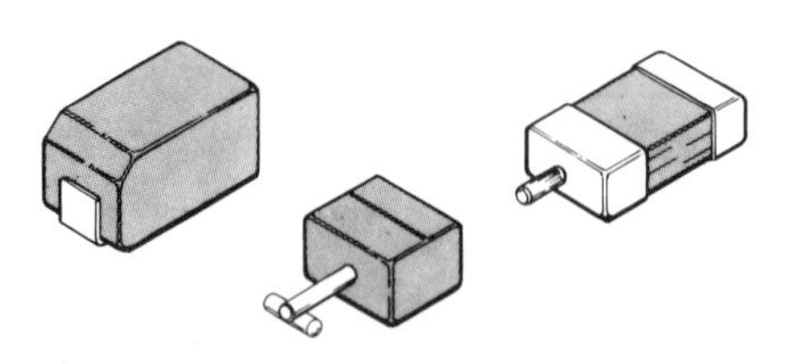

Fig. 2-5: The many styles and sizes of Tantalum Capacitors—each with excellent mechanical and electrical characteristics—help make the designer's job much easier.

Accepted levels of electrical and mechanical performance are required with each configuration and, at the same time, must be compatible with the automated placement equipment.

The standards proposed to the EIA include two families of capacitors, each covering a specific value and performance range. The current standard range includes values from 0.10 μF through 68.0 μF and the extended range provides capacitance value through 330 μF as illustrated on the following pages.

Resistors for SMT

Surface mounted resistors are available from several sources in all values. As with capacitors, some sizes are more common than others and higher wattage will have a limited value selection.

Chip resistors have a thick film element on an alumina substrate. The endcap terminations have a tin/lead solder plating over a nickel barrier for compatibility with SMT assembly and solder systems. The resistor element is covered with a resin coating and the 1206 size (and greater) is available with values clearly marked on the surface for easy inspection. See Fig. 2-6.

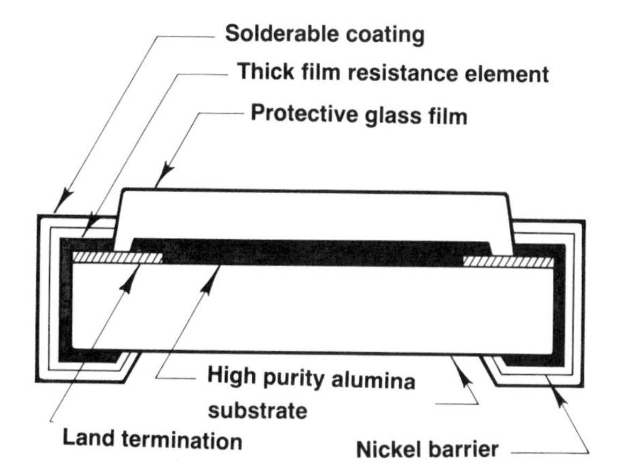

Fig. 2-6: Characteristics for a typical SMD resistor.

The most common chip size in use today is the 1206. This size is usually rated at ⅛ watt by the majority of the manufacturers. However, it is recommended for ¼ watt applications with specific temperature-range limits.

Other sizes are available as well—the 0805 for miniature applications with 1/10 watt rating and the 1210 device for ¼ watt applications. one-half watt and 1 watt are available in larger sizes, but value selection is limited.

Resistors are also supplied in the MELF package, a tubular shaped device with length and width dimensions similar to the ceramic chip components. The MELF is very popular for wave solder applications when the component is held to the solder side of the substrate with special epoxy. Before selecting the tubular shaped packages, verify the compatibility of the component with the assembly equipment to be used.

Resistor Networks

Resistor networks are available in many configurations. Choose packages that can be handled with automatic placement equipment. When possible, find a style that is mechanically the same from two or more sources. See Fig. 2-7.

The above networks are available from Dale Electronics and RCD components in a .300 inch overall width—SO-14 and SO-16 with a .250 inch overall width.

Be aware of the subtle differences when choosing your sources. If an incorrect substitution is made in purchasing, it could result in unwanted delays.

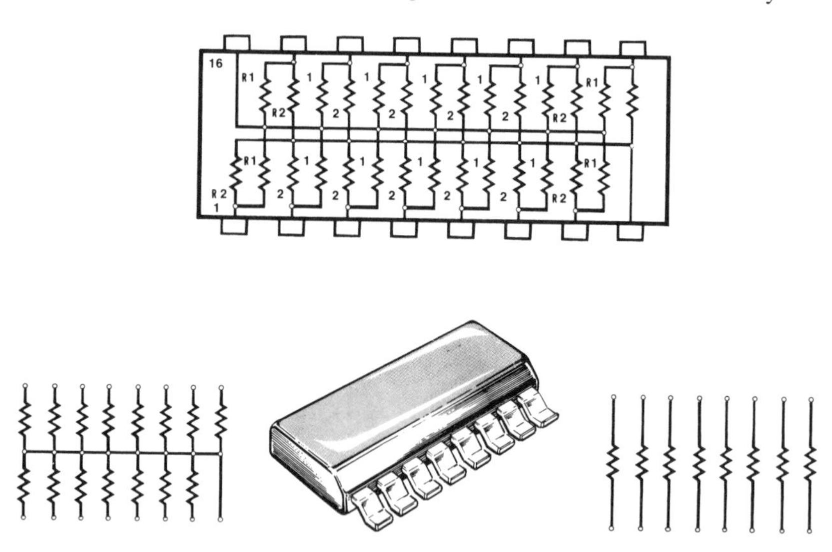

Fig. 2-7: Resistor networks are available in many configurations; choose one that works best with automated placement equipment.

Potentiometer for SMT

There is a multitude of single-turn cermet element posts on the market for PC boards requiring a variable resistor. As you research sources for these parts, you will find that they differ from one manufacturer to another.

Parts are available in open and sealed body types. It is important to choose one that is packaged for automated assembly equipment, as well as being usable in all types of solder reflow processes.

The devices, shown in Fig. 2-8, are designed for flat and right-angle mounting. Many manufacturers will offer a similar package with contacts on .050 inch spacing with open/closed body construction.

A small amount of adhesive may be added when mounting posts to enhance the mechanical integrity of a part requiring frequent adjustments.

Active Devices

The most common package types used for Bipolar and Field Effect transistors are the SOT-23, SOT-24 and the SOT-89 cases. The SOT-23 and SOT-24 are smaller packages with three or four leads respectively. See details in Fig. 2-9.

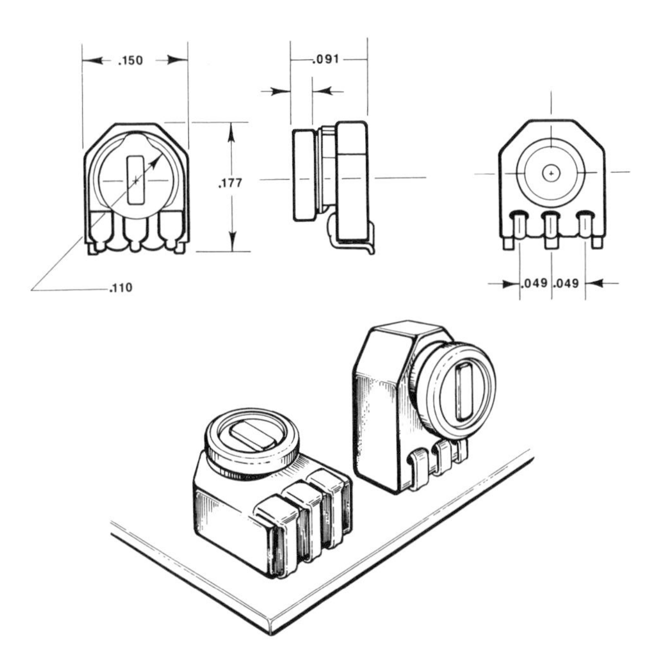

Fig. 2-8: Potentiometers designed for flat and right-angle mounting.

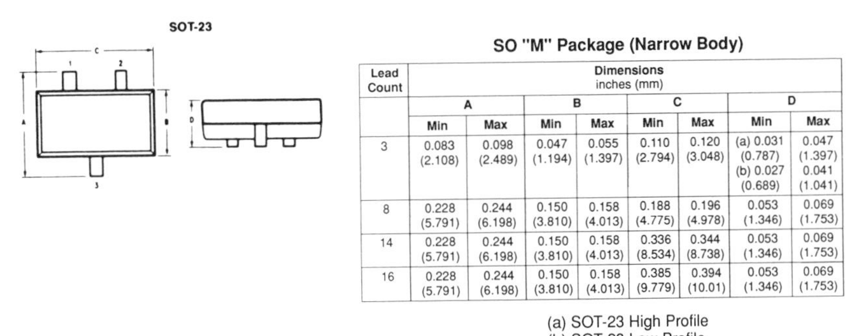

SO "M" Package (Narrow Body)

Lead Count	Dimensions inches (mm)							
	A		B		C		D	
	Min	Max	Min	Max	Min	Max	Min	Max
3	0.083 (2.108)	0.098 (2.489)	0.047 (1.194)	0.055 (1.397)	0.110 (2.794)	0.120 (3.048)	(a) 0.031 (0.787) (b) 0.027 (0.689)	0.047 (1.397) 0.041 (1.041)
8	0.228 (5.791)	0.244 (6.198)	0.150 (3.810)	0.158 (4.013)	0.188 (4.775)	0.196 (4.978)	0.053 (1.346)	0.069 (1.753)
14	0.228 (5.791)	0.244 (6.198)	0.150 (3.810)	0.158 (4.013)	0.336 (8.534)	0.344 (8.738)	0.053 (1.346)	0.069 (1.753)
16	0.228 (5.791)	0.244 (6.198)	0.150 (3.810)	0.158 (4.013)	0.385 (9.779)	0.394 (10.01)	0.053 (1.346)	0.069 (1.753)

(a) SOT-23 High Profile
(b) SOT-23 Low Profile

Fig. 2-9: Specifications of SOT-23 transistor.

The pinning of the contact is not the same as the leaded type device that you may be familiar with, and the pinning can vary or be optional with some manufacturers. The detail in Fig. 2-10 is typical for one type of common transistors, but FET devices vary a great deal, even with the same manufacturer. Study the specifications carefully before you design the PC board. An error here, at this stage of the project, will have predictably negative results.

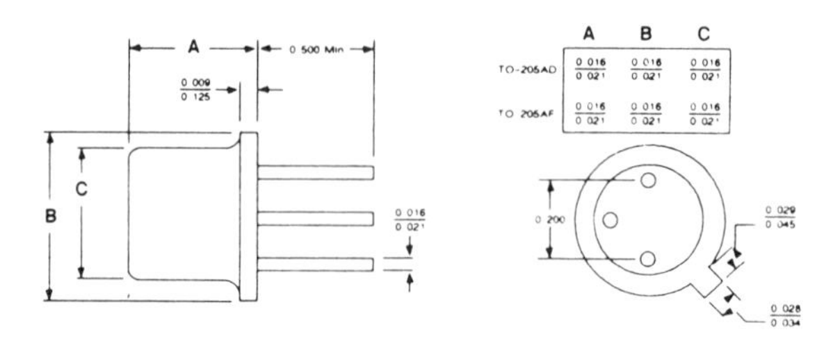

Fig. 2-10: Typical detail of a common transistor used with SMT.

Most general-purpose diodes and many Zener types are now available in the SOT-23 package. Pin assignment of diodes to the three-leaded package is optional from many suppliers, but to encourage standardization, use the more common configuration shown in Fig. 2-11 "A".

Dual diodes are also available in this package and in several direction options. To reduce space on your assembly, use two dual diodes packages to create a bridge network as shown in Fig. 2-12 "B" and "C".

Transistors and diode arrays are offered in other packages, but not always from multiple sources. Choose products that are furnished in JEDEC packages to insure standard size and compatibility. A package often used for single diodes is the MELF or tubular shaped devices. Size will vary with power rating and manufacturer. See Figure 2-13.

The round surface of the MEL is also difficult to handle on some of the assembly equipment. Further details on contact patterns and mounting methods will be furnished in the Designer Section.

ICs for SMT

Surface mounted ICs are available in several package styles: Small Outline (SO); Plastic Chip Carrier (PCC); and Quad Flat Pack. While all manufacturers do not conform to any one standard configuration, standards have been set for the SO and PCC packages. In all cases, these surface mounted ICs will be smaller than their leaded through-hole counterpart.

The JEDEC registered IC packages available today have .050 inch space between lead centers. However, specific contact geometry for these components are recommended by the component manufacturers. See Fig. 2-14.

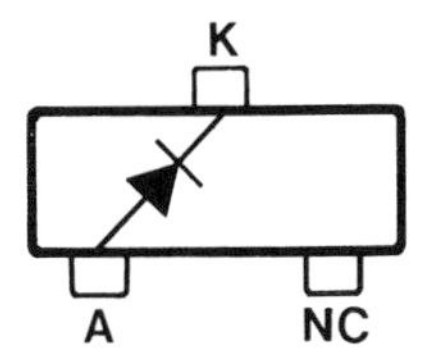

Fig. 2-11: The most common general-purpose diode.

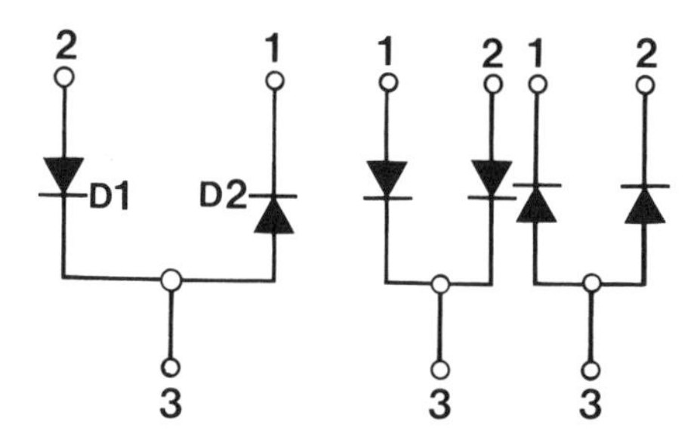

Fig. 2-12: Dual diodes are also available in the above configuration; use two dual diode packages to create a bridge network.

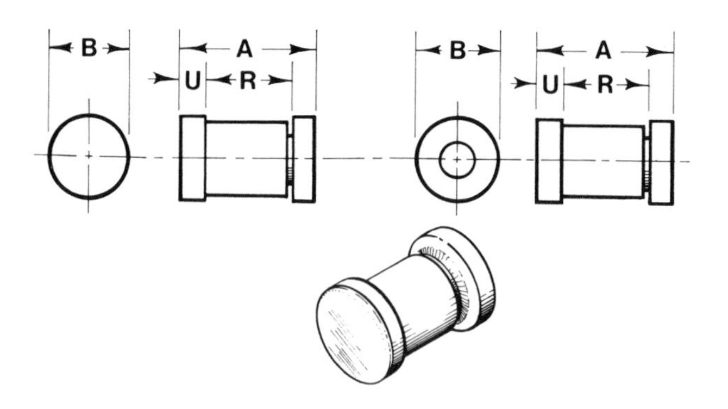

Fig. 2-13: The MELF (tubular-shaped devices) is often used for single diodes in SMT.

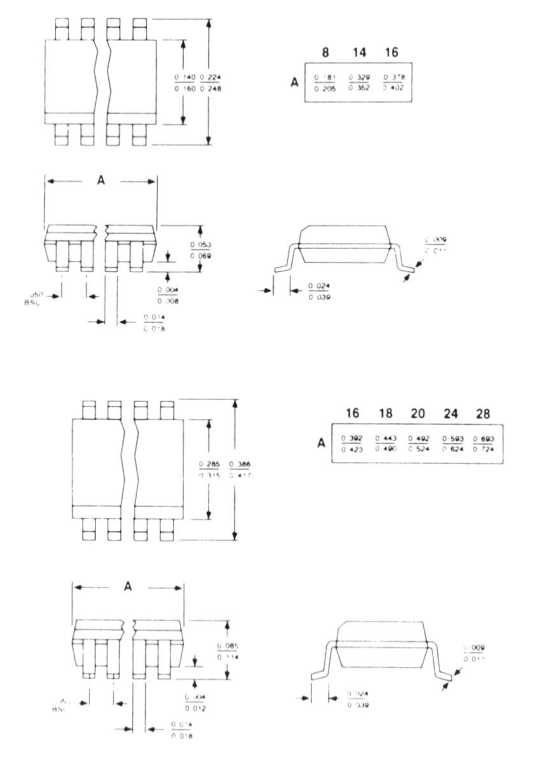

Fig. 2-14: Recommended dimensions for SO ICs.

Small Outline (SO) ICs

The JEDEC registered Small Outline or SOIC contact pattern is easily adaptable by the designer. The two parallel rows of contacts have the same pin assignments as the dual in-line through-hole IC they replace. For most logic devices, the spacing between contact row centers will be consistent.

Depending on the manufacturer, the same IC may be packaged in different packages for SMT. The designer needs to be aware of this, as the pinouts will vary. An example of the same device packaged in PCC-20 and SO-14 packages are shown in Fig. 2-15.

In comparing these packages, the designer will note that the SOIC has the same pinout as the conventional DIP, while the PCC-20 has a very unique pinout with non-connected pins dispersed on each of the four sides.

The selection of the style of component is an important factor in planning the layout. Compare the SOICs footprint to the DIP that performs the same electrical function. First, the component profile above the surface of the board is

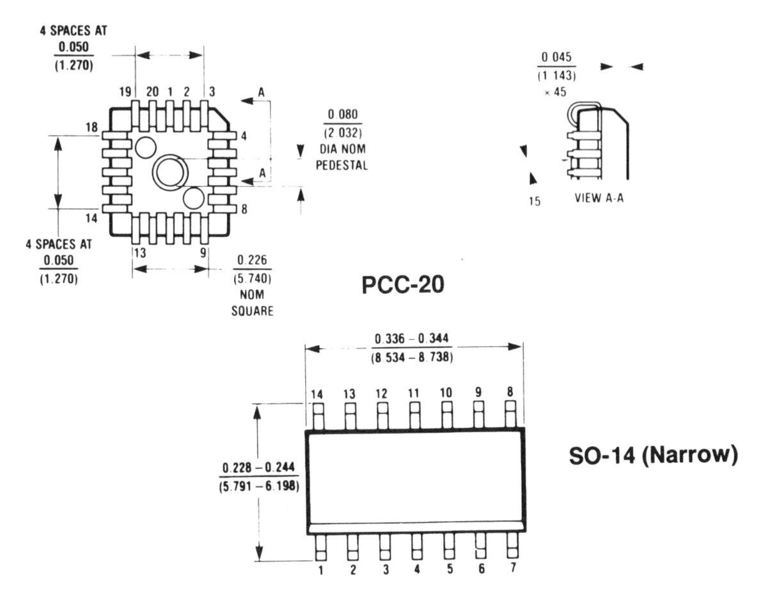

Fig. 2-15: The same IC may be packaged in different configurations for SMT; the designer needs to be aware of this.

much lower. Second, where component height is restricted, the SOIC offers a greater advantage over the DIP.

Some 16-pins require a wider body size to provide for the die size or heat dissipation. A PC board may have a mix of the SO-16 and SO-16L (wide) devices. Always check the component manufacturer's specifications. SOICs greater than 16 leads will be in the wider body size.

You will also have a choice in packaging. Tape and reel may be preferred by many companies with medium to high volume production, while tube magazine is recommended for lower volume production, or when inhouse testing of each IC is necessary.

Plastic Chip Carriers and Quad Packages

The most successful area of standardization has been in the plastic molded commercial grade devices. Several package styles registered with JEDEC comply to size and lead design that is common to many domestic, European and Japanese component manufacturers.

The plastic chip carrier (PCC) has equal amounts of contacts on each of its four sides. The contact is formed in a "J" shape, back under the package, to further decrease the surface area required for each device.

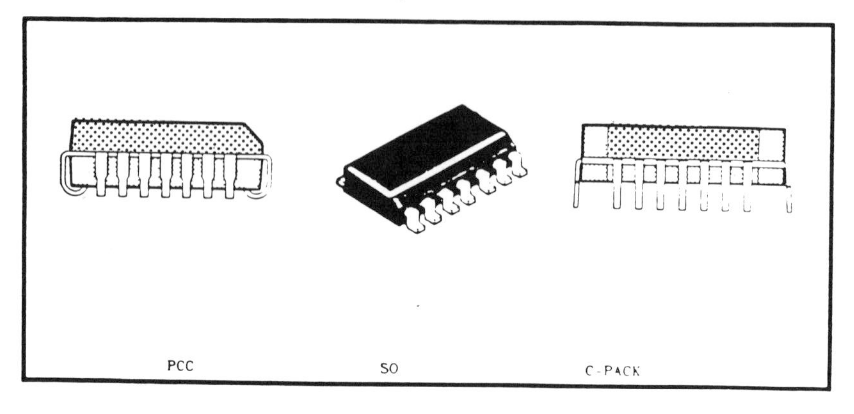

Fig. 2-16: Some plastic chip carrier will require a unique footprint for mounting.

Another style of package similar to the PCC is the C pack. The leads are spaced on.050 inch or less spacing on four sides, but instead of the "J" bend, the leads are bent straight down to the board surface. This style and the gull wing flat will require a unique footprint for mounting. See Fig. 2-16.

The plastic chip carrier (PCC) is a commercial, high-volume package resembling the early ceramic leadless chip carrier (LCC). The PCC has an excellent contact design and its symmetrical shape works well with most assembly equipment. The clearance under the PCC allows for more efficient cleaning after solder reflow.

The PCC footprint is similar to the LCC, but the lead design is much better for current production, reflow solder process and a larger selection of substrate materials. The uniform sizes and J-lead design of the PCC are well received by the users of these components.

IC manufacturers have attempted to standardize package and lead design. The JEDEC committee has approved the plastic (leaded) chip carrier or PCC in package configurations of 18, 22, 28, 32, 44, 52, 68, 84, 100 and 124 leads. See Fig. 2-17.

Some memory products are furnished in a PCC-18 package. Be aware that the PCC-18 is supplied in two sizes and each has a rectangular shape with four leads on each end and five leads per side. The width of the IC is consistent, but the length will vary more than .060 inch. (Check the manufacturer's specifications before board layout.)

The 18-lead device is now available in two case sizes: 18AA and 18AB.

The AB style allows for the component supplier to furnish 18- and 22-lead devices with the same lead frame and plastic mold tooling. The designer may select the AA footprint pattern for a particular position, but the second or third

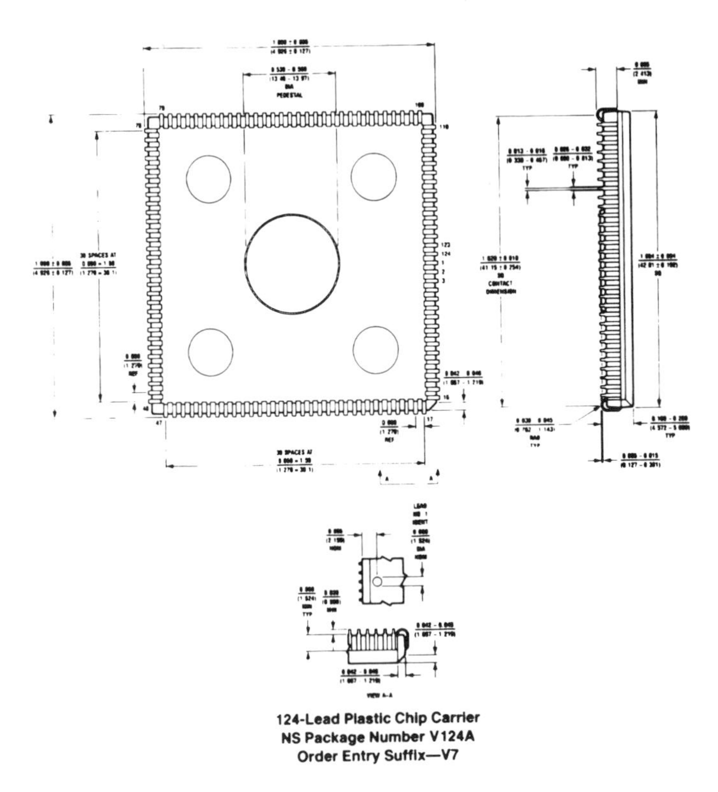

Fig. 2-17: Specifications of typical PCC devices.

source for the same function may use the AB outline.

In addition to the JEDEC registered devices, other component styles are supplied from several Asian manufacturers with a 1.0 mm and 0.8 mm lead spacing. These devices usually require special handling fixtures for efficient assembly processes.

Ceramic ICs

Due to the various size differences of IC packages offered by the IC manufacturers a multiple source may not be possible for applications requiring a ceramic body. In the case of leaded ceramic flat packs, there are some similarities. However, most will have lead spacing of .050 inch with a lead length that is generally excessive and extends far beyond the substrate surface. Therefore, leads on the ceramic flat packs, as shown in Fig. 2-18, must be modified before mounting the IC to the substrate surface. For further details on modification, refer to manufacturer's specifications.

CLT028

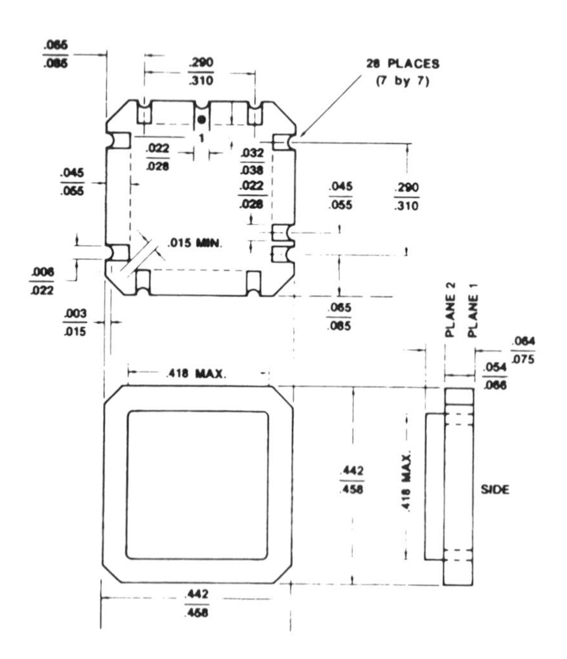

PID # 07703B

Fig. 2-18: Specifications for typical Ceramic ICs.

Connectors and Interface for SMT Assembly

Connector manufacturers are introducing several choices for surface mounted applications. In addition to the more prominent names in connectors, we are beginning to see several specialty and off-shore suppliers of very unique interconnection systems which will offer more choices.

Selecting a connector that will mate with existing connector families is preferred, since this obviously provides flexibility and, in many cases, multiple sources. See Fig. 2-19.

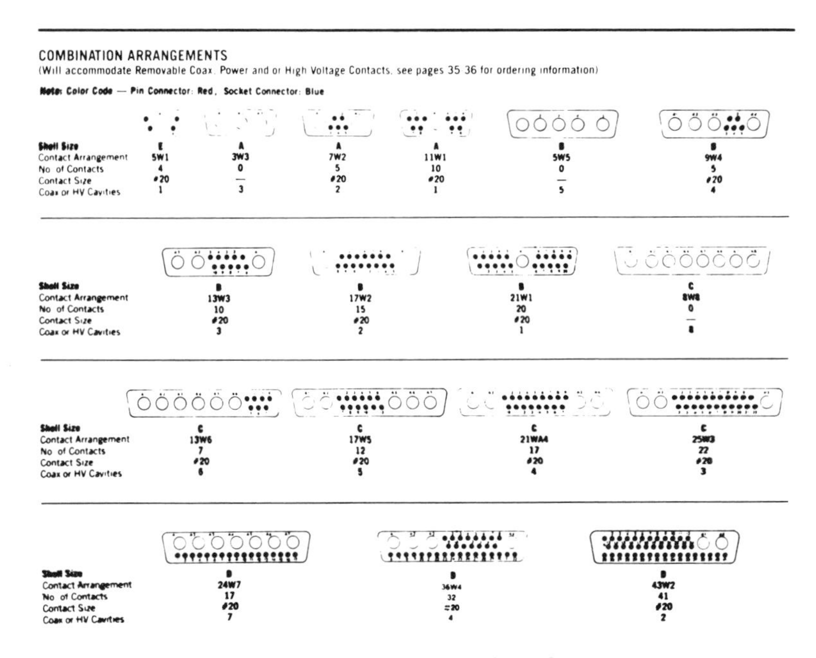

Fig. 2-19: SMT designers have a wide choice of connectors and interface. Selecting those that will mate with existing connector families is preferred.

For those users who have asked for additional mechanical support beyond just the solder connection, the connector manufacturers have furnished retention or stress relief mounting table, bosses, etc. The techniques for mechanically retaining the connector will vary somewhat from one manufacturer to another.

Heat Seal

Other methods for interfacing assemblies are the heat/pressure seal flat cable and compression contact connectors (HSC). The heat/pressure seal flexible cable is a polyester film base material with parallel rows of graphite/silver conductives covered by an insulating layer.

The ends of this flat cable are free of insulation and, when heat and pressure are applied, the electrical connection is made to the substrate.

Heat seal is a popular technique for mating the PC board with liquid crystal display (LCD) components. Refer to the HSC flexible cable manufacturers for specifications on compatible mating contact material and environmental limitations. See details in Fig. 2-20.

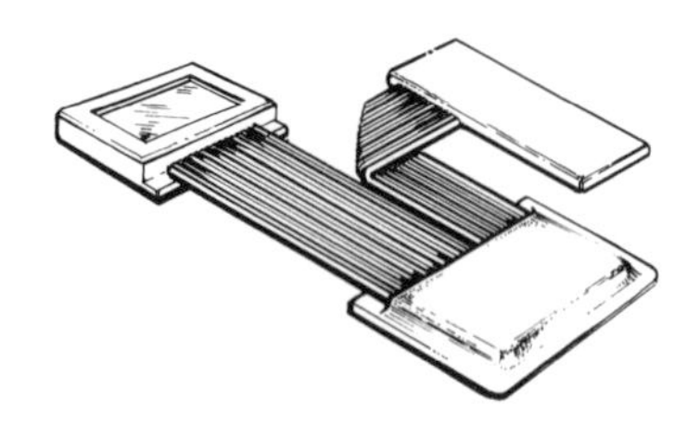

Fig. 2-20: Heat seal is a popular technique for mating the PC board with liquid crystal display (LCD) components.

Compression

Compression connectors are designed to mate two or more parallel substrates that are matched by contact areas on each surface as shown in Fig. 2-21. The substrate must be mechanically retained because this is usually how the connector material is captured within the assembly. If a more reliable method of retention is desired, a small amount of epoxy adhesive is placed at the ends of the compression material on one substrate or the other. Avoid getting adhesive on the contact areas.

After completing the component selection and the conversion of all available ICs and passive devices into surface-mount technology, a space study should be conducted. In general, the space study is a pre-planning procedure to assist the designer in the organization of component orientation, interconnection and density factors. This consists of establishing the component arrangement and the circuit pad and trace patterns necessary. For this, one needs to determine the device dimensions and tolerances, as these will combine with the machine placement accuracy and the etched or printed circuit accuracy tolerances to define a set of acceptable pad layout arrangement standards.

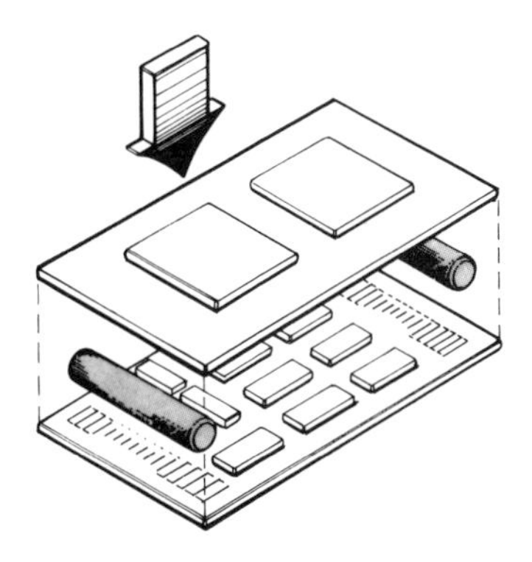

Fig. 2-21: Compressor connectors are designed to mate two or more matching parallel substrates.

There are many sources of suggested pad and grid, or trace, layout data available, both in the trade literature and from private publishers. All of these will provide you with guidance and a variable degree of help. Figure 2-22 is an indication of what you will find for typical surface mounted components. The data generally will show the lead and body dimensions and suggested pad dimensions and locations. What is finally used, however, is highly dependent on the particular combination of components you select, the placement medium, the planned placement machines, the remaining process combinations and materials which you select for your application.

Before the final mounting pad arrangement standards and medium material decisions are considered acceptable, one should go through a series of controlled tests using these candidate designs, plus minor variations as well, and the proposed medium material to evaluate the soundness of the choice of design.

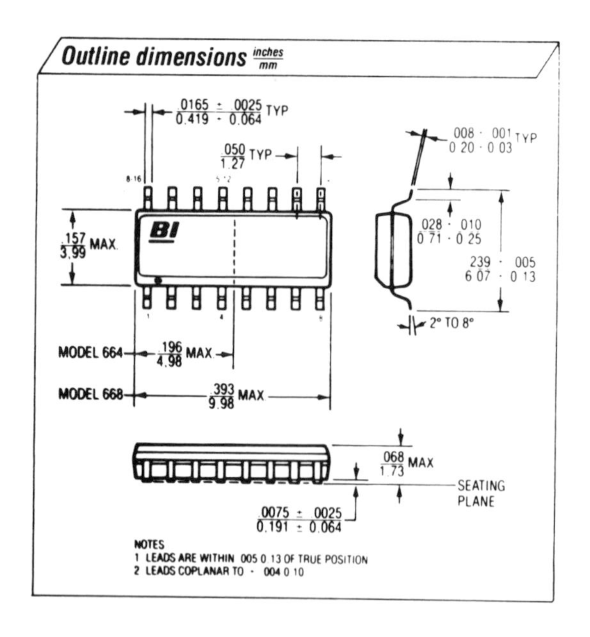

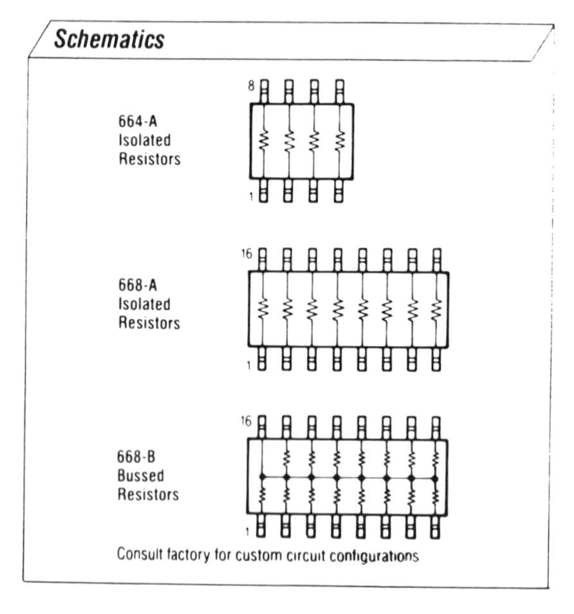

Fig. 2-22: A small sampling of suggested SMD pad-and-grid, or trace, layout data.

These tests should include several soldering profiles, including flux and solder temperature evaluations. After acceptable soldering parameters have been established, thermal cycling should be done to and beyond the degree to which the design will be exposed in normal use.

With printed circuit boards, all techniques used in the past would be applicable to any design, including double-side and multi-layer. In both cases, "via's", or plated-thru-holes, can be used for the layer-to-layer interconnection. This technique may be used with or without the additional use of thru-hole inserted components. It does require a drilling operation, however.

The major goal in the pre-planning stage is to allow adequate space to facilitate the manufacturing of the assembly. When component density increases on the circuit board, some compromises must be addressed:

1. Conductor width must be reduced.
2. Additional circuit layers must be added.
3. Inspection, testing and rework become more difficult.

Assembly Considerations

During the planning stage, specific assembly methods should be considered.

Medium overall size and shape: This is primarily a function of the specific selected machine(s) and/or any other process limitation that may be imposed. Generally there is a maximum and a minimum size beyond which a system cannot handle.

For manual medium handling, shape is less of a consequence so long as accurate locating points are used; holes, other precision surfaces, or, with the use of "vision", at least two acceptable and suitable located fiducials. These should be placed as shown in Fig. 2-23. For automatic medium handling, there also needs to be a rectangular shape involving parallel edges and squareness. For typical suggested values see the table in Fig. 2-24 and the supporting illustration in Fig. 2-25.

Areas which cannot be populated: Referring again to Figures 2-23 and 2-25, note the shaded portions which represent unpopulatable areas due to the edge handling means. If the edges themselves are to be used for medium location, a ceramic substrate for example, then the same three points must be used for all operations on that side where precision location is a requirement.

If the medium is located by means of tooling holes on a PCB, for example, the hole-locating and clearance recommendations shown in Figures 2-24 and 2-25 should be followed.

Preferred coordinate reference system: First-quadrant dimensioning is preferred, as most machines use positive X and positive Y positioning. If the medium is a PCB with drilled holes used for both locating and insertion, the

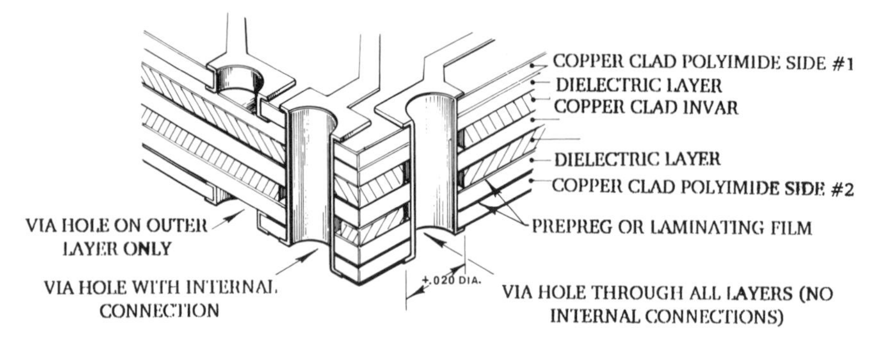

Fig. 2-23: Blind vias will add cost to the circuit board fabrication.

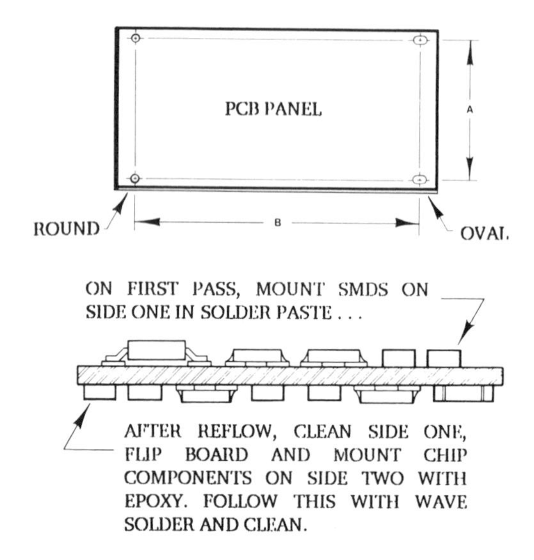

Fig. 2-24: This tooling-hole pat- tern allows for placement of SMDs on both sides of the board.

two reference lines should include the centers of both reference holes in one axis, and the other axis should intercept the center of one of these holes. This approach is much preferred to having an offset or "phantom zero," as a reference coordinate system because this approach can allow two sets of tolerances between the reference holes and the insertion holes and/or the placement pad locations. By making the reference holes coincide with the coordinate reference system, only one tolerance is allowed to the insertion holes or placement pads.

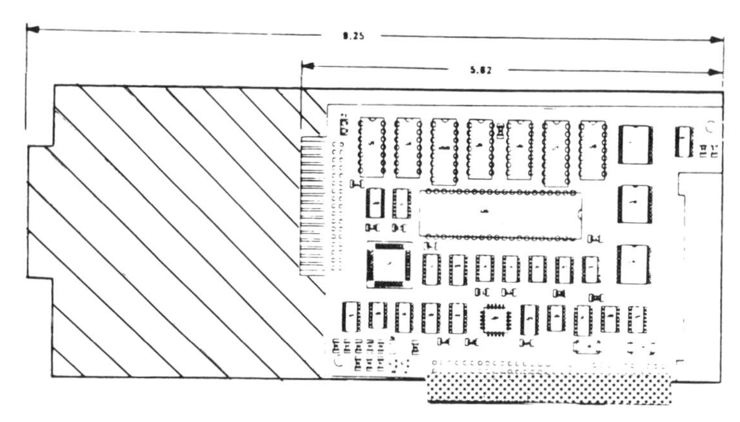

Fig. 2-25: PCB hole-locating and clearance recommendations.

Estimating Total Component Area

The entire footprint dimensions of the SMD must be included in the preliminary planning of the board. The component body dimensions alone do not furnish accurate information which is vital for determining board space. See Fig. 2-26.

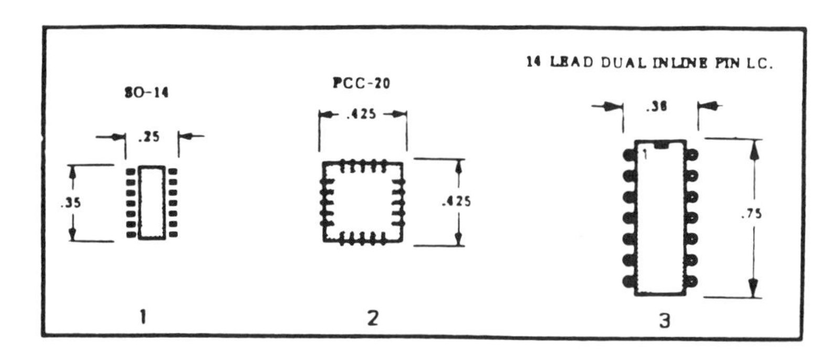

Fig. 2-26: The entire footprint dimensions of the SMD must be included in the preliminary planning.

A percentage multiplier is then added to the footprint area to allow for conductor traces and feedthrough on via hole pads. For example, detail No. 1 in Fig. 2-26 has a x 2 multiplier; No. 2 is multiplied by 1.7; and No. 3 has a multiplier of 1.5 (and on complex assemblies will probably require a multilayer).

Choosing components that maximize efficient use of space will best facilitate interconnection. Planning the areas required for the assembly can be estimated by using the space analysis sheet in Fig. 2-27.

SPACE ANALYSIS SHEET

Density factor: 1-comp. area x 2.00 2- comp. area x 1.75 3- comp. area x 1.50

	Component Footprint	Basic Area	Comp. Ft. Print W/Clearance	Full Area	Item Qty.	Total Area	x	
SOT-23	.120 x .125	.015	.145 x .150	.022				
SOT-89	.180 x .220	.040	.205 x .245	.050				
0805-C	.060 x .140	.008	.085 x .165	.014				
0805-R	.050 x .140	.007	.075 x .165	.012				
1206-C	.072 x .186	.013	.102 x .211	.022				
1206-R	.062 x .186	.012	.087 x .211	.018				
1210-C	.110 x .186	.020	.135 x .211	.028				
1812-C	.130 x .250	.032	.155 x .275	.043				
1005-C	.060 x .160	.010	.085 x .185	.016				
1005-R	.050 x .160	.008	.075 x .185	.014				
MELF-34	.067 x .206	.014	.092 x .231	.021				
MELF-41	.100 x .265	.027	.125 x .290	.036				
	x		x					
SO-8	.260 x .200	.052	.310 x .250	.078				
SO-14	.260 x .350	.091	.310 x .375	.116				
SO-16	.260 x .400	.104	.310 x .450	.140				
SO-16L	.450 x .410	.185	.500 x .460	.230				
SO-20	.450 x .510	.230	.500 x .560	.280				
SO-24	.450 x .610	.275	.500 x .660	.330				
SO-28	.450 x .710	.320	.500 x .760	.380				
	x		x					
	x		x					
PCC-18A	.350 x .500	.175	.400 x .550	.200				
PCC-18B	.350 x .570	.200	.400 x .620	.248				
PCC-20	.425 x .425	.181	.475 x .475	.226				
PCC-28	.525 x .525	.276	.575 x .575	.331				
PCC-44	.725 x .725	.526	.775 x .775	.601				
PCC-52	.825 x .825	.681	.875 x .875	.766				
PCC-68	1.025 x 1.025	1.051	1.075 x 1.075	1.156				
	x		x					
DIP-8	.360 x .400	.144	.460 x .500	.230				
DIP-14	.360 x .755	.272	.460 x .855	.393				
DIP-16	.360 x .785	.282	.460 x .885	.407				
DIP-24	.660 x 1.250	.825	.760 x 1.350	1.026				
DIP-28	.660 x 1.440	.950	.760 x 1.540	1.170				
DIP-40	.660 x 2.090	1.379	.760 x 2.190	1.664				

Fig. 2-27: A space analysis sheet will facilitate planning areas for assembly.

Optimizing Component Placement

During this initial design phase, the optimum location of components should be planned for efficient interconnection of related devices. The interconnection of components on the component side of the PC board can present a challenge to the designer. It is this interconnection factor which requires that careful attention be given to the relationship, the orientation and the placement of the SMDs.

The arrangement of components in a functional position to each other is only one consideration in the planning of a SMD PC board. The left layout in Fig. 2-28 may have the most direct interconnection possible. However, it becomes a costly assembly due to the arrangement of the components. Remember that assembly time increases each time the direction changes on a particular component.

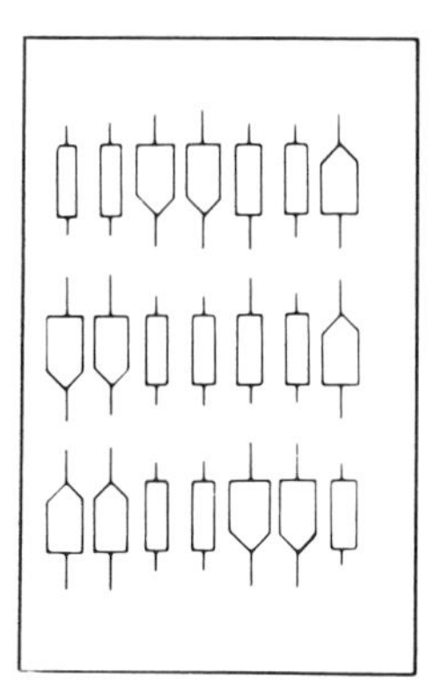

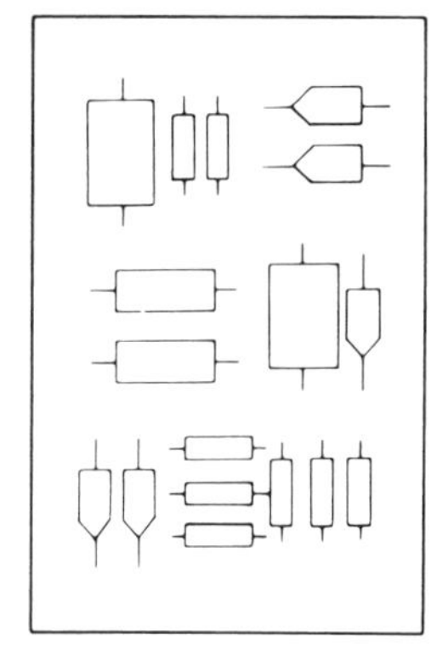

Fig. 2-28: The layout on the left may have the most direct interconnections possible, but assembly would be costly. The layout on the right is planned for cost savings.

Component Orientation

The proper component orientation is vital to maintaining an easy, cost-effective assembly. Both component orientation and signal paths are planned to take advantage of the surface area for the majority of the interconnections. Interfacing connectors usually determine the placement of the ICs in order to reduce excessive crossover of signal traces. Related or interactive devices are then grouped into functional clusters to make the most direct interconnection possible.

The correct orientation of components will speed up the assembly process, a cost-saving factor that must be considered at all times during the design phase of the project. Some low and medium volume placement equipment will allow for a 360 degree rotation of parts, which gives the designer flexibility in design placement.

Changing direction on some high speed placement equipment does, however, create additional cycle time during the process—since the assembly must be withdrawn, turn 90 degrees and placed back into the machine before continuing.

Utilizing Both Sides of Substrate

Mixing two assembly technologies on the same board has become common practice for some time-consuming discrete components. This has typically involved placing simple components like chip resistors, chip capacitors and SOT transistors on the reverse side of a standard insertion mount board. See Fig. 2-29. The assembly process is as follows:

1. Apply adhesive for surface mount components.
2. Place surface mount components.
3. Stuff insertion mount components.
4. Preheat.
5. Foam flux.
6. Wave solder.
7. Clean.

When it is necessary to mix surface mount and insertion mount components on the same board, but the surface mount components cannot be readily wave soldered, one solution is to mount both types of components on the top of the board and use both solder paste reflow and wave soldering. The assembly process is as follows:

1. Apply solder paste for surface mount components.
2. Place surface mount components.
3. Reflow solder paste.
4. Clean.
5. Stuff insertion mount components.
6. Preheat.
7. Foam flux.
8. Wave solder.
9. Clean.

Note that the surface mount solder joints typically will not reflow during the wave soldering. If they do, the components will stay in position due to the surface tension of the solder.

A third method sometimes used to mix surface mount and insertion mount components is to socket the surface mount components using a solder tail socket. This method is typically used only when a few high leadcount devices that only come in surface mount packages are to be mixed with standard insertion mount components.

The "rat's nest" technique of drawing lines to connecting ICs is a good way to refine component placement on complex logic circuits. Keep in mind that when leaded components are to be used, your assembly will generally be passed through a wave solder process. By mounting the majority of the chip components on the wave solder side, you can free up more surface area on the component side. The chip components are generally no higher off the board surface than the leaded ends. The resistors and capacitors mounted on the wave solder side are held in place by adhesive or UV cured epoxy. Locating surface-mounted ICs under leaded devices will maximize the area and the volume of your PC board as shown in Fig. 2-30.

Note: Beware of an increased occurrence of solder bridging between IC pins when wave soldered. Some touch-up will be required even with advanced wave solder techniques and secondary hot air processes.

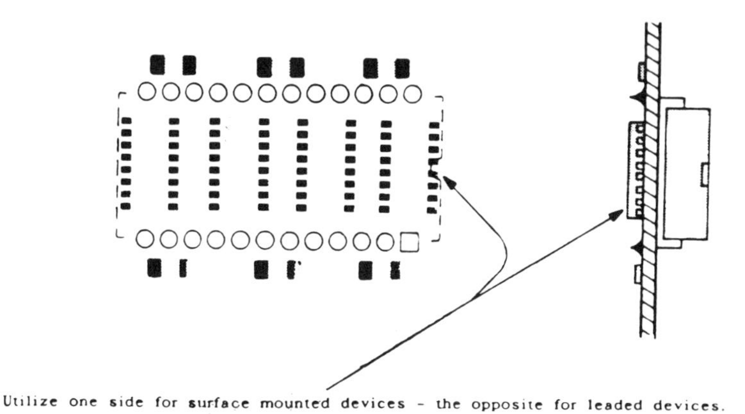

Fig. 2-30: Locating SMDs under leaded devices will maximize the work area.

When it becomes necessary to mix SMDs with leaded devices, one option is the two-sided assembly. The majority of the active components will be mounted on side one of the module, while chip components will be mounted on the opposite side—the solder side of the assembly.

Careful placement of chip components will take advantage of the small spaces not available to their bulky counterparts. By placing surface mounted parts in close proximity to the related leaded parts, the connection is kept short and allows for the more efficient use of the top side of the PC board. This also leaves the opposite side of the PC board free for interfacing with other leaded components. Details are shown in Fig. 2-31.

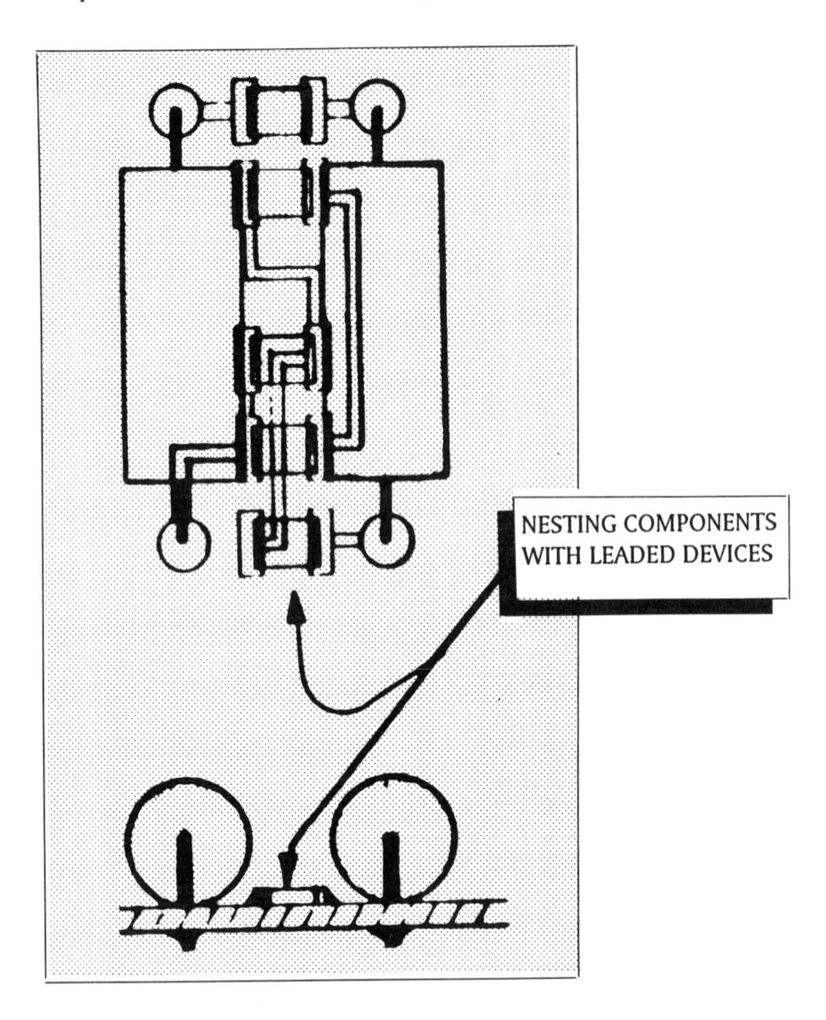

Fig. 2-31: Place SMDs as close as possible to leaded components; this will keep the connections short and also allow for more efficient use of the top side of the PCB.

Mixing leaded components on the SMT assembly will, of course, add additional steps to the soldering process and could require the use of more complex assembly equipment in an automated line.

Thermal Considerations

Thermal characteristics of integrated circuit (IC) packages have always been a major consideration to both designers and users of electronics products. This is because an increase in junction temperature (T_J) can have an adverse effect on the long term operating life of an IC. The advantages realized by miniaturization can often have trade-offs in terms of increased junction temperatures. Some of the *variables* affecting T_J are controlled by the *producer* of the IC, while others are controlled by the *user* and the *environment* in which the device is used.

With the increased use of SMT, management of thermal characteristics remains a valid concern because not only are the SMD packages much smaller, but the thermal energy is concentrated more densely on the printed wiring board (PWBP). For these reasons, the designer and manufacturer of surface mounted assemblies (SMAs) must be more aware of all the variables affecting T_J.

Power Dissipation

Power dissipation (P_D), varies from one device to another and can be obtained by multiplying V_{CC} Max by typical I_{CC}. Since I_{CC} decreases with an increase in temperature, maximum I_{CC} values are not used.

Thermal Resistance

The ability of the package to conduct this heat from the chip to the environment is expressed in terms of thermal resistance. The term normally used is Theta JA (θ_{JA}). θ_{JA} is often separated into two components: thermal resistance from the junction to case, and the thermal resistance from the case to ambient. θ_{JA} represents the total resistance to heat flow from the chip to ambient and is expressed as follows:

$$\theta_{JC} + \theta_{CA} = \theta_{JA}$$

Junction Temperature (T_J)

Junction temperature (T_J) is the temperature of a powered IC measured at the substrate diode. When the chip is powered, the heat generated causes the T_J to rise above the ambient temperature (T_A). T_J is calculated by multiplying the power dissipation of the device by the thermal resistance of the package and adding the ambient temperature to the result.

$$T_J = (P_D \text{ x } \theta_{JA}) + T_A$$

Factors Affecting θ_{JA}

There are several factors which affect the thermal resistance of any IC package. Effective thermal management demands a sound understanding of all these variables. Package variables include the leadframe design and materials, the plastic used to encapsulate the device, and to a lesser extent other variables such as the die size and die attach methods. Other factors that have a significant impact on the θ_{JA} include the substrate upon which the IC is mounted, the density of the layout, the air-gap between the package and the substrate, the number and length of traces on the board, the use of thermally conductive epoxies, and external cooling methods.

Package Considerations

Studies with dual-in-line plastic (DIP) packages over the years have shown the value of proper leadframe design in achieving minimum thermal resistance. SMD leadframes are smaller than their DIP counterparts (see Figures 2-32a and 2-32b). Because the same die is used in each of the packages, the die-pad, or flag, must be at least as large in the SO as in the DIP.

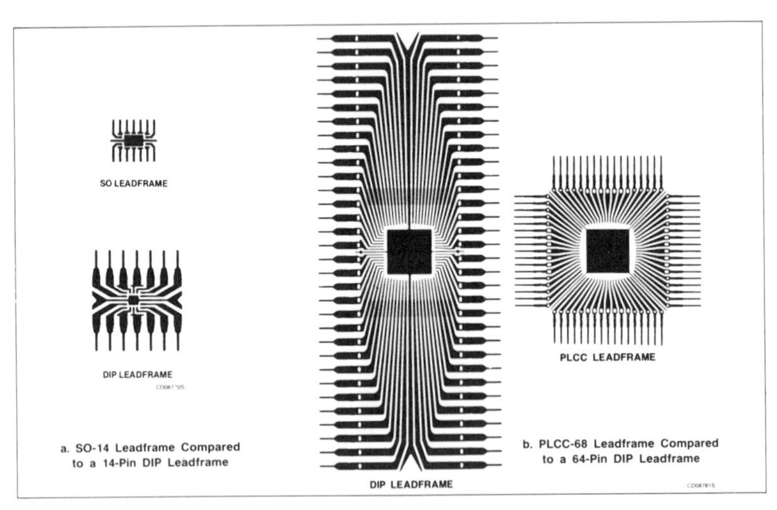

Fig. 2-32: Comparison of SO, SO-14 and PLCC-68 leadframes to a DIP leadframe.

While the size and shape of the leads have a measurable effect on θ_{JA}, the design factors that have the most significant effect are the die-pad size and the tie-bar size. With sign constraints caused by both miniaturization and the need to assemble packages in an automated environment, the internal design of an SMD is much different than in a DIP. However, the design in one that strikes a balance between the need to miniaturize, the need to automate the assembly of the package, and the need to obtain optimum thermal characteristics.

Lead Frame Material is one of the more important factors in thermal management. For years the DIP leadframes were constructed out of Alloy-42. These leadframes met the producers' and users' specifications in quality and reliability. However three to five years ago, the leadframe material of DIPs was changed from Alloy-42 to Copper (CLF) in order to provide reduced θ_{JA} and extend the reliable temperature-operating range. While this change has already taken place for the DIP, it is still taking place for the SO package. This change to CLF will produce dramatic results in the θ_{JA} of SO packages. All PLCCs are assembled with copper leadframes.

The Molding Compound is another factor in thermal management. The compound most often used is the same high purity epoxy used in DIP packages (at present, HC-10, Type II.) This reduces corrosion caused by impurities and moisture.

Other Factors often considered are the die size, die attach methods, and wire bonding. Tests have shown that die size has a minor effect on θ_{JA}.

While there is a difference between the thermal resistance of the silver-filled adhesive used for die attach and a gold silicon eutectic die attach, the thickness of this layer (1-2 mils) is so small as to make the difference insignificant.

Gold wire bonding in the range of 1.0 to 1.3 mils does not provide a significant thermal path in any package.

In summary, the SMD leadframe is much smaller than in a DIP and, out of necessity, is designed differently; however, the SMD package offers an adequate θ_{JA} for all moderate power devices. Further, the change to CL will reduce the θ_{JA} even greater margin of reliability.

Thermal Resistance Measurements

The graphs illustrated on pages to follow show the thermal resistance of Signetics' SMD devices. These graphs give the relationship between θ_{JA} (junction-to-ambient) or θ_{JC} (junction-to-case) and the device die size. Data is also provided showing the difference between still air (natural convection cooling) and air flow (forced cooling) ambients. All θ_{JA} tests were run with the SMD device soldered to test boards (See the Test Ambient section for details). It is important to recognize that the test board is an essential part of the test environment and the boards of different sizes, trace layouts or compositions may give different results from this data. Each SMD user should compare their system to the following test system and determine if the data is appropriate or needs adjustment for their application.

Test Method

One method of testing is commonly called the TSP (temperature sensitive parameter) method. This method meets MIL-STD 883C, Method 1012.1. The basic idea of this method is to use the forward voltage drop of a calibrated diode to measure the change in junction temperature due to a known power dissipation. The thermal resistance can be calculated using the following equation:

$$\theta_{JA} = \Delta T_J/P_D = T_J - T_A/P_D$$

TSP Calibration

The TSP diode is calibrated using a constant temperature oil bath and constant current power supply. The caliber temperatures used are typically 25°C and 75°C and are measured to an accuracy of ±0.1°C. The calibration current must be kept low to avoid significant junction heating, data given in this report used constant currents of either 1.0mA or 3.0mA. The temperature coefficient (K-Factor) is calculated using the following equation:

$$K = T_2 - T_1/V_{F2} - V_{F1}$$
$$I_F = \text{Constant}$$

Where:

- K = Temperature Coefficient (°C/mV)
- T_2 = Higher Test Temperature (°C)
- T_1 = Lower Test Temperature (°C)
- V_{F2} = Forward Voltage at I_F and T_2
- V_{F1} = Forward Voltage at I_F and T_1
- I_F = Constant Forward Measurement Current

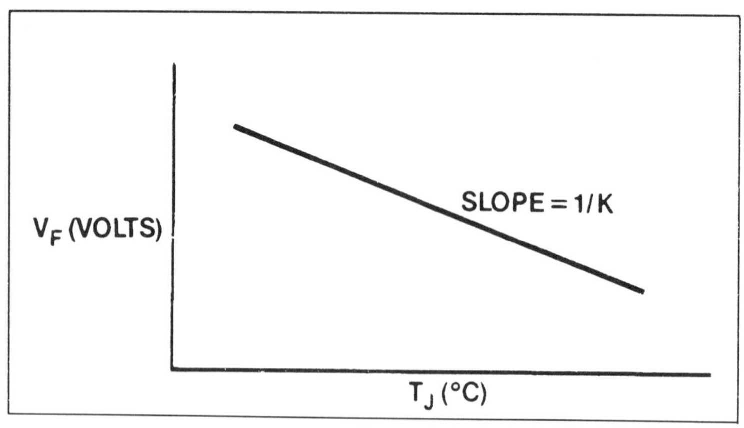

Fig. 2-33: Forward voltage — Junction Temperature characteristics of a semiconductor junction operating at a constant current. The K factor is the reciprocal of the slope.

Thermal Resistance Measurement

The thermal resistance is measured by applying a sequence of constant current and constant voltage pulses to the device under test. The constant current pulse (same current at which the TSP was calibrated) is used to measure the forward voltage of the TSP. The constant voltage pulse is used to heat the part. The measurement pulse is very short (less than 1% of cycle) compared to the heating pulse (greater than 99% of cycle) to minimize junction cooling during measurement. This cycle starts at ambient temperature and continues until steady-state conditions are reached. The thermal resistance can be calculated using the following equation:

$$\theta_{JA} = \Delta T_J/P_D = K\,(V_{FA} - V_{FS})/V_H \times I_H$$

Where:

V_{FA}	=	Forward voltage of TSP at ambient temperature (mV)
V_{FS}	=	Forward voltage of TSP at Steady-State Temperature (mV)
V_H	=	Heating Voltage (V)
I_H	=	Heating Current (A)

Test Ambient

All θ_{JA} test data collected here was obtained with the SMD devices soldered to either Philips SO Thermal Resistance Test Boards or Signetics PLCC Thermal Resistance Test Boards. SO devices are set at 8 - 9 mil stand-off and SO boards use one connection pin per device lead. PLCC boards generally use 2 - 4 connection pins regardless of device lead count. Fig. 2-34 shows a cross-section of an SO part soldered to a test board.

The still air tests were run in a box having a volume of 1 cubic foot of air at room temperature. All devices were soldered on test boards and held in a horizongal test position. The test boards were held in a Textool SIF socket with 0.16″ stand-off. Fig. 2-35 shows the air flow test setup.

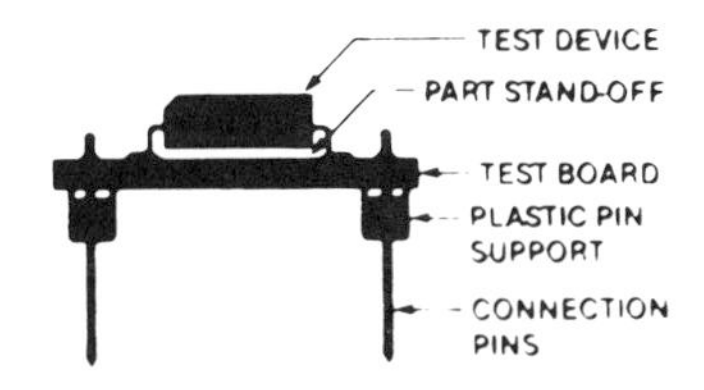

Fig. 2-34: Cross-section of test device soldered to test board.

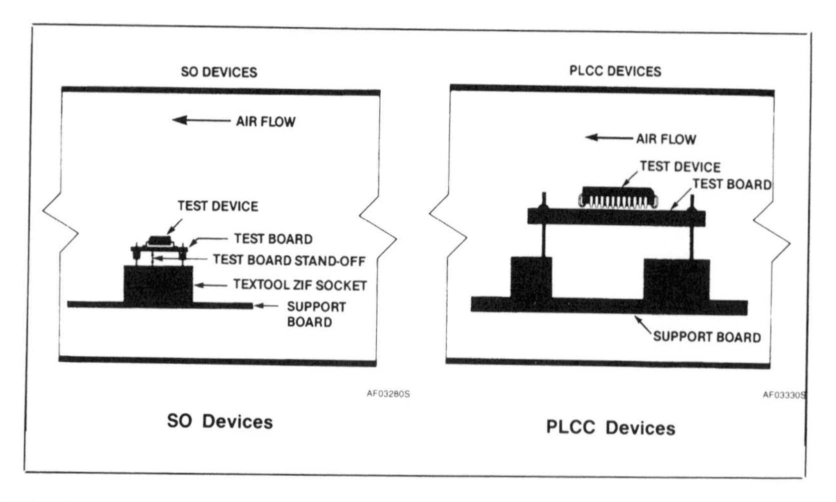

Fig. 2-34: Air flow test setup.

θ_{JC} Tests

The θ_{JC} test is run by holding the test device against an infinite heat sink (water cooled block approximately 4 inch x 7 inch x 0.75 inch) to give a θ_{CA} (case-to-ambient) approaching zero. The copper heat sink is held at a constant temperature (20°C)and monitored with a thermocouple (0.040″ diameter sheath, grounded junction type K) mounted flush with heat sink surface and centered below die in the test device. Figure 2-35 shows theθ_{JC} test mounting for a PLCC device.

SO devices are mounted with the bottom of the package held against the heat sink. This is achieved by bending the device leads straight out from the package body. Two small wires are soldered to the appropriate leads for tester connection. Thermal grease is used between the test device and heat sink to assure good thermal coupling.

PLCC devices are mounted with the top of the package held against the heat sink. A small spacer is used between the hold-down mechanism and PLCC bottom pedestal. Small hook up wires and thermal grease are used as with the SO setup. Figure 2-35 shows the PLCC mounting.

Data Presentation

The data presented here was run at constant power dissipation for each package type. The power dissipation used is given under Test Conditions for each graph. Higher or lower power dissipation will have a slight effect on thermal resistance. The general trend of thermal resistance decreasing with increasing power is common to all packages. Figure 2-36 shows the average effect of power dissipation of SMD θ_{JA}.

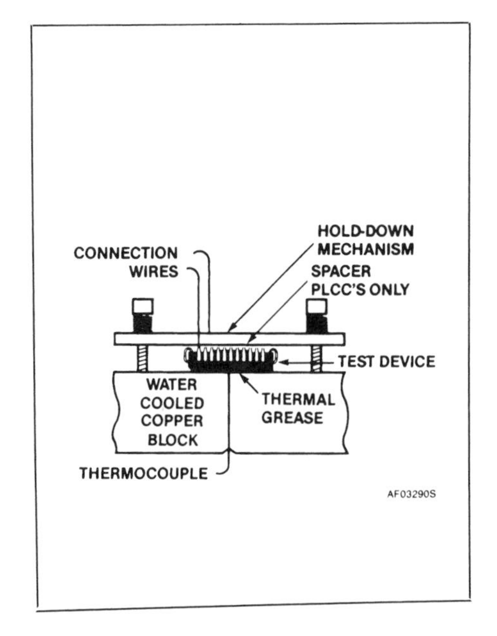

Fig. 2-35: Test setup with PLCC device.

Thermal resistance can also be affected by slight variations in internal leadframe design such as pad size. Larger pads give slightly lower thermal resistance for the same size die. The data presented represents the typical leadframe/die combinations with large die on large pads and small die on smallpads. The effect of leadframe design is within the ± 15% accuracy of these graphs.

SO devices are currently available in both copper or alloy 42 leadframes; however, many firms are converting to copper only. PLCC devices are only available using copper leadframes.

The average lowering effect of air flow on SMD θ_{JA} is shown in Figure 2-37.

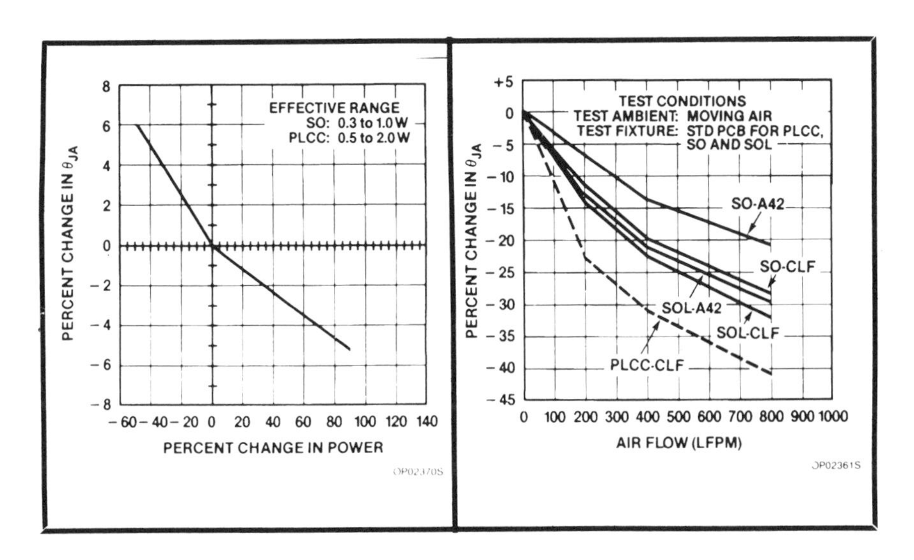

Fig. 2-36: Average effect of power dissipation on SMDs.

Fig. 2-37: Average effect of air flow on SMDs.

Thermal Calculations

The approximate junction temperature can be calculated using the following equation:

$$T_J = (\theta_{JA} \times P_D) + T_A$$

Where:

T_J = Junction Temperature (°C)
θ_{JA} = Thermal Resistance Junction-to-Ambient (° C/W)
P_D = Power Dissipation at a T_J
T_A = Temperature of Ambient(degree C)

Example: Determine approximate junction temperature of SOL-20 at 0.5W dissipation using 10,000 sq. mil die and copper leadframe in still air and 200 LFPM air flow ambients. Given T_A = 30°C.,

1. Find θ_{JA} for SOL-20 using 10,000 sq. mil die and copper leadframe from typical θ_{JA} data—SOL-20 graph.
   Answer: 88°C/W @ 0.7W

2. Determine θ_{JA} @ 0.5W using Average Effect of Power Dissipation on AMD θ_{JA}, Figure 2-36.

   Percent change in Power =

   $$0.5W - 0.7W/0.7W \times 100 = -28.6\%$$

   From Figure 2-36:
   28.6% change in power gives 3.5% increase in θ_{JA}
   Answer:
   88°C/W + (88 x 0.035) = 91°C/W @ 0.5W

3. Determine θ_{JA} @ 0.5W in 200 LFPM air flowfrom Average Effect of Air Flow on SMD θ_{JA}, Figure 2-37.
   From Fig. 2-37:
   200 LFPM air flow gives 14% decrease in θ_{JA}
   Answer:
   91°C/W - (91 x 0.14) = 78°C/W

4. Calculate approximate junction temperature
   Answer:
   T_J(still air) = (91°C/W x 0.5W) + 30 = 76°C
   T_J (200 LFPM) = (78 degrees C/W x 0.5W) + 30 = 69 degrees C

System Considerations

With the increases in layout density resulting from surface mounting with much smaller packages, other factors become even more important. *The user is in control of these factors.*

One of the most obvious factors is the substrate material on which the parts are mounted. Environmental constraints, cost considerations and other factors come into play when choosing a substrate. The choice is expanding rapidly, from the standard glass epoxy PWB materials and ceramic substrates to flexible circuits, injection molded plastics, and coated metals. Each of these has its own thermal characteristics which must be considered when choosing a substrate material.

Studies have shown that their air gap between the bottom of the package and the substrate has an effect on θ_{JA}. The larger the gap, the higher the θ_{JA}. Using thermally conductive epoxies in this gap can slightly reduce the θ_{JA}.

It has long been recognized that external cooling can reduce the junction temperatures of devices by carrying heat away from both the devices and the board itself. Manufacturers have done studies on the effects of external cooling on boards with SO packages. Sample results are shown in Figure 2-38.

The designer should avoid close spacing of high power devices so that the heat load is spread over as large an area as possible. Locate components with a higher junction temperature in the cooler locations on the PCBs.

The number and size of traces on a PWB can affect θ_{JA} since these metal lines can act as radiators, carrying heat away from the package and radiating it to the ambient. Although the chips themselves use the same amount of energy in either a DIP or an SO package, the increased density of a Surface Mounted Assembly concentrates the thermal energy into a smaller area.

It is evident that nothing is free in PWB layout. More heat concentrated into a smaller area makes it incumbent on the system designer to provide for the removal of thermal energy from his system.

Large conductor traces on the PCB conduct heat away from the package faster than small traces. Thermal vias from the mounting surface of the PCB to a large area ground plane in the PCB reduces the heat buildup at the package.

In addition to the package's thermal considerations, thermal management requires one to at least be aware of potential problems caused by mismatch in thermal expansion.

The very nature of the SMD assembly, where the devices are soldered directly onto the surface, not through it, results in a very rigid structure. If the substrate material exhibits a different thermal coefficient of expansion (TCE) than the IC package, stresses can be setup in the solder joints when they are subjected to temperature cycling (and during the soldering process itself) that may ultimately result in failure.

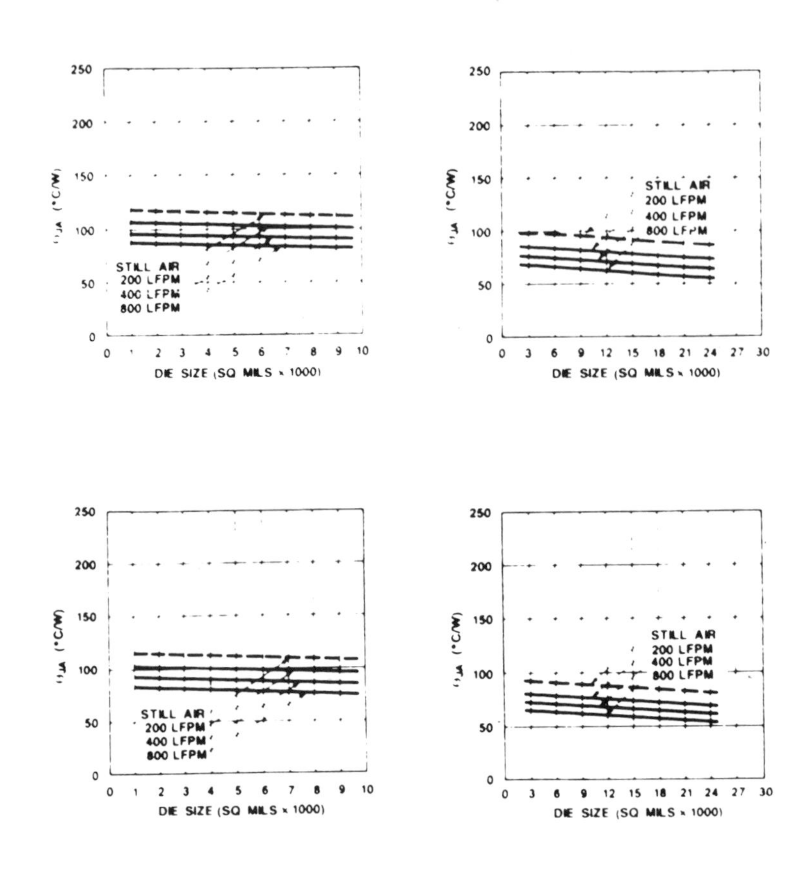

Fig. 2-38: Results of air flow on θ_{JA} on various SMDs with copper leadframe. (a) SO-14; (b) SOL-16; (c)SO-16; (d)SOL-20.

Because some of the boards assembled will require the use of Leadless Ceramic Chip Carriers (LCCCs), TCE must be understood. As will be seen below, TCE is less of a problem with the commercial SMD packages with leads.

Take the example of a leadless ceramic chip carrier with a TCE of about 6 x 10^{-6} power/K soldered to a conventional glass-epoxy laminate with a TEC in the region of 16 x 10^{-6}/K. This thermal expansion mismatch has been shown to fracture the solder joints during thermal cycling. Substrate materials with

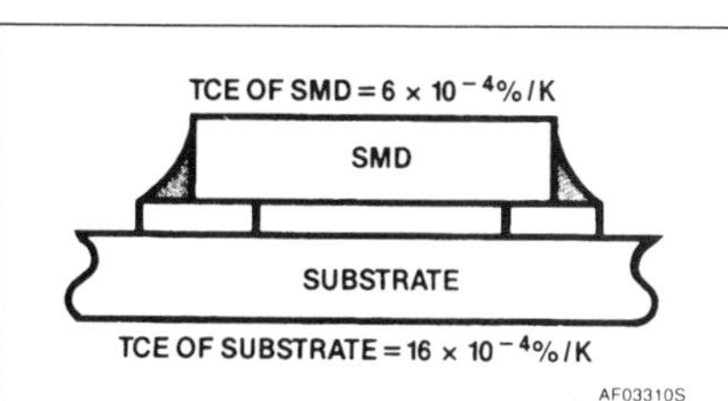

Figure 2-39: The basic problem of thermal expansion mismatch is that the substrate and component may each have different thermal coefficients of expansion.

NOTE: Data provided by N.V. Philips

matched TCEs should be evaluated for these SMD assemblies to avoid problems caused by thermal expansion mismatch.

The stress level associated with thermal expansion and contraction of small SMDs such as capacitors and resistors, where the actual change in length is small, are normally rather low. However, as component sizes increase, stresses can increase substantially.

Thermal expansion mismatch is unlikely to cause too many problems in systems operating in benign environments; but, in harsher conditions, such as thermal cycling in military or avionic applications, the mechanical stresses setup in solder joints due to the different TCEs of the substrate and the component are likely to cause failure.

The basic problem is outlined in Figure 2-39. The leadless SMD is soldered to the substrate as shown, resulting in a very rigid structure. If the substrate material exhibits a different TCE from that of the SMD material, the amount of expansion for each will differ for any given increase in temperature. The soldered joint will have to accommodate this difference, and failure can ultimately result. The larger the component size, the higher the stress levels so that this phenomenon is at its most critical in applications requiring large LCCCs with high pin-counts. To address this problem, three basic solutions are emerging. First, the use of leadless ceramic chip carriers can sometimes be avoided by using leaded devices; the leads can flax and absorb the stress. Second, when this solution is not feasible, the stresses can be taken up by inserting a compliant elastomeric layer between the ceramic package and the epoxy glass substrate. Third, TCE values of component and substrate can be matched.

Using Leaded Devices (SO, SOL & PLCC)

The current evolution in commercial electronics includes the adoption of the commercial SMD packages, i.e. SO with gull-wing leads or the PLCC with rolled-under J-leads, rely on the compliance of the leads themselves to avoid any serious problems of thermal expansion mismatch. At elevated temperatures, the leads flex slightly and absorb most of the mechanical stress resulting from the thermal expansion differentials.

Similarly, leaded holders can be used with LCCCs to attach them to the substrate and thus absorb the stress.

Unfortunately, using a lead does not always ensure sufficient compliancy. The material from which the lead is made, and the way it is formed and soldered can adversely affect it. For example, improper soldering techniques, which cause excess solder to over-fill the bend of the gull-wing lead of an SO can significantly reduce the lead's compliancy.

Compliant Layer

This approach introduces a compliant layer onto the interface surface of the substrate to absorb some of the stresses. A 50um thick elastomeric layer is bonded to the laminate. To make contacts, carbon or metallic powders are introduced to form conductive stripes in the nonconductive elastomer material. Unfortunately, substrates using this technique are substantially more expensive than standard uncoated boards.

Another solution is to increase the compliancy of the solder joint. This is done by increasing the stand-off height between the underside of the component and the substrate. To do this, a solder paste containing lead or ceramic spheres which do not melt when the surrounding solder reflows, thus keeping the component above the substrate can be used.

Matching TCE

There are two ways to approach this solution. The TCE of the substrate laminate material can be matched to that of the LCCC either by replacing the glass fibers with fibers exhibiting a lower TCE (composites such as epoxy-Kevlar or polyimide-Kevlar and polyimide-quartz), or by using low TCE metals (such as Invar, Kovar, or molybdenum).

This latter approach involved bonding a glasspolyimide or a glass-epoxy multilayer to the low TCE restraining core material. Typical of such materials are copper-Invar-copper, Alloy-42, copper-molybdenum-copper, and copper-graphite. These restraining-core constructions usually require that the laminate be bonded to both sides to form a balanced structure so that they will not warp or twist.

This inevitably means an increase in weight, which has always been a negative factor in this approach. However, the SMD substrate can be smaller and components more densely packed in many cases overcoming the weight disadvantages. On the positive side, the material's high thermal conductivity helps to keep the components cool. More over, copper-clad-Invar lends itself readily to moisture-proof multilayering for the creation of ground and power planes and for creation of ground and power planes and for providing good inherent EMI/RFI shielding.

Kevlar is lighter and widely used for substrates in military applications; but, it suffers from a serious drawback which, although overcome to a certain extent by careful attention to detail, can cause problems, The material, when laminated, can absorb moisture and chemical processing fluids around the edges. Thermal conductivity, machinability and cost are not as attractive as for copper-clad Invar.

For the majority of commercial substrates however, where the use of ceramic chip carriers in any quantity is the exception rather than the rule, and when adequate cooling is available, the mismatch of TCEs poses little or no problem. For these substrates traditional FR-4 glass-epoxy and phenolic-paper will no doubt remain the most widely used materials.

Although FR-4 epoxy-glass has been the traditional material for plated-through professional substrates, it is a phenolic-paper laminate (FR-2) which finds the widest use in consumer electronics. While it is the cheapest material, it unfortunately has the lowest dimensional stability, rendering it unsuitable for the mounting of the LCCCs.

Substrate Types

FR-4 glass-epoxy substrates are the most commonly used for commercial electronic circuits. They have the advantage of being cheap, machinable, and lightweight. Substrate size is not limited. On the negative side, they have poor thermal conductivity and a high TCE, between 13 and 17 x 10^{-6}/K. This means they are a poor match to ceramic.

Glass polyimide substrates have a similar TCE range to glass-epoxy boards, but better thermal conductivity. They are, however, three to four times more expensive.

Polyimide Kevlar substrates have the advantage of being lightweight and not restricted in size. Conventional substrate processing methods can be used and its TCE (between 4 and 8), matches that of ceramic. Its disadvantages are that it is expensive, difficult to drill and is prone to resin microcracking and water absorption.

Polyimide quartz substrates have a TCE between 6 and 12 making them good match for LCCs. They can be processed using conventional techniques, although drilling vias can be difficult. They have good dielectric properties and compare favorably with FR-4 for substrate size and weight.

Alumina (ceramic) substrates are used extensively for high-reliability military applications and thick-film hybrids. The weight, cost, limited substrate size and inherent brittleness of alumina means that its use as a substrate material is limited to applications where these disadvantages are outweighed by the advantage of good thermal conductivity and a TCE that exactly matches that of LCCCs. A further limitation is that they require Thick-film screening processing.

Copper-clad Invar substrates are the leading contenders for TCE control at present. It can be tailored to provide a selected TCE by varying the copper-to-Invar ration. Figure 2-40 shows the construction of a typical multilayer substrate employing two cores providing the power and ground planes. Plated-through holes provide and integral board-to-board interconnection. The low TCE of the core dominates the TCE of the overall substrate, making it possible to mount LCCCs with confidence.

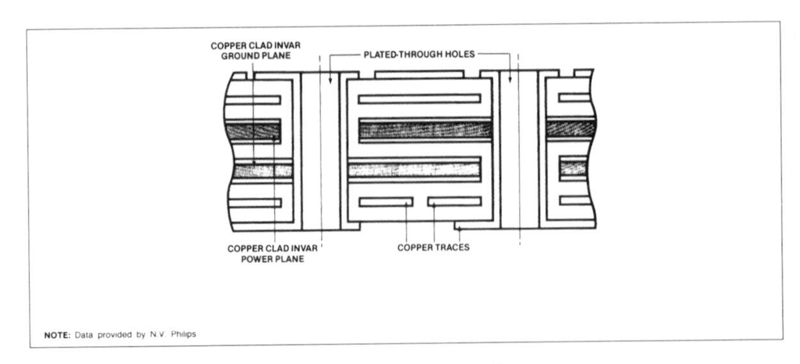

Fig. 2-40: Section through a typical multilayer substrate incorporating copper-clad invar ground and power planes, interconnected via plated-through holes.

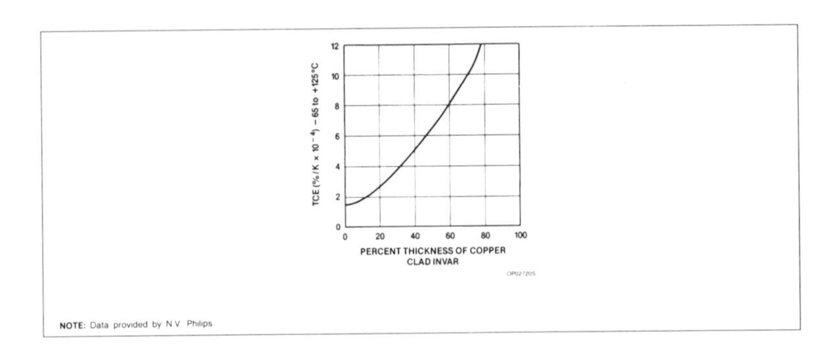

Fig. 2-41: The TCE range of coppper-clad invar as a function of copper thickness.

Because the TCE of copper is high, and that of Invar is low, the overall TCE of the substrate can be adjusted by varying the thickness of the copper layers. Figure 2-41 plots the TCE range of the copper-clad Invar as a function of copper thickness, and shows the TCE range of each several other materials to which the clad material can be matched. For example, if the TCE of Alumina is to be matched, then the core should have about 46% thickness of copper. When this material is used as a thermal mounting plane, it also acts as a heatsink.

Conclusion

Thermal management remains a major concern of producers and users of ICs. The advent of SMD technology has made a thorough understanding of the thermal characteristics of both the devices and the systems they are used in mandatory. The SMD package, being smaller, does have a higher θ_{JA} than its standard DIP counterpart...even with Copper Lead Frames. That is the major trade-off one accepts for package miniaturization. However, consideration of all the variables affecting IC junction temperatures will allow the user to take maximum advantage of the benefits derived from the use of this technology.

chapter 3

SPECIFYING MATERIALS FOR SUBSTRATES

For many years, glass-reinforced resin laminate or FR-4 has been the primary substrate material for etched copper circuit boards.

This has been ideal for two-sided and multilayer PC boards using plated through-holes to interface one layer with another. The fire-retardant material has become the industry standard for communications, instrumentation, computers and nearly all electronic products requiring high quality performance.

Of course, many alternatives to FR-4 are used in electronic applications. Low-cost consumer products use several of the paper-based laminates on the market. Many etched circuits with very simple functions may require only copper on one surface. Holes for leaded components could be punched rather than drilled through the substrate as shown in Fig. 3-1.

Common Substrate Materials

FR-2: Low cost, flame-retardant paper, the workhorse of consumer electronics when electrical and mechanical properties are not overly demanding.

FR-6: Glass-mat, reinforced polyester. Low cost favorite of the automotive industry.

CEM-1: Paper-glass composite. Best laminate property for price ratio. Can be used to replace other laminate grades to improve yields, cut laminate costs or obtain better process. (Example: replaces single-sided FR 4).

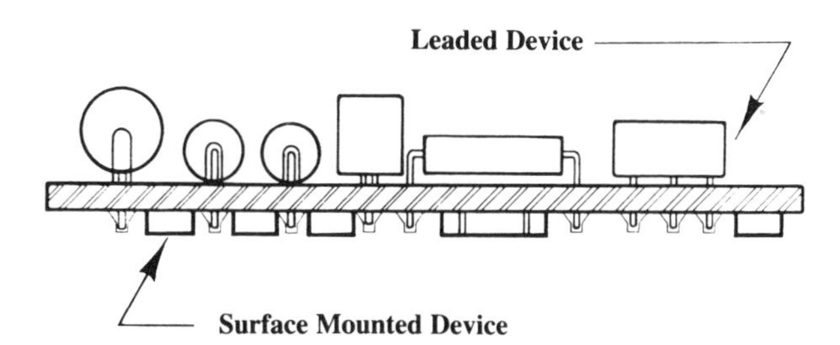

Fig. 3-1: At the designer's option, holes for leaded components may be either punched or drilled.

CRM-5: Composite material using a polyester resin system to bind together a sandwich of random glass mat core and woven glass fabric surfaces.

CEM-3: All glass composite. A new entry into the laminate spectrum. Cheaper than FR-4 with almost equivalent properties. Entering automotive and appliance applications.

FR-4: Fire retardant glass laminate, more expensive than the above materials, but very good. Widely used in computer and telecommunications applications.

In addition to the commercial materials above, the engineer can meet more demanding environments with some of the "high-tech" materials now available.

For hi rel, military, or other special applications, there are complex substrate materials with unique characteristics suitable for the more exotic functions or environments.

Teflon and ceramic substrates are used in many RF applications requiring a very stable dielectric characteristic. The teflon materials are furnished copper clad as are the more conventional laminates and are processed in very much the same way as FR-4.

Ceramic substrates are fabricated using much different techniques. The substrate area is limited and conventional drilling or routing of this material is impossible. The ceramic substrate can be punched and profiled in its green or unfired state; but, after firing of the organic material, holes and additional alterations to its shape require laser technology. The circuit traces are added to the surfaces of the ceramic layers and conductive fillers are drawn into the laser drilled via holes to connect one layer to the other.

Ceramic substrates are used mainly for hybrid circuits. In many cases, IC chips are attached directly to the surface of the ceramic and are terminated using wire bonding techniques similar to those used in IC fabrication. The advantage in using a hybrid circuit is the choice the engineer has in mixing several integrated circuits, transistors and miniature surface mounted devices onto the substrate. The result is a customized, electronic function in a small package. The finished module may have leads attached directly for mounting into a larger PC board or, as in hi rel applications, sealed in a metal housing with glass insulated leads. See Fig. 3-2.

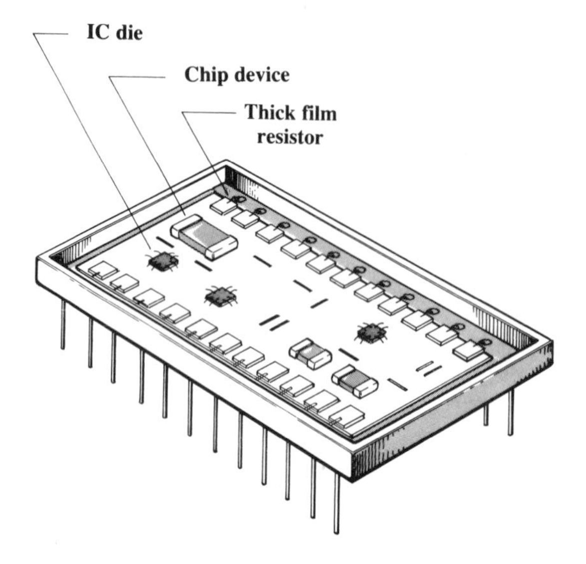

Fig. 3-2: Typical ceramic hybrid substrate.

Other alternatives to the more common laminates are polyamide glass and polyamide quartz materials. More stable than FR-4, the polyamide products can be processed with most of the same techniques used in standard board fabrication. Because the polyamide has a more stable thermal coefficient of expansion than FR-4, it is popular for chip on board applications similar to those used in hybrid circuits. Polymides can be laminated to metal core layers to provide an even more thermally stable substrate surface. Fig. 3-3.

Fabrication and Material Planning

Most commercial products will continue to use epoxy-glass laminate, while paper epoxy will be confined to consumer products.

Substrate materials are furnished to the board fabricator in 36 X 48 inch sheets. The fabricator then cuts the material into smaller panels for processing,

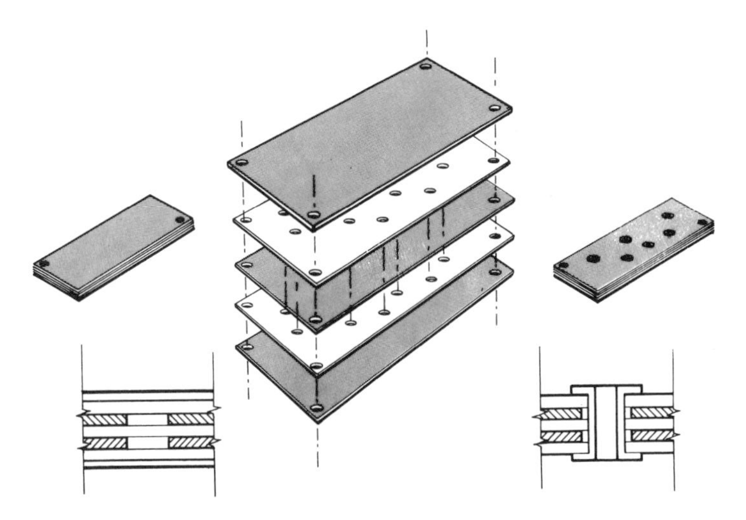

Fig. 3-3: Polyimides can be laminated to metal core layers to provide an even more thermally stable substrate surface.

i.e. 18 X 24 inch, 12 X 18 inch. From these panels, your PC board would be processed into one, two or more PC units per panel. The fabricator will plan material usage carefully to minimize waste. If the designer of the board prepares film in multiple image format, two areas are to be considered:

1. The maximum size board the assembly equipment can handle.
2. Minimizing waste of the board material. See Fig. 3-4

When the board is designed, furnish at least two holes for tooling location pins on each unit. In the case of panelization, two additional tooling holes are required, generally on a break-away tab. The break-away tab and holes will be discarded after assembly processes are completed. The tooling holes in each board will be used in test fixtures on the finished assembly. The tooling hole pattern shown in Fig. 3-5 allows for placement of SMDs on both sides of the board. An optional elongated hole allows the assembly to be extracted from the tooling pins without binding.

Routing to profile the board is one of the last steps in the fabrication process, but planning must be done in advance to insure the best results.

Methods of profiling the individual board shape include rotational cutters, high pressure hydro routing, laser profile, bevel scoring and shearing after assembly. For the latter, adequate clearance to component bodies are necessary for the straight cuts.

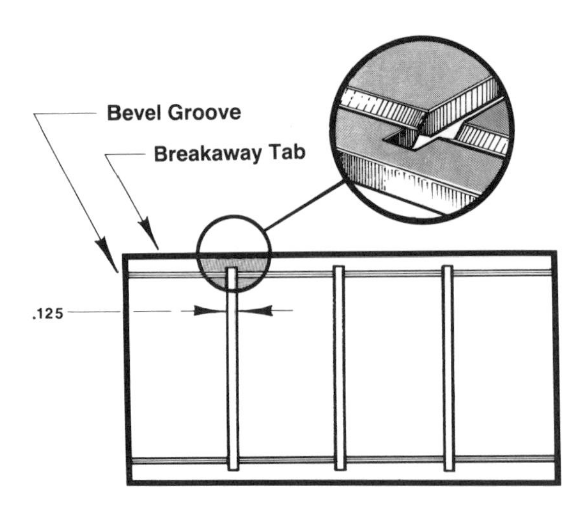

Fig. 3-4: Details of substrate panel. Careful planning is necessary to minimize waste.

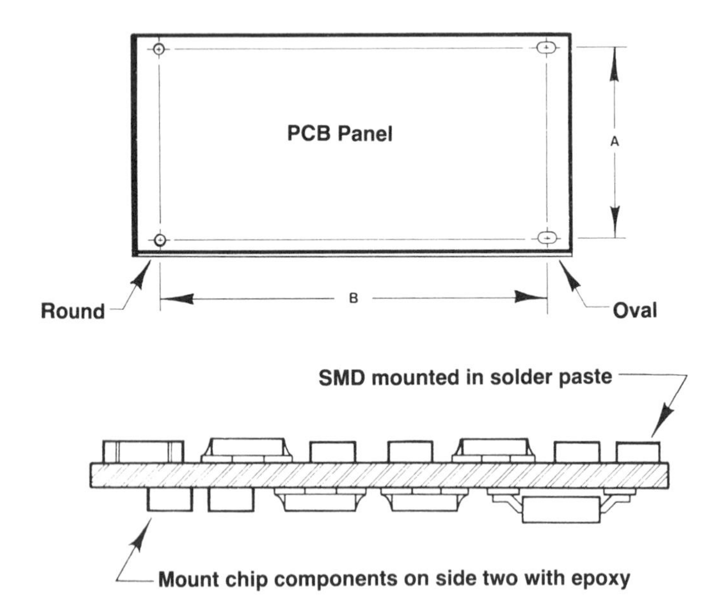

Fig. 3-5: This tooling-hole pattern allows for placement of SMDs on both sides of the board.

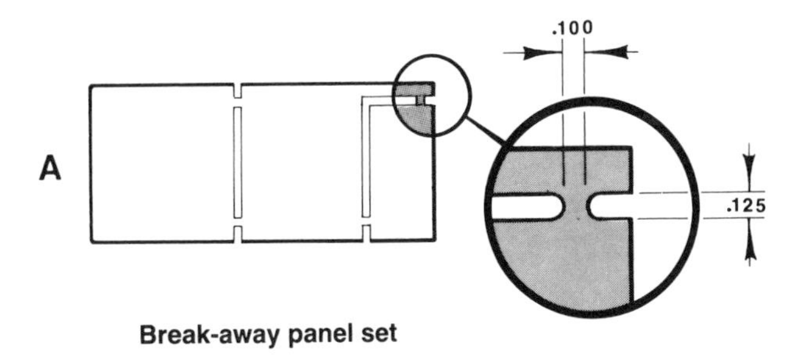

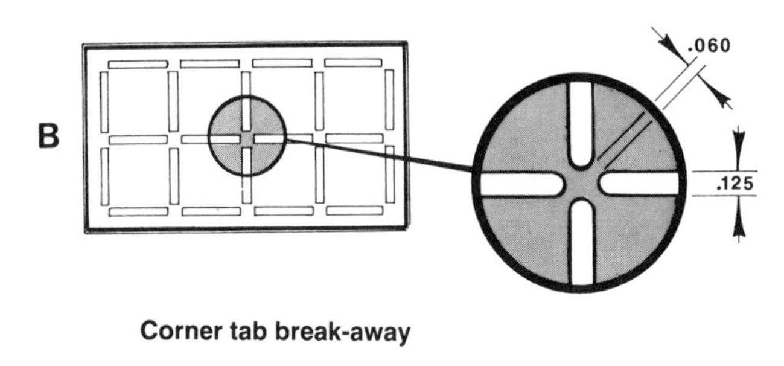

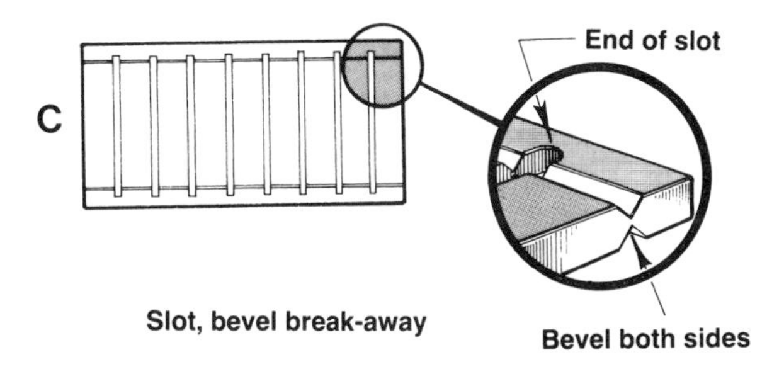

Fig. 3-6: Several variations are used for connecting tabs on PCBs.

The slot separating the break-away tab and the individual boards on a panel should be a minimum of .125 inch wide. Generally a .100 inch diameter router tool is used to create the .125 inch wide slot in two passes to insure greater tooling hole to board edge accuracy. The tabs that retain the board to the panel are usually .100 inch wide and are spaced to adequately support the panel as a unit through the assembly processes. There are several variations used for the connecting tabs. Figure 3-6 illustrates a few.

Die cutting or punching methods are normally reserved for higher volume products and have proven to be efficient means of fabrication. While glass laminates are less than ideal for this method, due to the abrasive nature of the glass fiber material, the paper-base laminates are excellent. Punchpress profiling of the board (Fig. 3-7) may be configured similar to a routed panel, but processed with greater speed and economy.

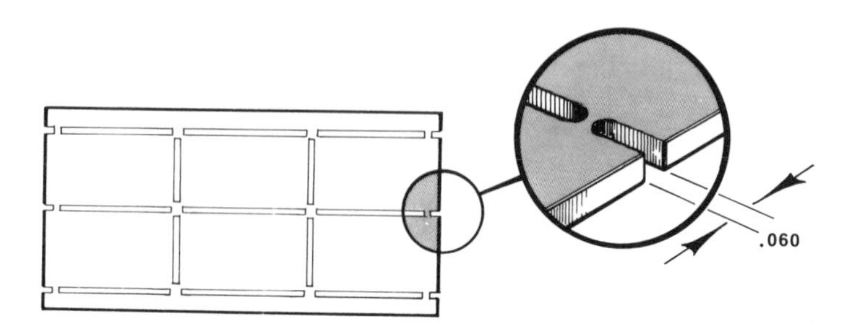

Fig. 3-7: Punch-press profiling of the board may be configured similar to a routed panel, but processed with greater speed and economy.

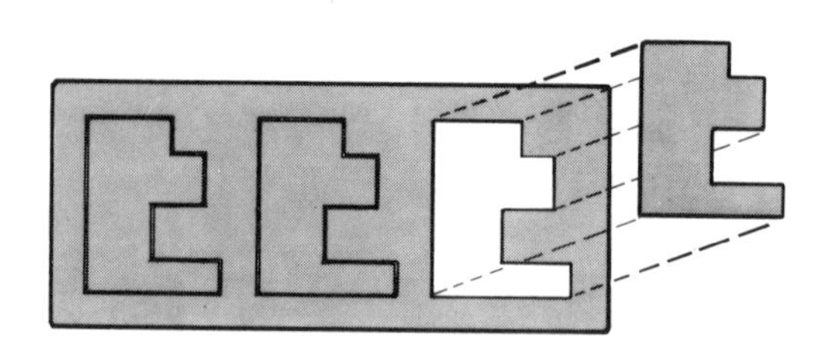

Fig. 3-8: Irregularly-shaped boards should be avoided if at all possible.

Panelizing small boards is a must for efficient assembly processing. Fig. 3-8 illustrates a typical example of an irregularly shaped board that would be awkward to handle. Small profiles like this can be blanked from a panel in one hit, like a cookie cutter, then each one is pressed back into the panel for efficient handling through all assembly operations. The final step after assembly requires simply pushing the finished module from the panel. This concept is referred to as a punch-back or punch-and-retain panel. Adequate clearance between each module must be provided for the male/female die set contact with the panel surface.

Providing for Assembly Automation

Before preparing the fabrication detail, the designer of the PC board must know the assembly equipment that will be used. Basic information would include maximum panel size, tooling-hole requirements and placement area for the surface mounted devices.

There are many equipment choices on the market and when the right combination is selected all assembly stations will be compatible. If your company uses one of the better contract manufacturers, you should be furnished with their specific guidelines which will assist in preparing products for assembly.

Before going to the expense of panelizing a product, a design review of the assembly by the manufacturing engineer would allow for improvements that may save the company time and money over the life of the product.

The maximum panel size and placement area for pick and place of surface mounted components will vary from one equipment manufacturer to another. Therefore, specifications from several SMD manufacturers should be consulted before finalizing the design. Listed below are a few examples:

EQUIPMENT	PLACEMENT AREA	MAX. BRD.
DYNAPERT MPS 500	16 X 18	18 X 20
QUAD QS-34	18 X 18	20 X 20
FUJE FP 60/90	9 X 13	10 X 14
FUJI CP 2	13 X 17	14 X 18
EXCELON CP30/40	14 X 16	16 X 18
AMISTER SM2001	16 X 20	16 X 16
UNIVERSAL 4621	14 X 16	16 X 18

Ideally, for long production runs, tooling holes on the board would be compatible with all machines in the assembly process from screen printing, pick and place and solder reflow to cleaning stations. However, since one size is not practical for all products, universal carrier pallets are often developed to transfer one particular assembly through each of the processes. This is usually the case when smaller boards are not panelized. Two or more substrates will be attached to a pallet and pass from one machine to another as a set. See Fig. 3-9.

The advantage to this concept is there is no downtime necessary for adjusting rails, tracks or belts.

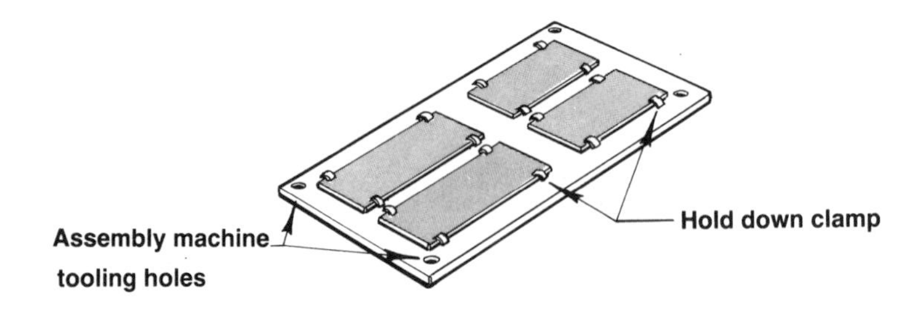

Fig. 3-9: Two or more substrates may be attached to a pallet and pass from one machine to another as a set.

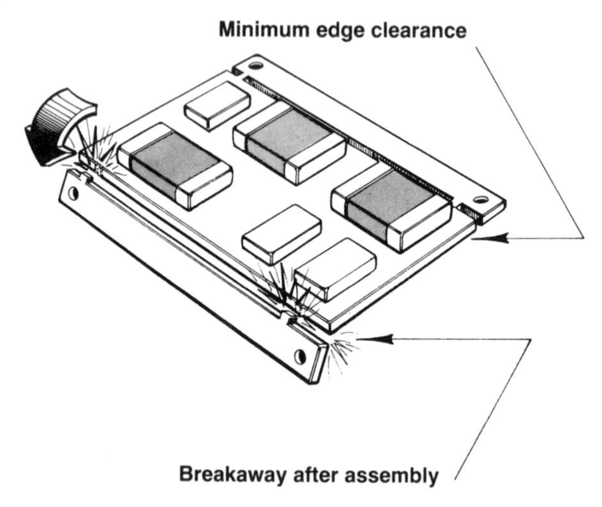

Fig. 3-10: Breakaway edge strips may be the best way to provide edge clearance on densely populated substrates.

Other factors in preparing for automation included component orientation, clearance, or spacing between devices.

Design Rules and Layout Guidelines.

Clearance of board edge to the component body is another point that must be addressed by the designer. If the assembly is to pass over a wave solder system, an adequate distance from components must be provided in order to eliminate interference with holding rails or pallet fixtures. On boards that are very dense, breakway edge strips may be the best way to provide edge clearance. See Fig. 3-10.

Plated Through-Holes

Leaded through-hole parts continue to be a viable part of the electronics industry. The guidelines for plated holes of the leaded parts are well established.

Component lead diameters will vary over a wide range. To reduce fabrication cost, PC board manufacturers advise standardization. For the greatest majority of through-hole components, the lead diameter falls in the range of .018 inch to .028 inches. A common hole size can be selected if allowances are made for the tolerance of lead-forming equipment, automatic insertion accuracy and wave solder characteristics. The industry has generally accepted the .040 inch diameter hole as a standard for auto insertion application. Smaller holes can usually be adapted for hand assembly.

The .040 inch size provides .010 inch to .012 inch clearance of the larger diameter leads supplied on resistors and capacitors, while at the same time compensating for the more difficult alignment of leaded DIP ICs.

Multilayer and Fine Line Construction

The industry is pressed to push the state of the art in fabrication technology as component density increased and circuit complexity evolves.

When the trace width and air gap are less than .010 inch, or plated through-hole diameters are less than .020 inch/.025 inch in diameter, the increased difficulty of manufacturing will be reflected in the board cost.

Before reverting to the more difficult and costly high tech method of board manufacturing the designer should make an effort to keep fabrication less complex. The board is usually the most expensive single component of the assembly; therefore, savings at this level will be reflected over the entire project.

The majority of manufacturers of fixed-space leaded components have complied with the .100 inch grid arrangement. The .100 inch grid pattern has been used for many years as a standardized form with PC board designers. It is a common practice to use a conductor trace between .015 inch to .012 inch for most signal-carrying applications. Routing a conductor trace between .100 inch spaced holes will further restrict the outside diameter of the contact pad. The conductor path must be spaced away from the non-related contact pads and conductors, maintaining the so called "air gap" or clearance. This air gap will allow for a clean etch during fabrication of the PC board and reduce the chance of solder bridging during soldering processes. See Fig. 3-11.

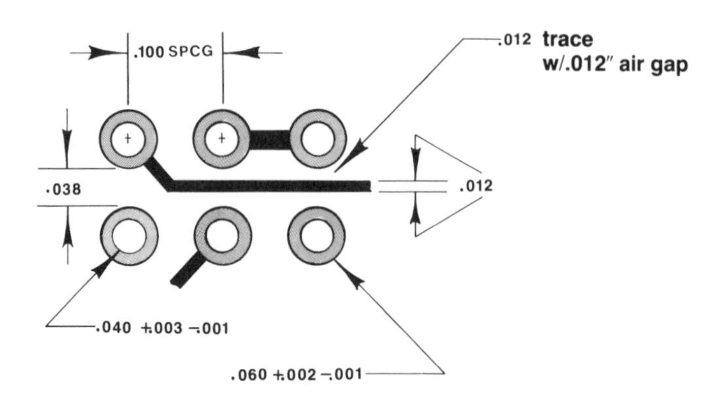

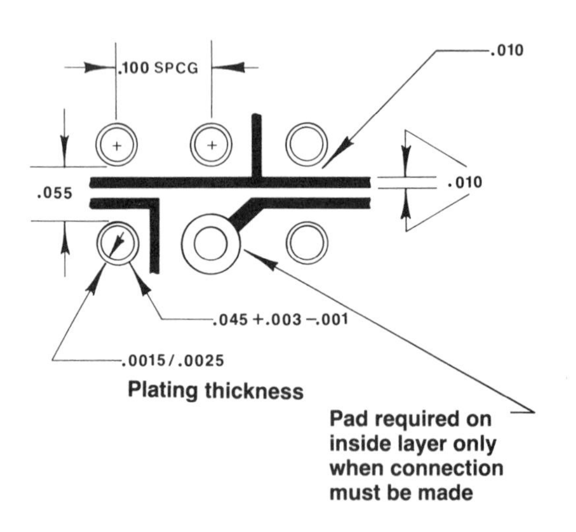

Fig. 3-11: An air gap will allow for a clean etch during fabrication of the PCB and reduce the chance of solder bridging during the soldering process.

The tolerance of a drilled and plated hole is typically +.003 inch and -.001 inch diameter. Keeping the above mentioned tolerances and limitations in mind, we recommend a .060 inch - .062 inch outside diameter contact pad size for leaded components requiring the .100 inch grid pattern.

Component density is further increased when applying multilater technology. Typically the contact pad will appear on the outside layers and on an inside layer only when a connection must be made. Conductor density can be increased because the plated through-holes do not have the annular ring on the inside layers. Two .010 inch wide conductor traces will provide .010 inch air gap on these inside or laminated layers. The importance of accuracy in pad location and conductor trace spacing cannot be over stressed. Many companies are relying heavily on CAD systems that are designed to alleviate these complex applications. The .100 inch grid pattern is also the primary location element in the auto-routing feature of more advanced systems. Fig. 3-12 provides an example.

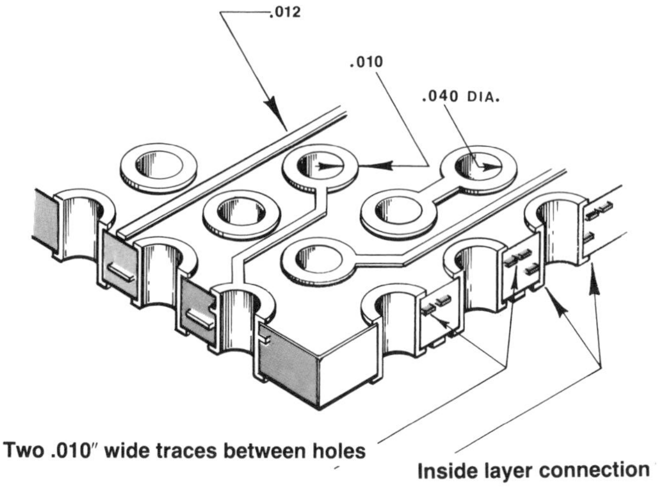

Fig. 3-12: The .100″ grid pattern is the primary locating element in the auto-routing feature of more advanced SMD manufacturing systems.

Surface Mount and Via Holes

The surface mounted component, of course, does not require a hole in the substrate. On a smaller, less complex surface mounted circuit, it is possible to design the PC board without using any holes.

As SMT circuits become more complex and component density is increased, the need to add feedthrough (via) pads in the substrate to maximize available space also increases. Because feedthrough holes will not have to clear a component lead, a much smaller drill size is possible. In hybrid technology using ceramic materials, via holes are smaller than .010 inch diameter. However, in a glass laminate fabrication, small drill size will add excessive cost to the fabrication of the PC board. To maximize drill speed and keep the drill breakage rate low, most board shops would prefer a minimum finished hole size of .020 inch diameter. This size hole allows for a reduction in pad size and increased conductor trace density. Standardizing on a .020 inch finished hole diameter will allow for a .040 inch outside diameter pad while maintaining the desired .010 inch annular ring. See Fig. 3-13.

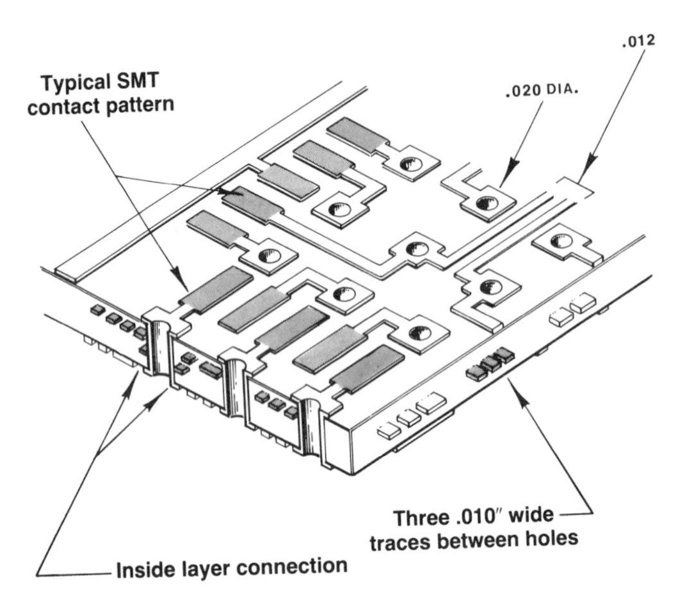

Fig. 3-13: To maximize drill speed and keep the drill breakage rate low, most PCB manufacturers prefer a minimum finished hole size of .020″ diameter.

To further provide for routing conductor traces and at the same time insure an acceptable air gap, you may choose to use a square pad of .035 inch to 1.037 for feedthrough holes. This square configuration will furnish more than enough metal in the diagonal corners of the pad to compensate for the reduced annular cross section at the sides of the square. The .035 inch -.037 inch square feedthrough pad can be spaced at .050 inch when necessary or on the more traditional .100 inch grid. When the .035 inch -.037 inch square pad is spaced at .100 inch, it is possible to route two conductor traces between pads. Originally, this density was possible only on internal layers of multilayer boards with the leaded, through-hole technology. Using multilayer for surface mounted applications will dramatically increase density possibilities, as shown in the two illustrations in Fig. 3-14.

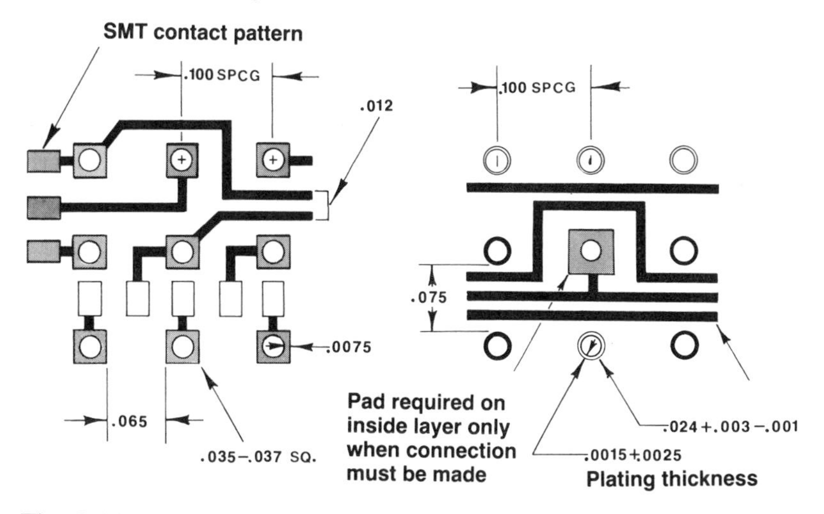

Fig. 3-14: Using multilayer for SMD application will dramatically increase density possibilities.

The reduced size of the plated feedthrough hole diameter will easily allow three conductor traces on internal layers without resorting to "fine line" (.006″ wide) traces. As component density increases further, it may be necessary to limit the outer layer surfaces to the component's contact patterns, feedthrough pads and a short conductor trace for connection. This is sometimes referred to as "pads only" - with all circuit traces buried on internal layers of the PC boards.

Computer Aided Design and Via Holes

Auto-routing for SMT, on most computers, required a via pad and a hole on all layers of the multilayer board.

To simplify this procedure, a standard contact or footprint pattern for each type of device would include a line or trace connection to the via pad. With this system, the footprint only appears on the surface and all the signal traces are buried on inside layers of the board. With all circuits on inside layers, it is more practical to use .008 inch, .006 inch or less trace width and air gap to interconnect the components.

Solder Mask on PC Boards Using SMT

Solder mask and plating specifications will always promote controversy. Some companies would like to eliminate solder mask altogether because of the extra cost and difficulty in registration on the small surface mounted footprint. Others insist on the coating to seal the exposed laminate surface and insulate signal traces from possible danger from foreign particles or contamination. In any case, except for very small (3″ x 4″) boards, the use of photo-imaged solder mask is recommended. Fig. 3-15.

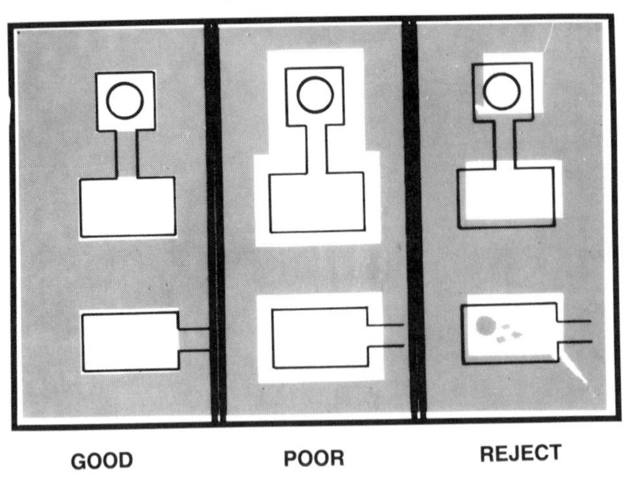

Fig. 3-15: In most cases, the use of photo-imaged solder mask is recommended.

With a photo-image solder mask process, zero clearance from footprint pattern is possible, but providing a clearance of up to .005 inch is usually more forgiving. What is to be avoided is solder mask overlap or solder mask overlap residue on the contact area itself. At the same time, if too much of the signal trace is exposed, solder paste will migrate away from the contact area during the reflow solder process.

The solder mask acts to contain the paste and to insure that each component contact receives an equal amount of solder. Contact areas are usually connected to a via pad with a narrow trace. This isolation is necessary to allow for a dam of solder mask to stop the solder from flowing to and down through the via hole.

A coating over the via hole is required, if the via pad is too close to the contact area to provide for an adequate solder mask barrier. Many companies choose to cover all via pads with mask to eliminate any possible problems.

NOTE: If you do cover vias, only cover one side or the other. If both sides are covered, air and gasses may expand during the reflow process and this can cause blow out, blistering or even dislodging a component from the contact pattern. In any case, don't cover pads or vias needed for test probe contact by auto test or bed of nails fixtures required for automatic testing.

It would not be appropriate to recommend a particular solder mask in this publication. We suggest that you discuss your needs with the board fabricator. There are several excellent products on the market, and this is a great opportunity to use your prototype to evaluate the various candidates.

Plating Process For SMT

The most process-compatible plating required on the contact area is tin/lead. Tin/lead is a customary plating for etched copper circuits, although tin/nickel is being frequently used. The advantage of the tin/nickel surface is the flatness and uniformity of the finished surface. Other fabricators, to achieve that flatness, will strip the tin/lead from the board after etching, leaving bare copper exposed over all traces.

To be compatible with the surface mounted devices and the solder and flux used, it will be necessary to plate the contact areas and any plated throughhole and pad mounting a leaded device with tin/lead. With the bare copper board, solder mask is applied, exposing only the contacts and holes. Dipping this board into a solder (tin/lead) bath with hot air leveling adds the proper alloy to the contact areas for process compatibility. The same technique can be used on the nickel plated board. If solder mask is to be eliminated, electroplating of tin or tin/lead with a selective process must be used on contact areas and through-holes for mounting leaded components. In all cases, the tin/lead should be reflow fused.

Once you have qualified you board fabricator, draft the finished board descriptions in cooperation with the supplier. This will insure that the finished PC board you are specifying can be fabricated at the most reasonable cost to your company.

A typical note for two-sided PC board with the solder mask over bare copper may read as follows:

NOTES: (Unless otherwise specified)

1. Material: Type FR-4, 1/2 oz. cu clad. .062 + .007 thick.
2. Location of all holes to be within +.003 in. of pad center. All holes diameters are after plating.
3. Copper plating in all holes to be .001 min.
4. Plating: Solder mask over bare cu, unmasked areas to be 1 to 2 mils of 60/40 tin-lead solder using: plating, hot air laveling or hydro squeegee technique. Solder covered areas to be flat and free of excess solder.
5. Coat both sides of P.C.B. with green solder mask or equivalent. Align or true position within .002 in. using tooling holes for reference. Application of solder mask should be in accordance with IPC-SN-840, Type B, Class II. (All pads and contact areas to be free of solder mask).

6. Space between trace and trace or trace and pad shall be within + .002 of paths shown on film.
7. Width of conductor paths shall be within + .002 of paths shown on film.
8. After baking in 150°C + 5°C oven for 15 minutes, board acceptability shall be based on IPC-A-600C, Class II.
9. Board twist and warp not to exceed .004 per linear inch.
10. No voids, opens, or shorts allowed in traces.
11. Vendor code and UL rating shall be shown on circuit side of P.C.B.
12. Silkscreen to be applied to component side of P.C.B. using whit epoxy ink. All pads must be free of silkscreen.
13. Gold flash or plate finger tabs .00005-.00008 thick over hark nickel.
14. Use artwork [specify here] .

High Tech Materials For Military Applications

Ceramic (alumina) materials have been traditionally used in military or extreme environmental applications. The need for larger substrate size and the less than uniform shapes of today's electronic applications has made it necessary to explore alternative materials. Metal core laminated construction of circuit boards using relatively new materials now being accepted for military and space application.

A metal core (Copper Clad Invar/Polyamide-glass or Polyamide/Kevlar) laminated substrate is being offered as an alternative material for applications using leadless ceramic chip carriers. Previously, the choice was limited to ceramic substrate in order to match the component's TCE (thermal coefficient of expansion).

The metal core laminate material has been approved for many applications and companies are specifying CCI in new products being developed. However, extensive testing, technical papers and reports have not resulted in either military specifications or applicable guidelines to assist the engineer and PC designer in implementing this process.

In this section we will present practical guidelines for the PCD board designer to assure the most producible finished product.

CCI = Copper clad invar is a nickel steel alloy core material with a layer of copper rolled on each surface.

Specifying Copper Clad Invar

In order to have a clear understanding of CCI fabrication guidelines, it is important to know the process steps. One method of adapting these more stable materials into you products is by laminating the finished PC boards to a layer of CCI to provide stability, ground plane and a thermal transfer medium. Figure 3-16.

Designs with unused signal layers should be avoided. If layers are uneven from one side of the CCI core to the other, the unused layer should be noted on the fabrication detail to reduce confusion for the fabricator. Adding a layer number on the working film will also reduce errors in fabrication.

Most of these recommendations for fabrication of the polyimides materials apply to conventional PC board construction as well.

The most basic metal core laminated substrate would have one or two CCI layers and one single sided copper clad polyamide sheet laminated to both outer layers as shown in Fig. 3-17.

This lamination would take place after clearance holes are drilled or etched in the CCI layers. This will prevent unwanted signal connection when holes are plated through.

Clearance holes must allow for insulation around via holes that are to be connected with CCI layers. The laminating adhesive film will flow into the clearance holes in the CCI to provide the dielectric separation of copper plated via holes. After lamination all holes are added to substrate, including those that will connect to CCI layers.

The entire substrate panel is plated with one ounce of copper, and simultaneously, holes are copper plated from one side to the other. Only holes not drilled for insulated clearance will connect to the internal layer, providing ground or plane connection. The circuit image is now plated to the outside surfaces and through the holes. This plating will act as a resist while exposed bare copper is etched away. If a tin/lead alloy is used as the plating medium, the etched panel is generally passed through reflow process to fuse the tin/lead with the copper, leaving a bright finish.

A solder mask coating can be applied to both surfaces of the finished substrate. Contact areas and via pads should be free of this mask material, as noted for conventional PC boards.

The advantages of coating the substrate are:

A. The solder paste used to attach the SMT components is contained on the land.
B. Solder bridging is reduced or eliminated.
C. The bare laminate and conductors are covered, reducing absorption of moisture and contamination.

If space permits, reference designations and component outlines are screen printed on the board surface in epoxy inks. This aids in the identification of each device as well as identifying the polarity and the orientation of the SMDs.

The fabrication procedure will vary for complex applications that require the stability of a metal core laminated substrate. This variable is dependent on the number of layers required to interconnect the circuit. Surface mounted components are often attached to both outside surfaces of the substrate.

It may be necessary to reduce the number of side-to-side plated through-holes to avoid excessive perforation of the metal core material. Many of the circuit interconnects can be made within the layer associated with each respective side. Only the via holes necessary to connect side #1 with side #2 and the power/ground CCI layers, will be drilled during this step.

After the final drilling, the fully laminated substrate is copper plated. Simultaneously, the copper plating connects through all via holes and the PC board is processed as described earlier in the basic substrate fabrication examples.

The process requiring blind vias-holes that do not continue through all layers will add cost to the circuit board fabrication. The continuity of the copper clad invar layers, however, will be maximized by reducing the number of holes in the CCI. This maintains the continuous layer, which is desirable to insure the most thermally stable finished substrate.

Most of the design rules recommended for surface mounted technology on conventional epoxy glass substrates will apply to copper clad invar/polyamide substrates. The assembly process control and solder selection may be more critical, but the components will be attached with reflow solder technology in the same procedure as normally used in SMT assembly.

Materials For Copper Clad Invar/Polyamide

Substrate Fabrication:

1. Standard material thickness for polyamide glass and polyamide Kevlar is .005 inch, .006 inch, .008 inch and .010 inch . Polyamide-Kevlar material is presently four times greater in cost compared to polyamide-glass. A careful selection process is recommended. Check with your supplier to learn the standard sheet sizes available.
2. In addition to the basic dielectric the copper thickness desired must also be chosen:

   .0007 Thk - ½ oz Copper Clad
   .0014 Thk - 1 oz Copper Clad
   .0028 Thk - 2 oz Copper Clad

Materials available: Bare; Copper Clad on one side, Copper Clad on two sides; or Copper Clad on two sides each having a different thickness. The latter is non-standard and will be more costly.

3. Prepeg or laminating film is used to bind the layers together and is furnished in a standard .0025 inch thickness. This material can be layered to "build up" to the final thickness specified for the finished board. Prepeg is supplied to the PC board fabricator in 38 inch wide rolls.

4. Copper Clad Invar material is available in several thicknesses—.006 inch is furnished in 24½ inch wide rolls. .010 inch, .020 inch, .030 inch, .050 inch and .060 inch thick materials are supplied in sheets.

Using standard material sizes whenever possible will keep the costs under control. However, all of these materials are presently more expensive than the conventional FR-4 epoxy glass laminates. (The additional cost is far less than multilayer ceramic substrate materials and mechanically the assembly will prove more durable.)

The Following Recommendations Will Help Assure A Successful Finished Product:

1. A symmetrical assembly that has equal stresses on both sides of the substrate at all times. The finished PC board should remain flat upon thermal cycling. Bonding the etched copper circuit polyamide layers on both sides of the copper clad invar also eliminates assembly flexing problems during the thermal cycling.
2. A conductor width of .010/.012 inch is preferred. .008 inch is the minimum conductor width on outside layers.
3. Total board thickness range is .062 inch to .100 inch with maximum of .125 inch.
4. Board thickness tolerance is 10% of nominal or .007 inch, whichever is greater.
5. .008 inch minimum is preferred for the thickness of the dielectric, but not less than .0035 inch.
6. An air-gap of .010/.012 inch between conductors is preferred, with a minimum of .005.
7. Finished board thickness to plated through-hole size—3:1 is preferred; 4:1 maximum.
8. Plated through-hole diameter of finished board thickness:

   .062 inch thickness or less: .020/.025 inch hole dia.
   .075 inch thickness or less: .025/.035 inch hole dia.
   .100 inch thickness or less: .035/.055 inch hole dia.

Hole sizes .015 inch or .018 inch diameter, if used, are to be used on outer layers only.

9. Annular plating around finished hole is .007/.010 inch -minimum of .005 inch is to be avoided.
10. Conductor width to edge of board:
    Internal layer of .100 inch is preferred, .031 inch min.
    External layer of .100 inch minimum.

Specifying standard materials will insure greater producibility, since this allows the fabricator to use "off the shelf" materials. Overall thickness can be specified including plating, but do not specify between the layers created by the prepeg or core thickness. If you must specify core and/or prepeg thickness when spacing is critical, give a loose tolerance on the overall thickness to maximize fabrication efficiency and yield.

Commercial SMD Packages

Commercial SMD packages fall into two general categories: Small Outline (both the 0.150 inch wide SO and the 0.300 inch wide SOL) and the Plastic Leaded Chip Carrier, also referred to as a PLCC or PCC. Both of these packages use the same materials and assembly technology as used for the standard plastic DIP. Both have 0.050 inch lead spacing. The SO/SOL is dual in-line with a Gull-Wing shaped lead bend, while the PLCC has leads on all four sides with a J-shaped lead bend. Both lead bends are designed for mounting directly to the surface of a PCB. The SO/SOL packages are for the smaller pin count devices; that is, those up to 28 pins, while the PLCC is for those parts with up to 84 pins. There is, however, some overlap due to the need to accommodate larger die in smaller pin counts as shown in the table in Fig. 3-18.

Pin Count	SO	SOL	PLCC	Philips Special Pkgs
8	X			SOL*
14	X			
16	X	X		
20		X	X	
24		X		
28		X	X	
32			X	
40				VSO*
44			X	
52			X	
56				VSO*
68			X	
84			X	
120				TAB PKG*

VSO Lead pitch for VSO-40 is 0.762 mm while that for VSO-56 is 0.75 mm. See special package outlines.

SOL Dimensions for SOL-8 do not fall within the JEDEC standard. See special package outlines.

*For special package outlines, see pages 21 through 23 (SOL-8, VSO-40, VSO-56).

Fig. 3-18: Available SMDs with pin count.

Military Packages

To support the implementation of defense and aero-space systems in SMT, most SMD manufacturers offer products processed to MIL-M-38510, DESC selected item drawings, and MIL-STD-88e Revision C including the following packages:

Ceramic leadless chip carriers; triple laminated with metal lid and solder coated terminals in 20 to 68 leads.

Ceramic flat packages; frit glass sealed alumina CER-PAC in 14 to 28 leads, and brazed leaded ceramic in 52 leads.

All military package case outlines and physical dimensions must conform with the current revision MIL-M-38510, Appendix C, except for package types which are not included in that specification. Also be aware that this specification may be revised frequently; always adhere to the latest specifications. For the latest information, contact Signetics Military Division, Sacramento California, (916) 924-6010.

The following gives specific dimensions of SMD packages. These dimensions, however, may vary slightly with individual manufacturers. Pinouts for devices in SO packages are the same as all standard DIPs except for a few linear devices.

D Package—Plastic (SO-8)

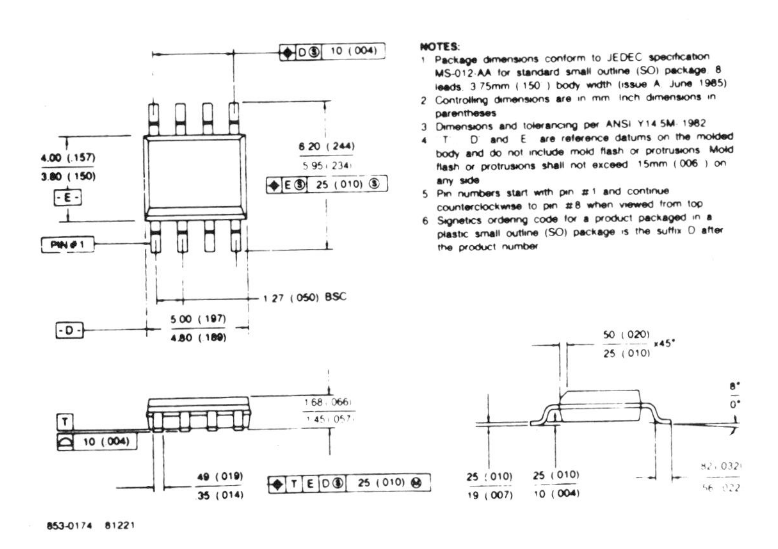

D Package—Plastic (SO-14)

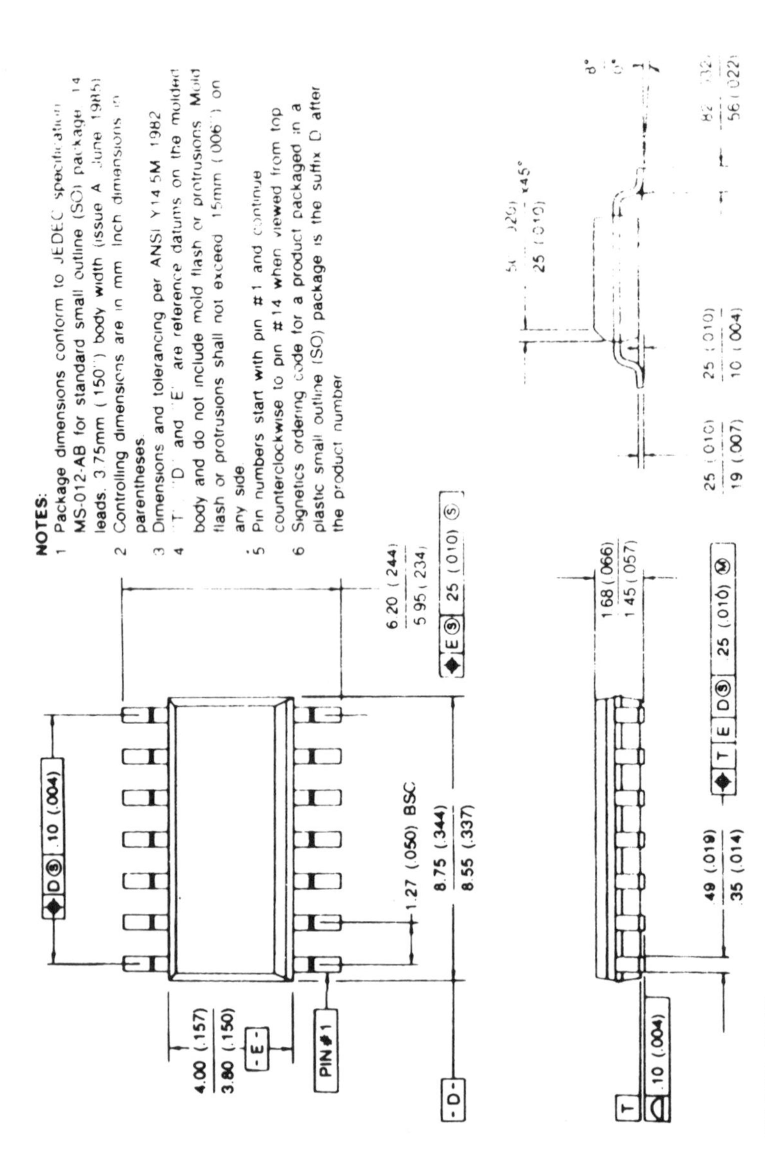

D Package—Plastic (SO-16)

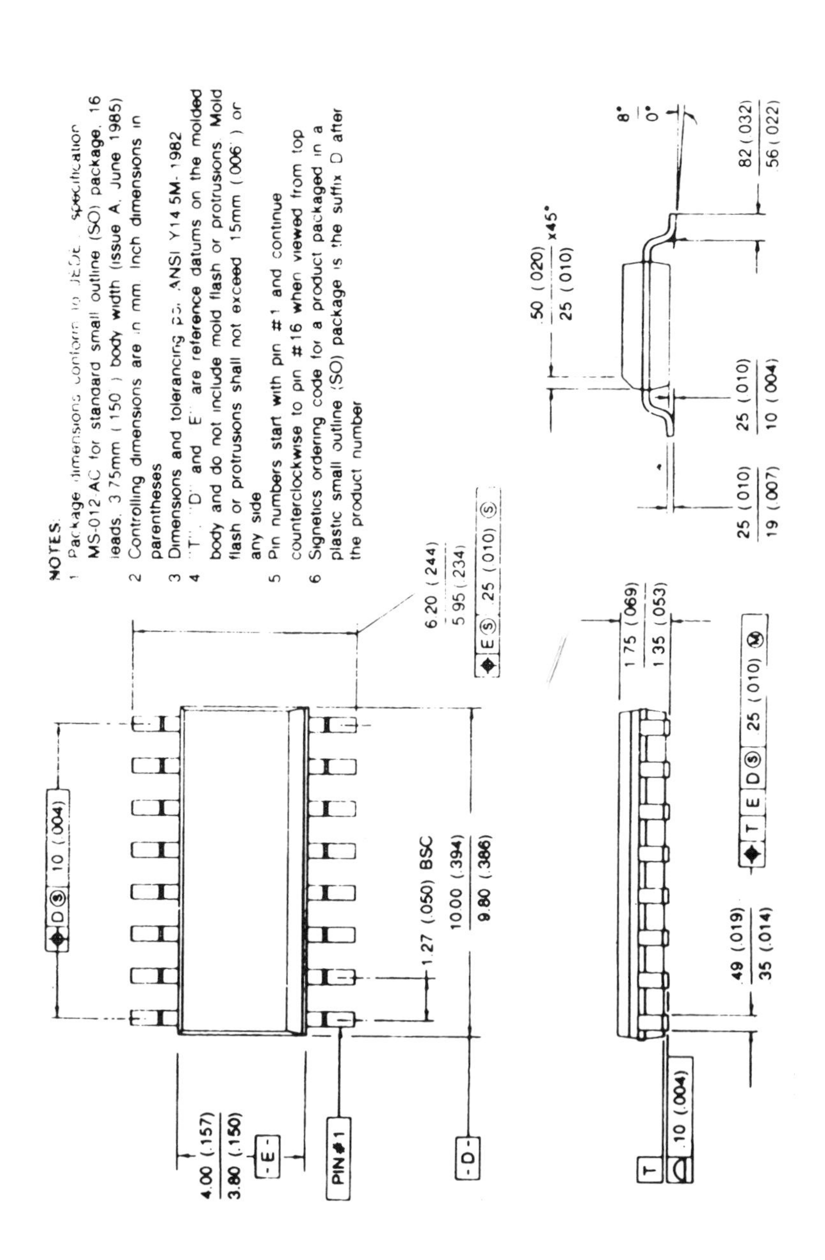

D Package—Plastic (SOL-16)

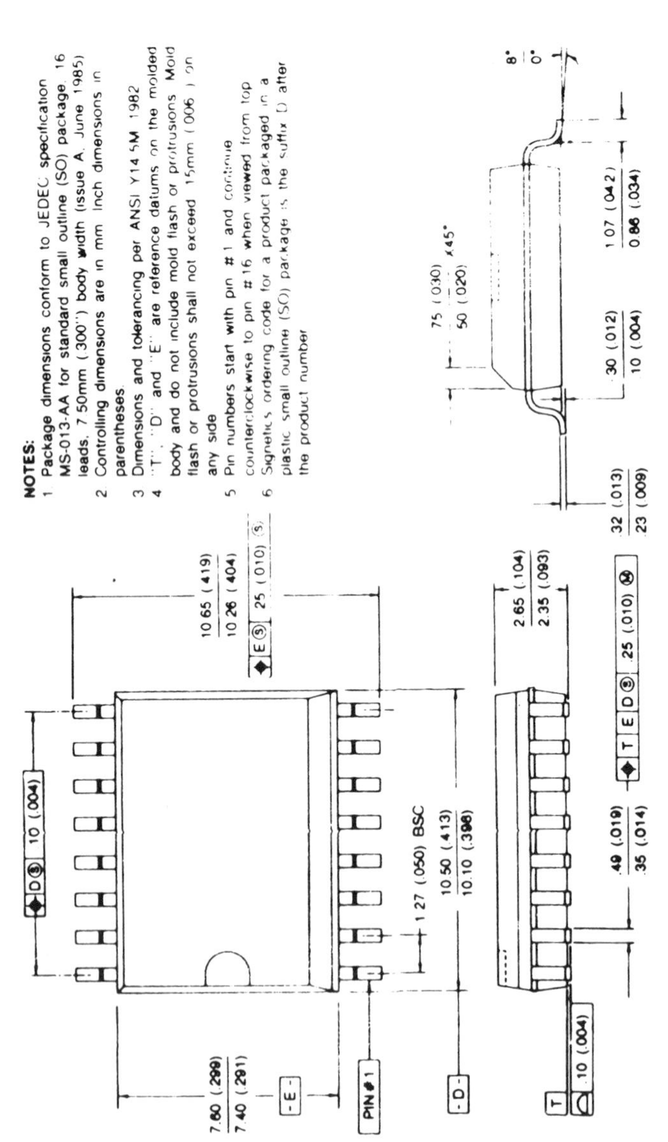

D Package—Plastic (SOL-20)

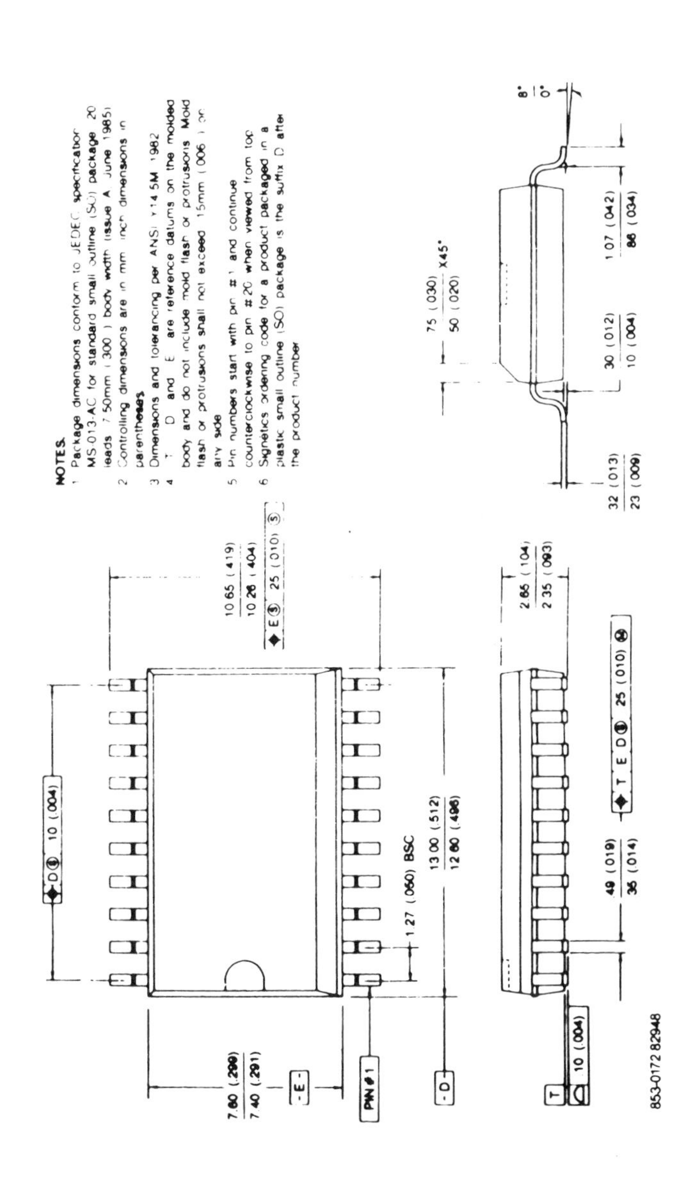

D Package—Plastic (SOL-24)

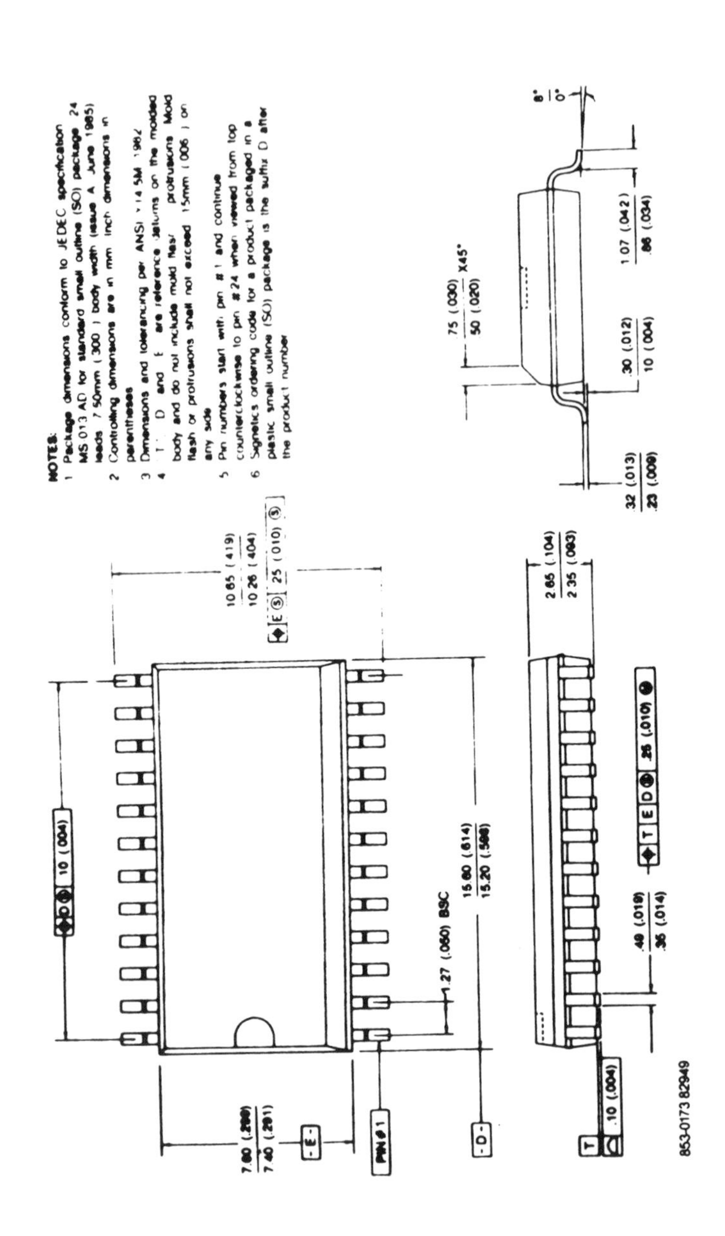

D Package—Plastic (SOL-28)

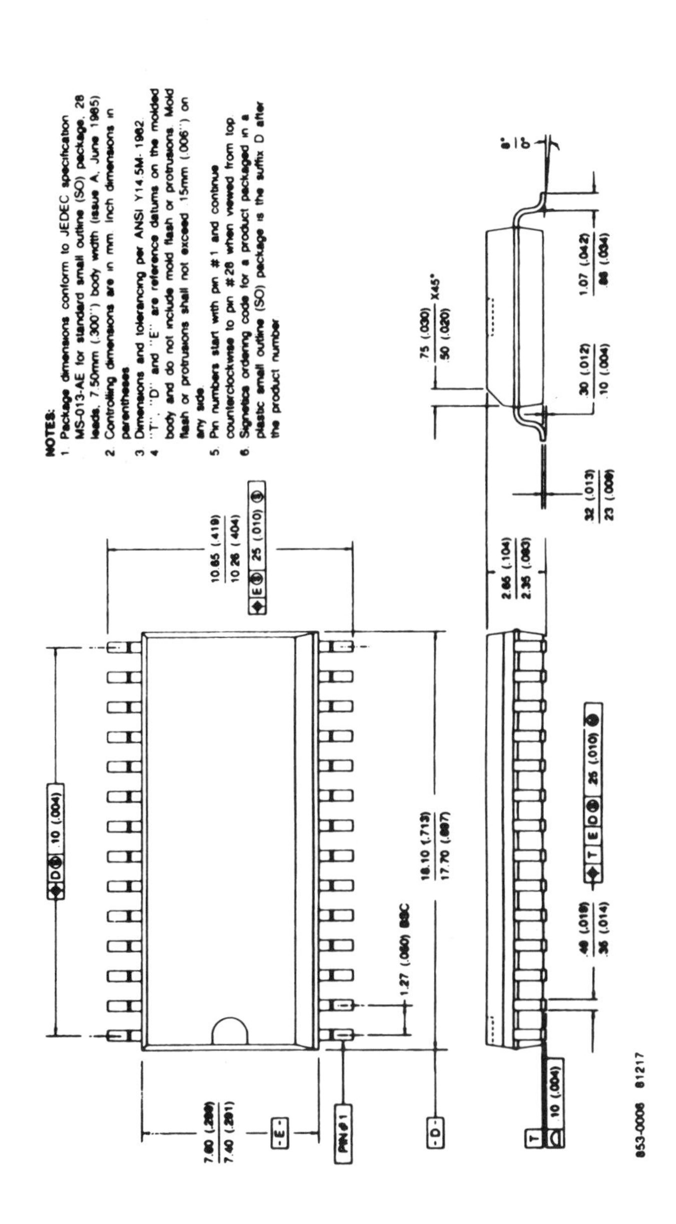

A Package (PLCC-20)

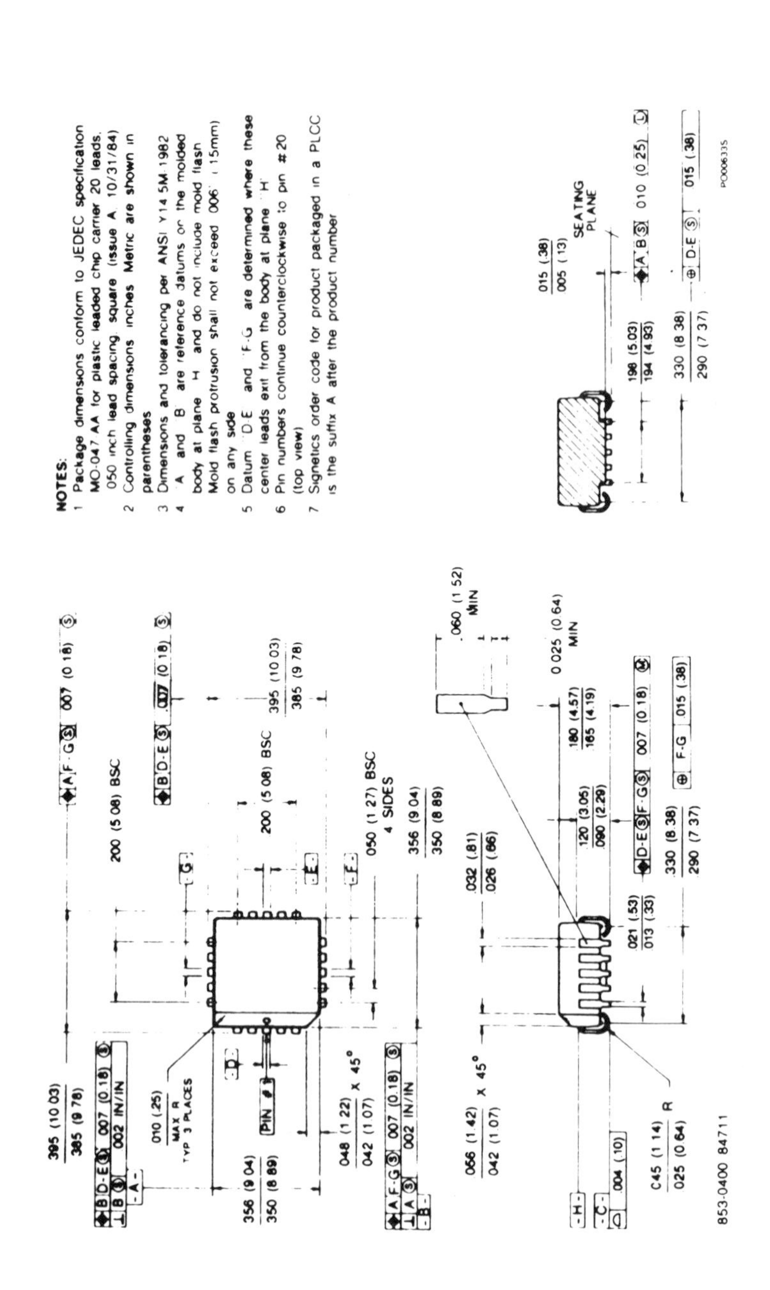

A Package (PLCC-28)

NOTES:

1. Package dimensions conform to JEDEC specification MO-047-AB for plastic leaded chip carrier 28 leads, .050 inch lead spacing, square (issue A, 10/31/84)
2. Controlling dimensions: inches. Metric are shown in parentheses
3. Dimensions and tolerancing per ANSI Y14.5M-1982
4. "A" and "B" are reference datums on the molded body at plane "H" and do not include mold flash. Mold flash protrusion shall not exceed .006" (.15mm) on any side
5. Datum "D-E" and "F-G" are determined where these center leads exit from the body at plane "H"
6. Pin numbers continue counterclockwise to pin #28 (top view).
7. Signetics order code for product packaged in a PLCC is the suffix A after the product number

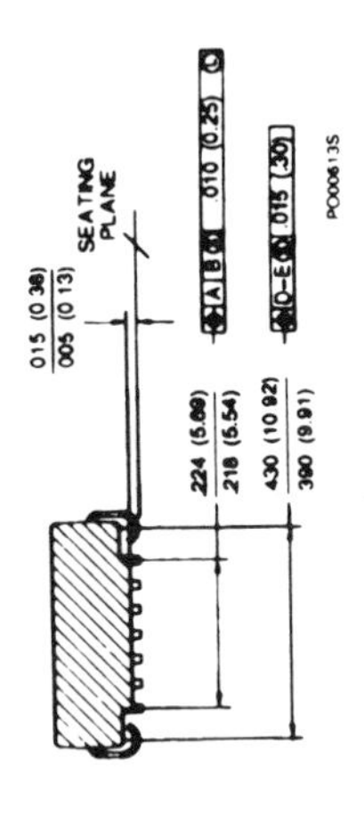

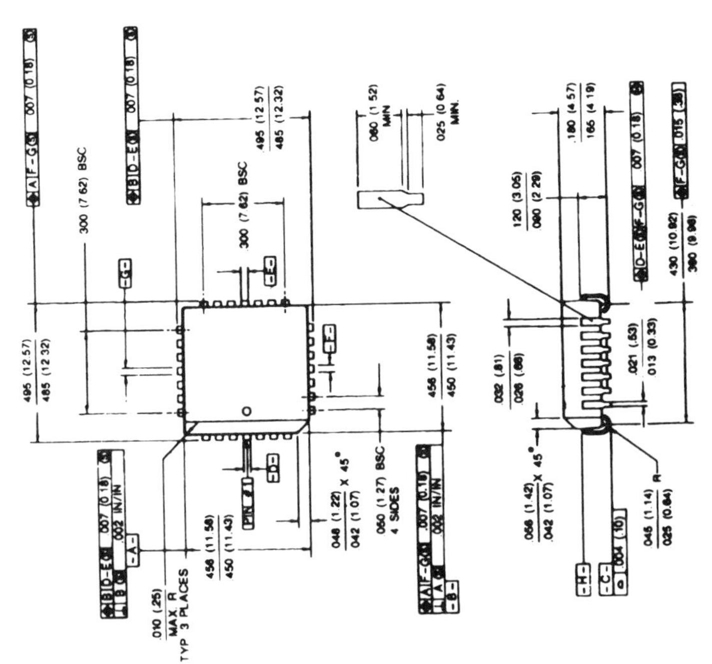

A Package (PLCC-32)

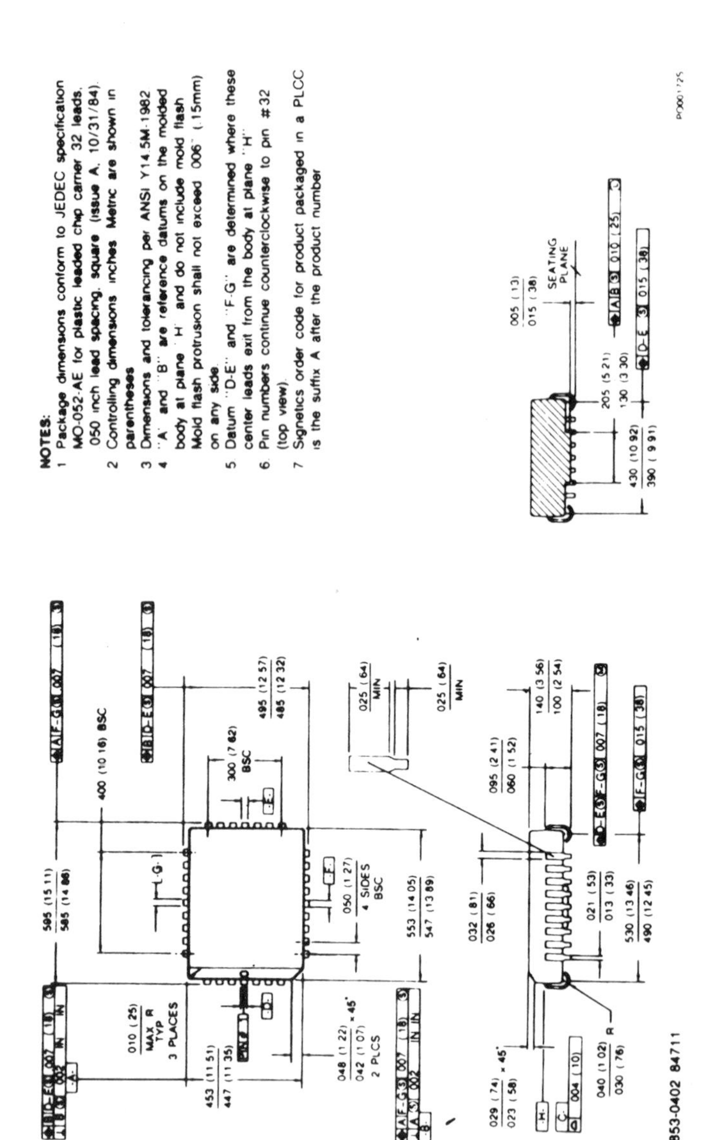

A Package (PLCC-44)

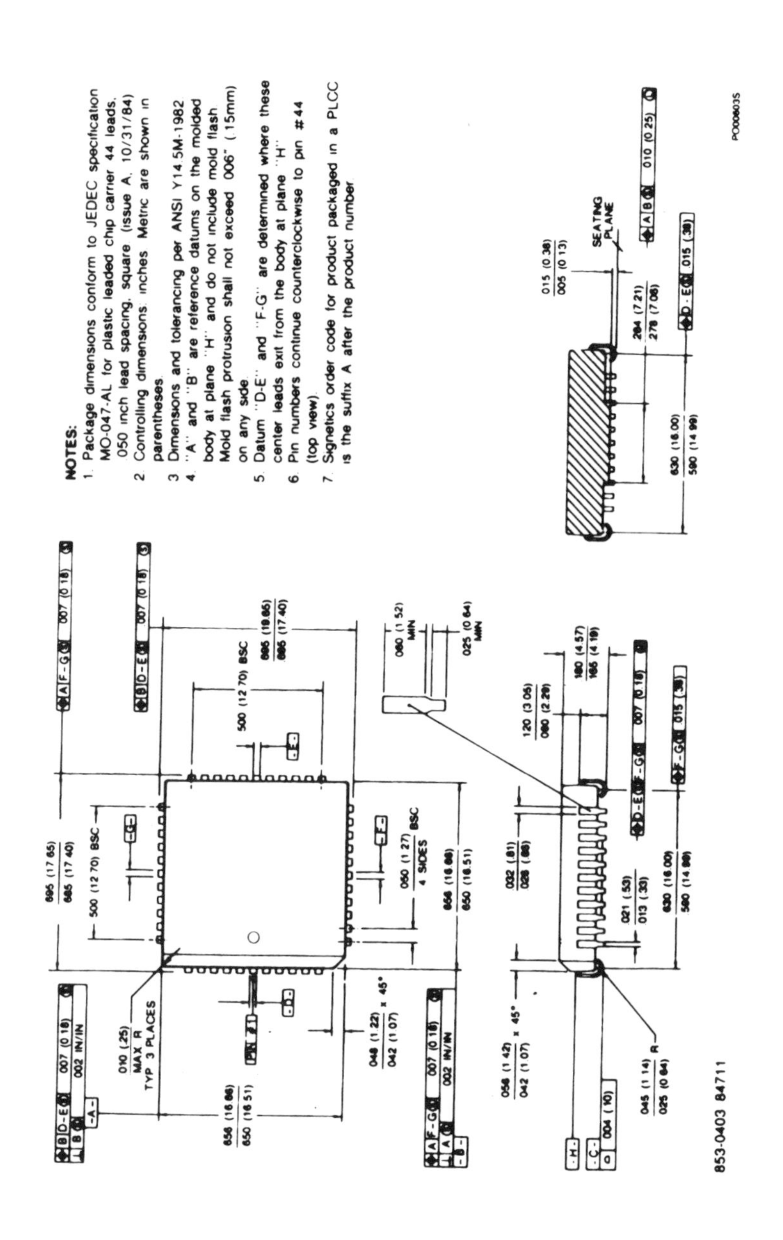

A Package (PLCC-52)

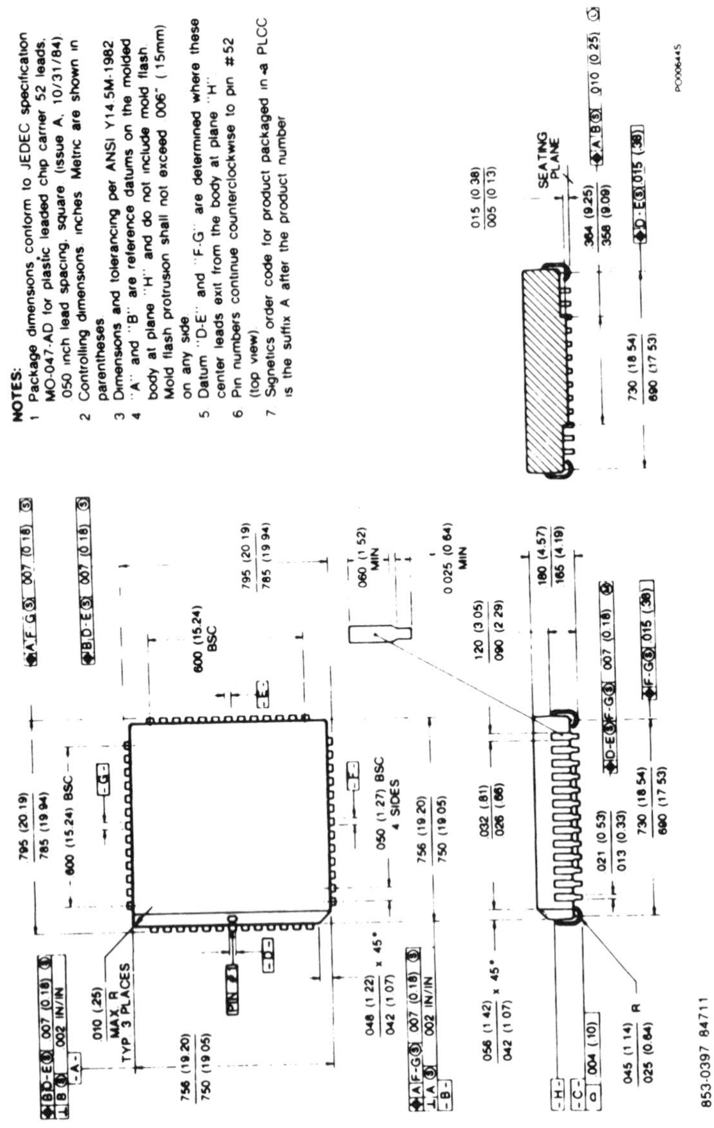

A Package (PLCC-68)

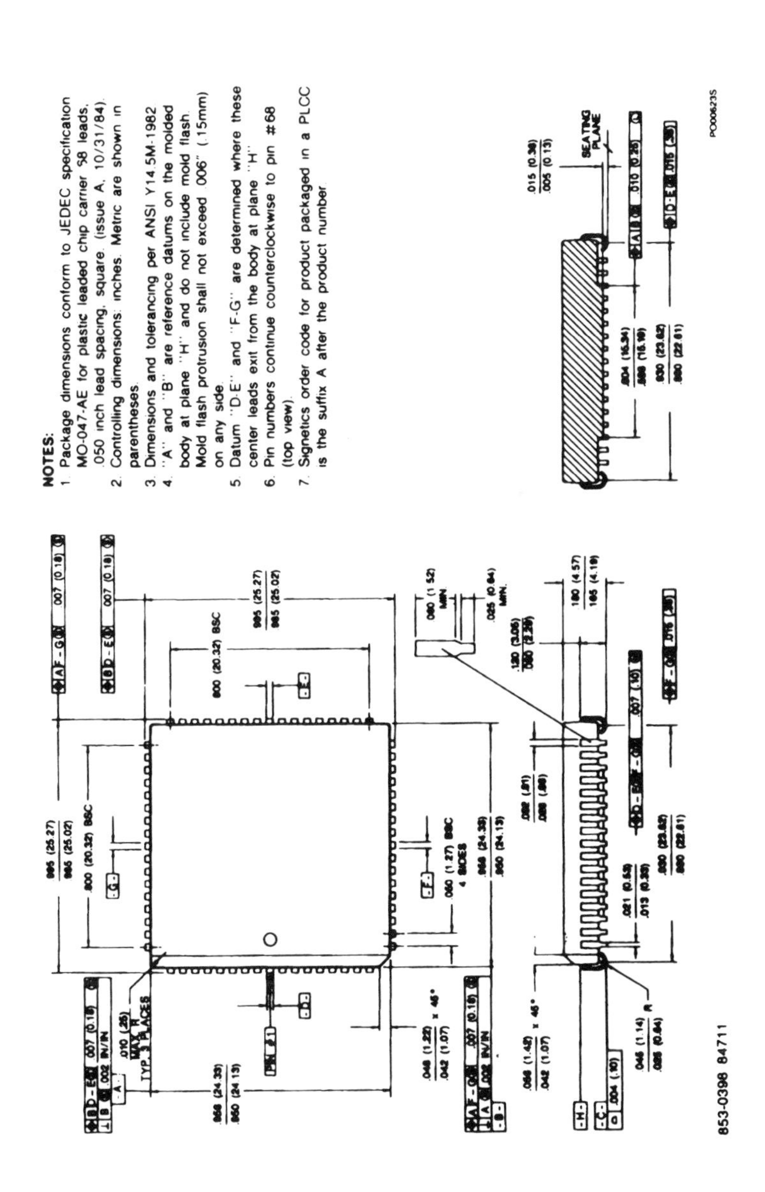

A Package (PLCC-84)

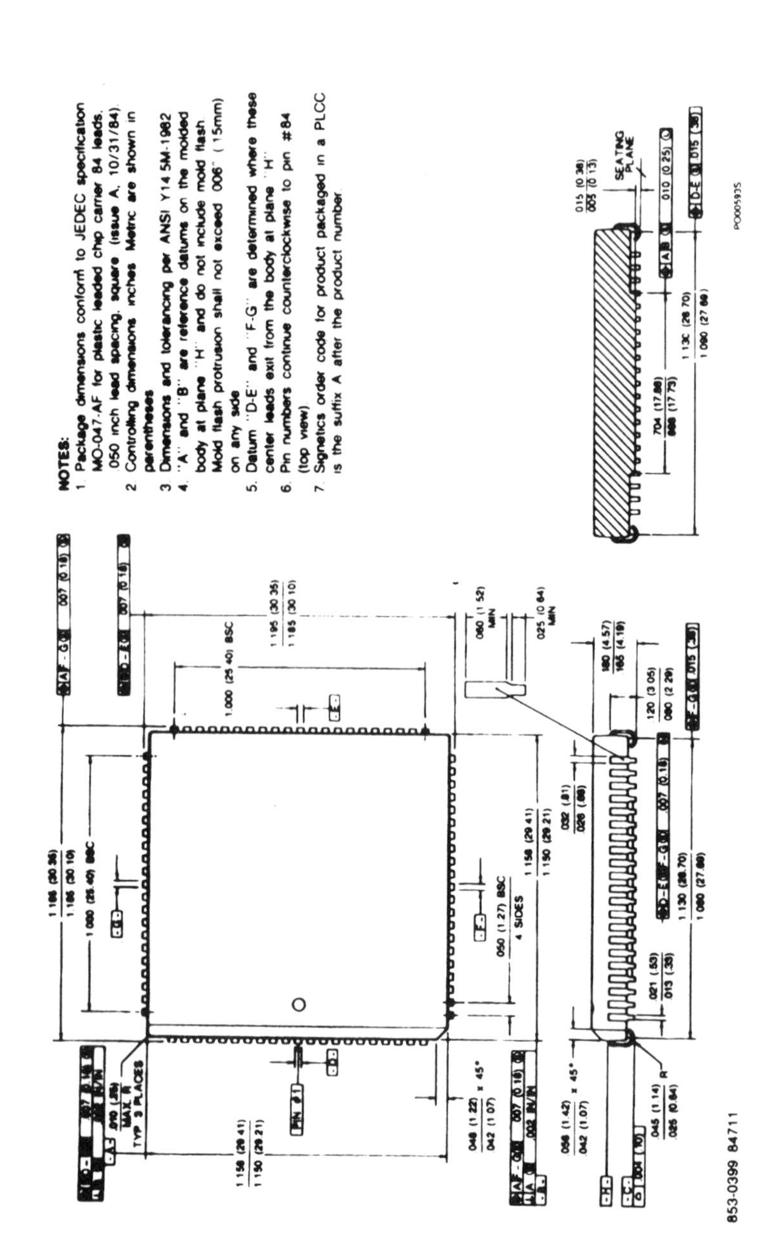

Special Package Outlines

The following package outlines are provided for certain part types, although these packages are not yet (as this goes to press) included under the JEDEC Standards.

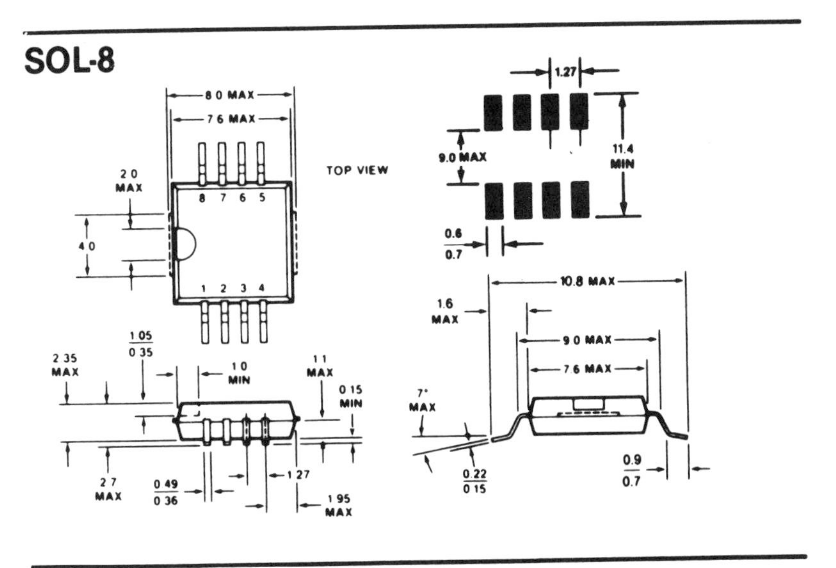

VSO-40

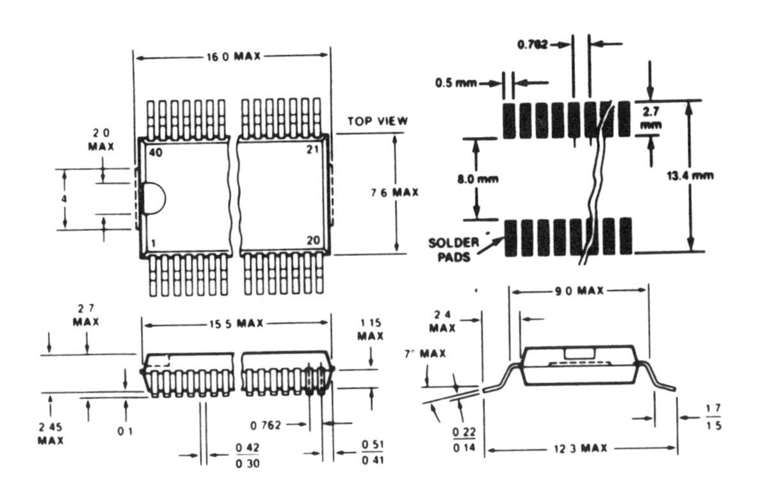

VSO-56

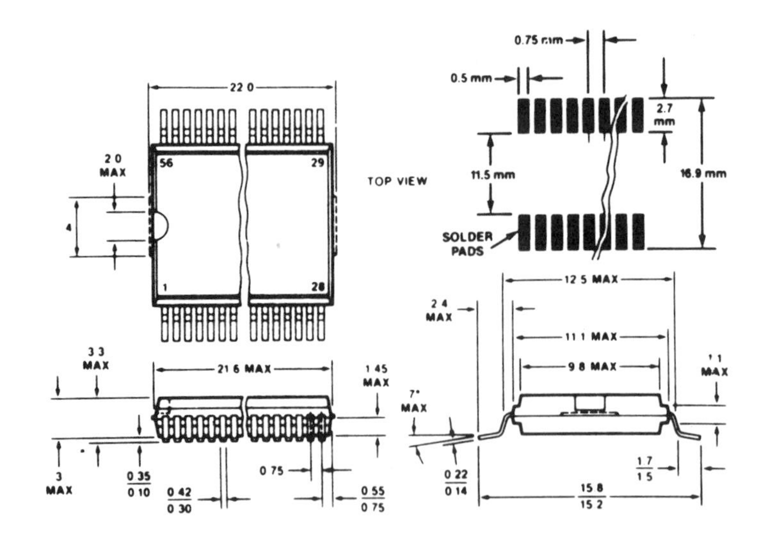
22 0
56
29
1
28
2 0
MAX
4
TOP VIEW
0.75 mm
0.5 mm
2.7
mm
11.5 mm
16.9 mm
SOLDER
PADS
12 5 MAX
2 4
MAX
11 1 MAX
9 8 MAX
3 3
MAX
21 6 MAX
1 45
MAX
7°
MAX
3
MAX
0 35
0 10
0 42
0 30
0 75
0 55
0 75
0 22
0 14
1 7
1 5
15 8
15 2

QFP-44

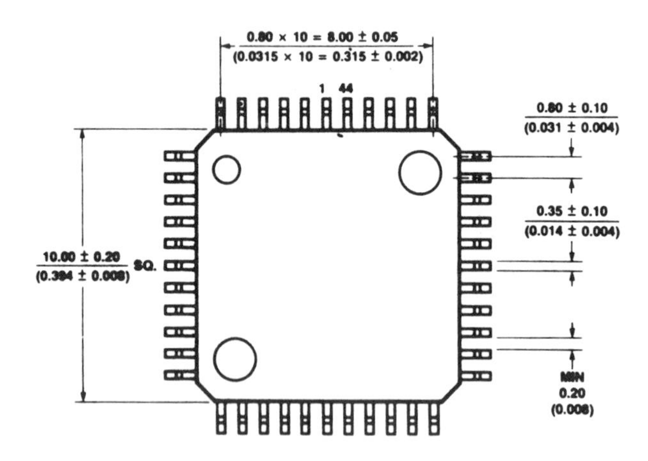
0.80 × 10 = 8.00 ± 0.05
(0.0315 × 10 = 0.315 ± 0.002)
1 44
0.80 ± 0.10
(0.031 ± 0.004)
0.35 ± 0.10
(0.014 ± 0.004)
10.00 ± 0.20
(0.394 ± 0.008)
SQ.
MIN
0.20
(0.008)

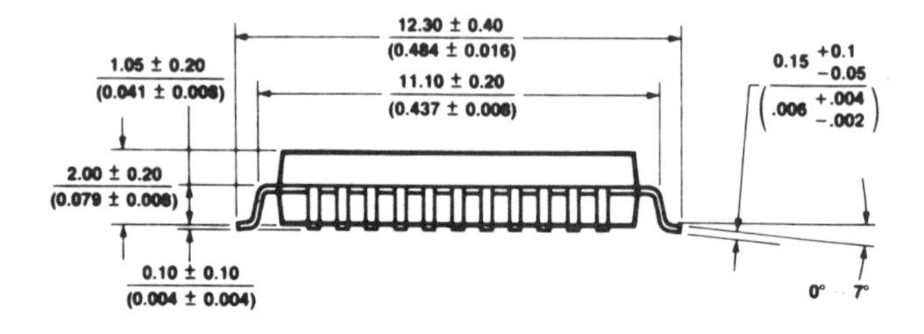
12.30 ± 0.40
(0.484 ± 0.016)
11.10 ± 0.20
(0.437 ± 0.008)
1.05 ± 0.20
(0.041 ± 0.008)
2.00 ± 0.20
(0.079 ± 0.008)
0.10 ± 0.10
(0.004 ± 0.004)
0.15 +0.1 −0.05
.006 +.004 −.002
0° 7°

Bar Code Labeling

Some SMD manufacturers offer bar coding on Tape & Reel products as a means to streamline identification and offer a simple, accurate, and cost effective approach to identifying components by machine-readable techniques.

Recommended by many industry organizations, as well as being specified for DoD use in MIL-STD-1189, Code 39 is the most widely used alphanumeric bar code.

Footprints

SMD manufacturer's footprints should be of concern to anyone working with SMT. Experienced designers stress that pad length can be important for SO packages. They should be long enough to provide a solder fillet of 0.005″ to 0.10″ after reflow, but not long enough to allow the device to float off centerline during the reflow operation.

For the PLCC, care must be taken not to run the footprint too far under the package, but rather, extend it out approximately 0.050″ to the outside. This will help reduce solder bridges under the package where they cannot be seen during visual inspection.

You will also find that SMD manufacturers have many new products under development; that is, they are not yet available in production quantities. However, most manufacturers will furnish preliminary technical data or engineering samples when requested by authorized designers.

The following footprints are typical of those currently in use by manufacturers and designers. All are quite readily adapted to CAD systems for timesaving designs. In fact, by the time this book goes to press, some CAD software suppliers will probably have many SMD footprints built into their systems whereas they may be retrieved merely by pressing a couple of keys on the computer keyboard.

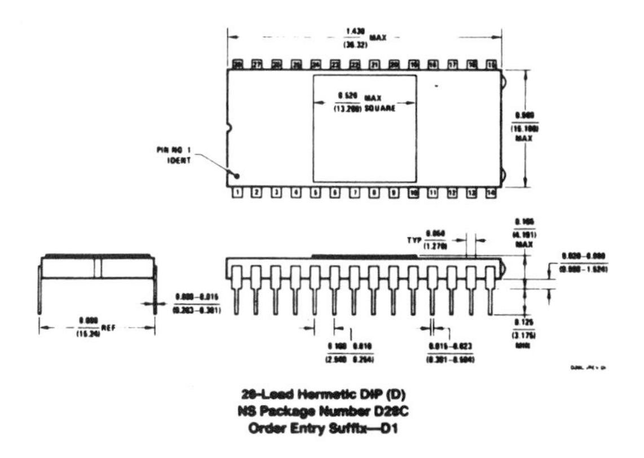

28-Lead Hermetic DIP (D)
NS Package Number D28C
Order Entry Suffix—D1

Footprints for PLCC

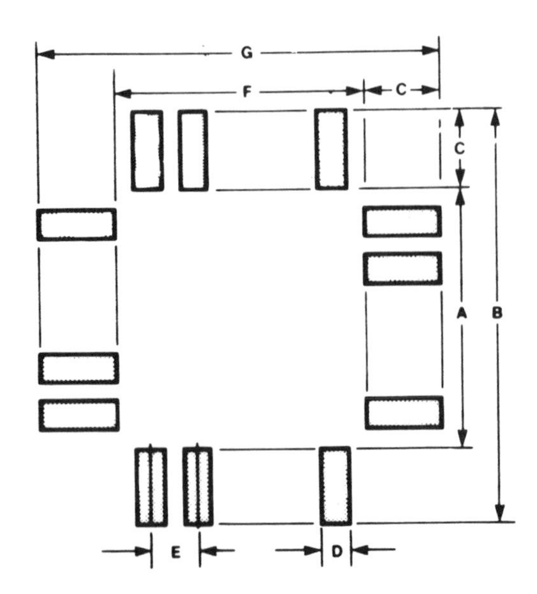

PACKAGE OUTLINE	INCHES						
	A	B	C	D	E	F	G
PLCC-20	0.260	0.440	0.090	0.024	0.050	0.260	0.440
PLCC-28	0.360	0.540	0.090	0.024	0.050	0.360	0.540
PLCC-32	0.360	0.540	0.090	0.024	0.050	0.460	0.640
PLCC-44	0.560	0.740	0.090	0.024	0.050	0.560	0.740
PLCC-52	0.660	0.840	0.090	0.024	0.050	0.660	0.840
PLCC-68	0.860	1.040	0.090	0.024	0.050	0.860	1.040
PLCC-84	1.060	1.240	0.090	0.024	0.050	1.060	1.240

Footprints for SO ICs

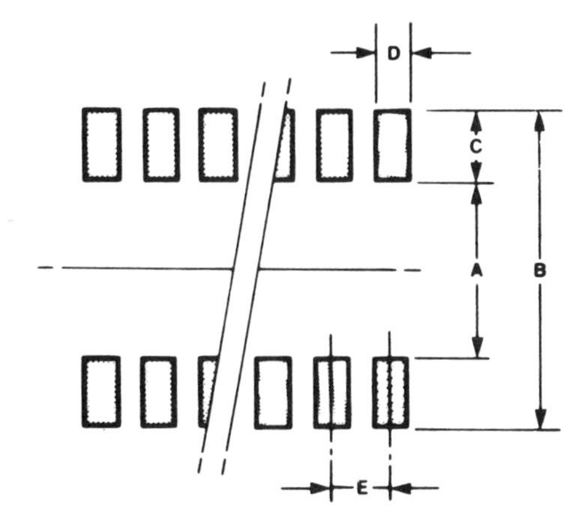

PACKAGE OUTLINE	INCHES A	B	C	D	E
SO-8, 14, 16	0.155	0.275	0.060	0.024	0.050
SOL-16, 20, 24, 28	0.310	0.450	0.070	0.024	0.050

PACKAGE OUTLINE	METRIC (mm) A	B	C	D	E
SO-8, 14, 16	4.0	7.0	1.5	0.6	1.27
SOL-16, 20, 24, 28	7.8	11.4	1.8	0.6	1.27

CERAMIC LEADLESS CHIP CARRIER (Cont'd.)

CLT052

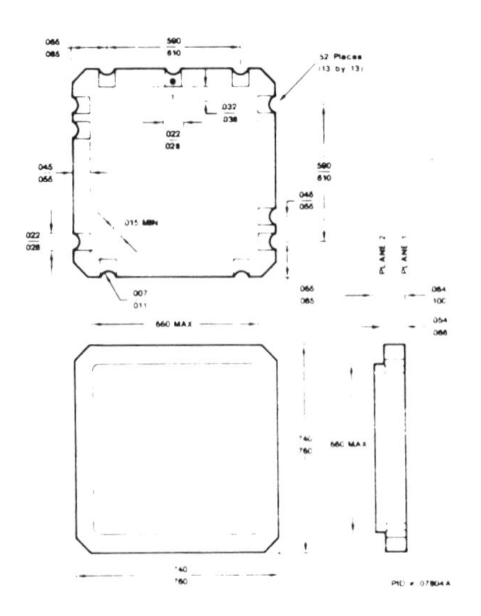

CL068

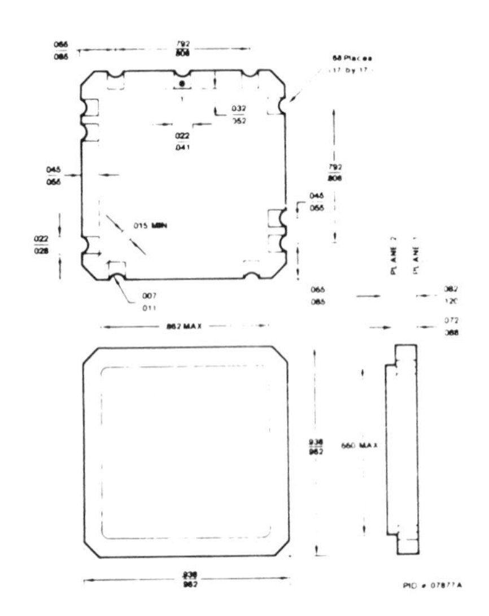

SOT-23

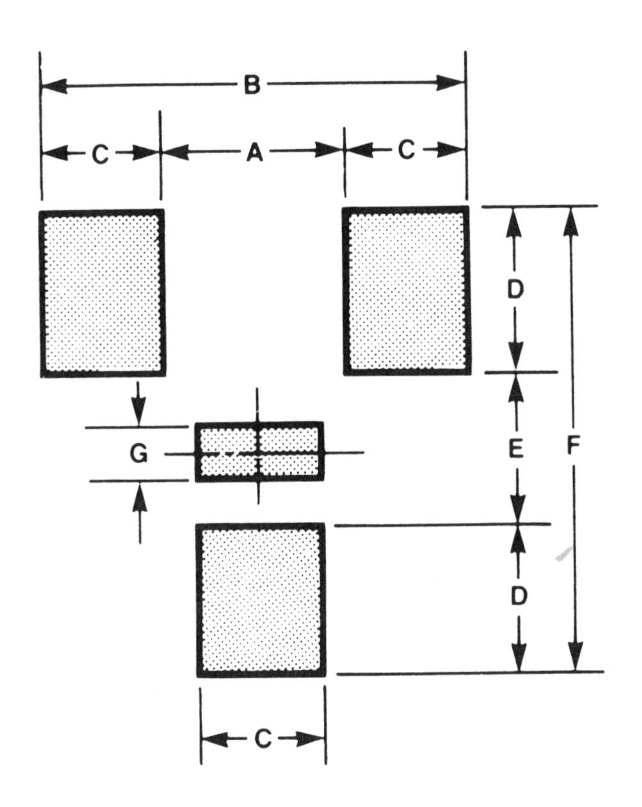

INCHES

SOT-23	A	B	C	D	E	F
Reflow	.048	.104	.028	.044	.104	—
Wave	.032	.136	.052	.052	.048	.152

METRIC (mm)

SOT-23	A	B	C	D	E	F
Reflow	1.2	2.6	.7	1.1	2.6	—
Wave	.8	3.4	1.3	1.3	1.2	3.8

PLASTIC LEADED CHIP CARRIER
(PLCC)

PL018

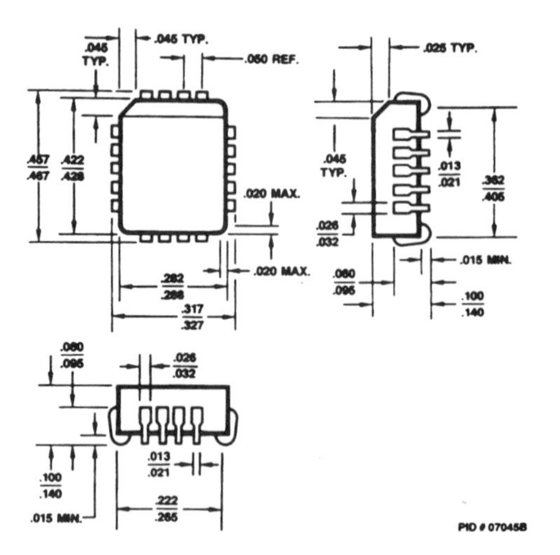

PLE018

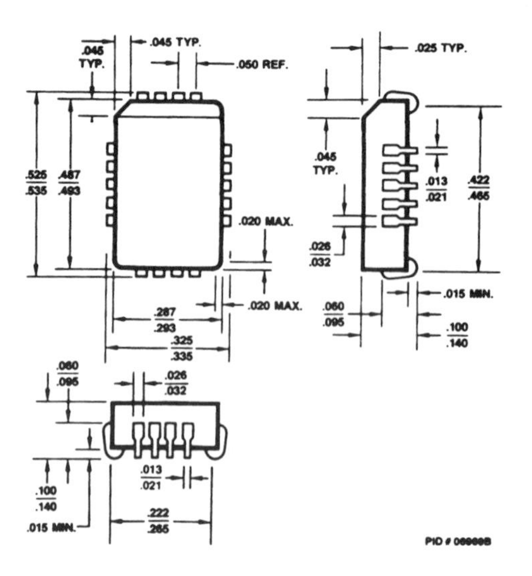

SOT-89

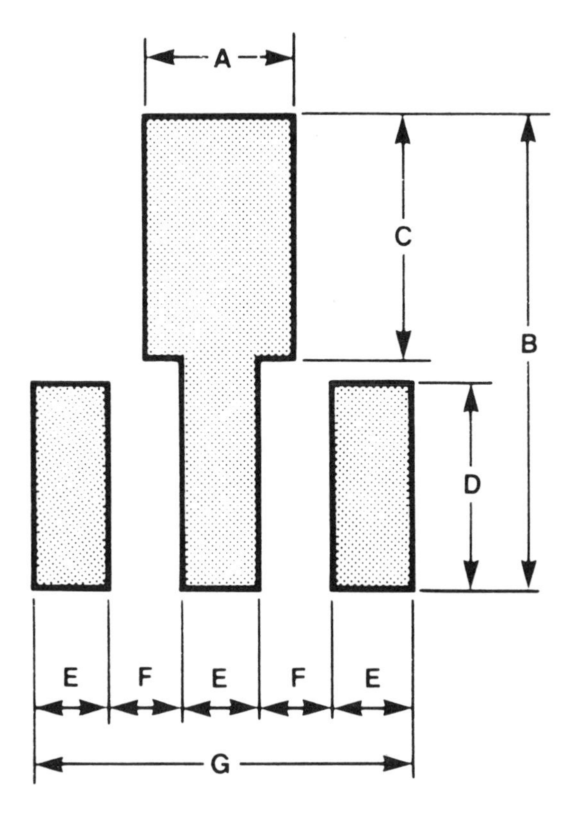

PACKAGE OUTLINE	INCHES A	B	C	D	E	F	G
SOT-89	.08	.[illegible]4	.104	.048	.032	.028	.152

PACKAGE OUTLINE	METRIC (mm) A	B	C	D	E	F	G
SOT-89	2.0	4.6	2.6	1.2	.8	.7	3.8

SOT-143

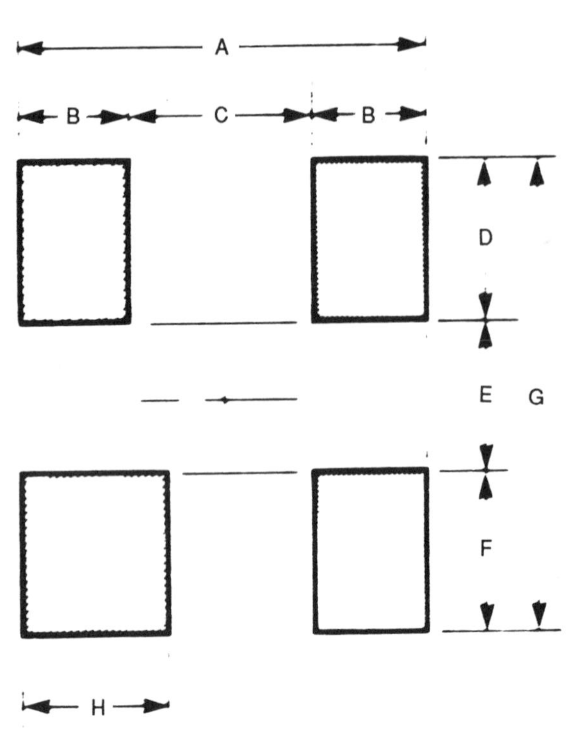

PACKAGE OUTLINE	INCHES A	B	C	D	E	F	G	H
SOT-143	.104	.028	.048	.036	.044	.036	.116	.044

PACKAGE OUTLINE	METRIC (mm) A	B	C	D	E	F	G	H
SOT-143	2.6	.7	1.2	.9	1.1	.9	2.0	1.1

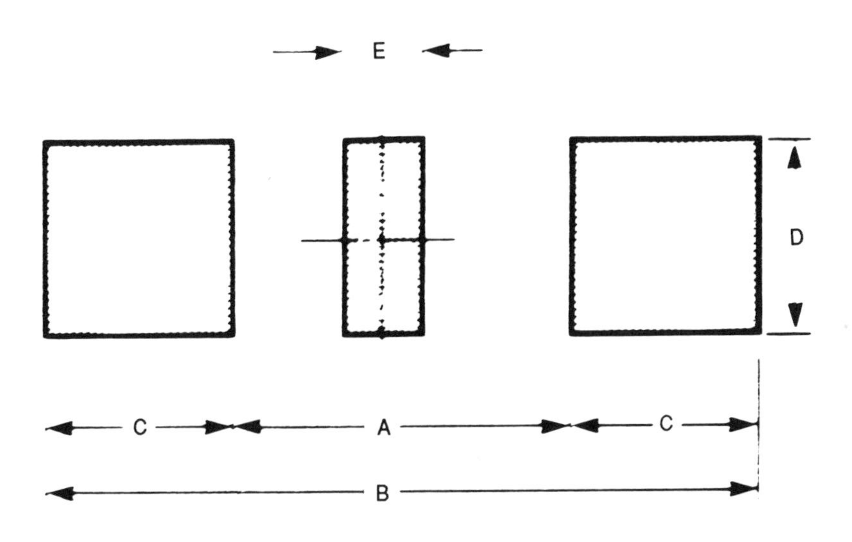

INCHES

CODE	SIZE	A	B	C	D	E
C0805	.08 x .05	.048	.144	.048	.048	.016
R/C 1206	.128 x .064	.08	.192	.056	.056	.020

METRIC (mm)

CODE	SIZE	A	B	C	D	E
C0805	2.0 x 1.25	1.2	3.6	1.2	1.2	0.4
R/C 1206	3.2 x 1.6	2.0	4.8	1.4	1.4	0.5

Footprints for wave soldered surface mounted resistors and ceramic multilayer capacitors.

SOD-80

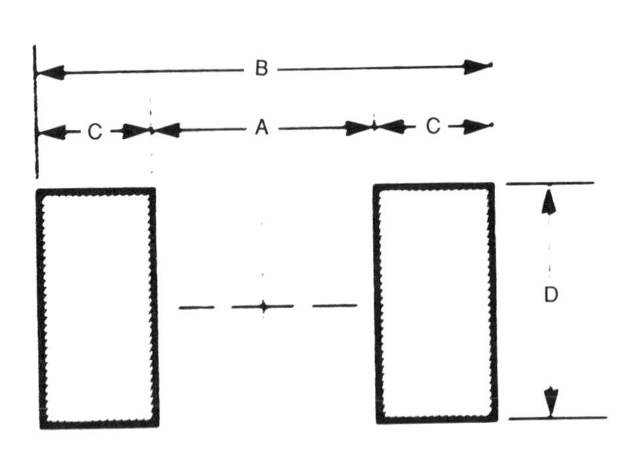

INCHES

SOD-80	A	B	C	D
Reflow	.096	.208	.056	.056
Wave	.10	.2	.05	.08

METRIC (mm)

SOD-80	A	B	C	D
Reflow	2.4	5.2	1.4	1.4
Wave	2.5	5.0	1.25	2.0

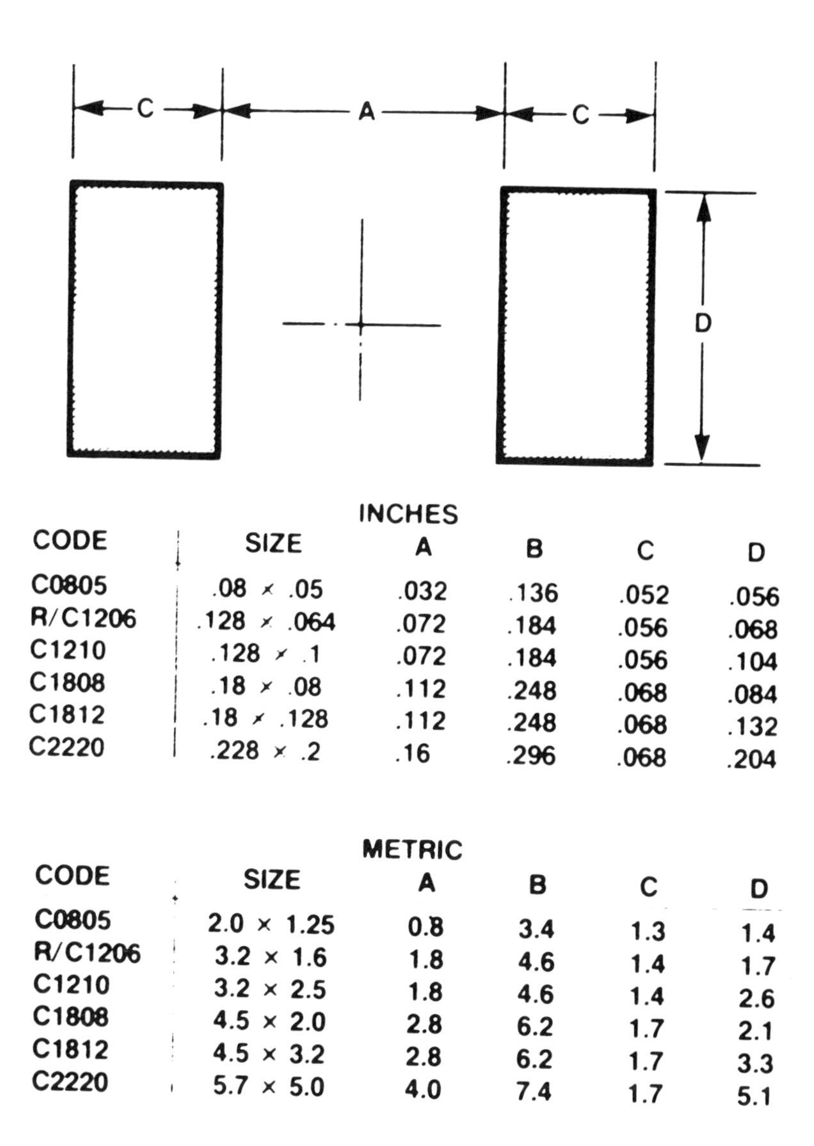

INCHES

CODE	SIZE	A	B	C	D
C0805	.08 × .05	.032	.136	.052	.056
R/C1206	.128 × .064	.072	.184	.056	.068
C1210	.128 × .1	.072	.184	.056	.104
C1808	.18 × .08	.112	.248	.068	.084
C1812	.18 × .128	.112	.248	.068	.132
C2220	.228 × .2	.16	.296	.068	.204

METRIC

CODE	SIZE	A	B	C	D
C0805	2.0 × 1.25	0.8	3.4	1.3	1.4
R/C1206	3.2 × 1.6	1.8	4.6	1.4	1.7
C1210	3.2 × 2.5	1.8	4.6	1.4	2.6
C1808	4.5 × 2.0	2.8	6.2	1.7	2.1
C1812	4.5 × 3.2	2.8	6.2	1.7	3.3
C2220	5.7 × 5.0	4.0	7.4	1.7	5.1

Footprints for reflow soldered surface mounted resistors and ceramic multilayer capacitors.

chapter 4

SMD ASSEMBLY PROCESS

Both the equipment and the processes of SMD assembly are quite different from the more common through-hole methods. While bench-top assembly is adequate for small assemblies, robotic and automatic assembly equipment will be more practical as the products move into higher volume.

The manufacturing and assembly process in SMDs today involves a large percentage of circuit assemblies that use both leaded IC components and surface mounted devices on the same PC board. Figure 4-1 shows a flow chart of a medium to high-volume assembly for mixed technology PC boards.

Solders for SMD Applications

Solder is an alloy made principally of tin and lead and is used to join metals. It provides reliable electrical connections and mechanically strong joints, when and if, the correct alloy is selected. Other metals such as antimony, silver, cadmium, indium and bismuth are alloyed with tin and lead to control certain physical and mechanical properties of the alloy; that is, melting range, tensile and shear strength, hardness and even corrosion resistance.

The tin/lead or "soft" solder alloys are the most widely used in electronic applications since their low melting temperatures make them ideal for rapid joining of most metals by conventional heating methods.

Care must be taken in specifying the proper alloy for each soldering job since each alloy has unique properties. When referring to the tin/lead alloys, tin is customarily listed first; for example, 60/40 refers to 60% tin, 40% lead.

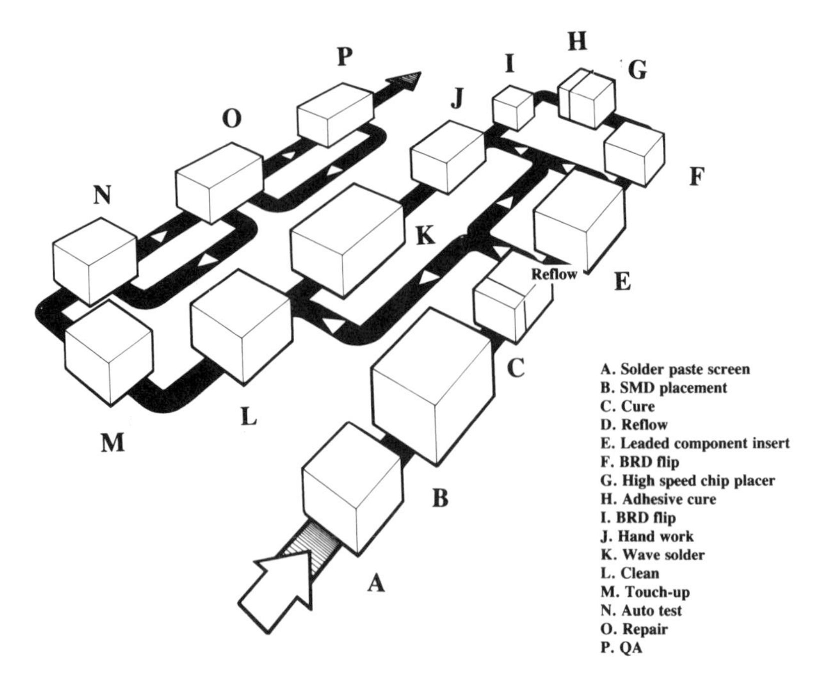

Fig. 4-1: Flow chart showing various operations of a medium-high volume SMD assembly set-up.

General-purpose solders include 40/60 and 50/50, which are typically used for plumbing and sheet-metal work as well as electrical applications. However, where minimum heat must be used during formation of the solder joint, as in electronic assemblies with heat-sensitive components and materials, higher tin-content alloys are required, such as 60/40 and 63/37.

Alloys of tin/lead with a small percentage of silver (63/37/2) are used to reduce the leaching of silver from the silver alloy end terminations of some passive components. These types of alloys are also ideal for soldering to thick-film silver alloy coatings on ceramics.

Bismuth-containing solders, which are frequently used as fusible alloys, may be used in applications where the soldering temperature must be below 381 degrees F (183°C). Indium alloys, also with low temperature melting ranges, are very ductile and are thus suitable for joining metals with greatly different coefficients of thermal expansion.

Table 4-1 gives the approximate temperature range required to melt soldering alloys:

TABLE 4-1
APPROXIMATE TEMPERATURE RANGE REQUIRED TO MELT SOLDERING ALLOYS:

SOLDER ALLOY	MELTING POINT
TIN/ZINC	400°F - 700°F
LEAD/SILVER	590°F - 690°F
TIN/LEAD	350°F - 620°F
TIN/ANTIMONY	480°F
TIN/LEAD/INDIUM	210°F - 420°F
TIN/LEAD/BISMUTH	100°F - 300°F

Fluxing and Cleaning

Populating a substrate involves the soldering of a variety of terminations simultaneously. In one operation, several different types of materials—each possessing varying degrees of solderability—must be attached to a common substrate using a single solder alloy.

It is for this reason that the choice of the flux is so important. The correct flux will remove surface oxides, prevent reoxidization, help to transfer heat from source to joint area, and leave non-corrosive, or easily removable corrosive residues on the substrate. It will also improve wettability of the solder joint surfaces.

The wettability of a metal surface is its ability to promote the formation of an alloy at its interface with the solder to ensure a strong, low-resistance joint.

However, the use of flux does not eliminate the need for adequate surface preparation. This is very important in the soldering of SMD substrates, where any temptation to use a highly-active flux in order to promote rapid wetting of ill-prepared surfaces should be avoided because it can cause serious problems later when the corrosive flux residues have to be removed. Consequently, optimum solderability is an essential factor for SMD substrate assembly.

Flux is applied before the wave soldering process, and during the reflow soldering process (where flux and solder are combined in a solder cream). By coating both bare metal and solder, flux retards atmospheric oxidization which would otherwise be intensified at soldering temperature. In the areas where the oxide film has been removed, a direct metal-to-metal contact is established with one, low energy interface. It is from this point of contact that the solder will flow.

Types of Flux

Electronic grade solder pastes are manufactured to meet the critical requirements of electronic component assembly. The composition of the pastes may vary as the individual requirements vary. A wide variety of compositions may be specified from suppliers which comply with recognized standards, like those of the ASTM (American Society for Testing and Materials).

While most solder pastes are pre-mixed compounds, the compound materials, namely, the metal powder and the flux binder, can be purchased separately. Mixing would take place in-house. One advantage of mixing solder paste in-house is that it is possible to mix just enough for immediate use. However, for best mixing results, all the powder in a packet should be mixed at the same time. For both pre-mixed and mix-it-yourself pastes, the remaining paste should be stored in small, tightly sealed containers.

There are two main characteristics of flux. The first is efficacy—its ability to promote wetting of surfaces by solder within a specified time. Closely related to this is the activity of the flux, that is, its ability to chemically clean the surfaces.

The second is the corrosivity of the flux, or rather the corrosivity of its residues remaining on the substrate after soldering. This is again linked to the activity; the more active the flux, the more corrosive are its residues.

Although there are many different fluxes available, and many more being developed, they fall into two basic categories: those with residues soluble in organic liquids, and those with residues soluble in water.

Flux Types

Flux used in solder pastes consists of the following primary components: (1) solvent or vehicle; (2) rosin/resin of synthetic (solid); (3) activator; (4) viscosity control additive.

1. The solvent selection is based on compatibility and ability to dissolve high concentrations of solids. Other important properties are: slow evaporation at room temperature; low moisture absorption; high flash point; and compatibility with supplemental activators and viscosity modifiers.
2. The solids used in solder pastes are selected based on their characteristics. These include: cleaning of surfaces to be soldered; oxidation protection to the solder powder and solder joint during heating; binder and surface protection after curing; stability at soldering temperatures; and ability to be completely removed by conventional cleaning solvents.
3. Activation levels of flux include: non-activated (R); mildly activated (RMA); activated (RA); and super-activated (RSA).
4. Organic thickeners are added to the flux systems to alter physical properties required for typical application. Thickness is selected based on the methods of dispensing—screening, stencil printing or other.

Solder-flux residues must be removed from the board surfaces after reflow. Both organic and inorganic activators are available. Organic systems are preferred for their ease of cleaning and environmental acceptability and safety.

Solder Applications

Solder pastes are dispensed in several ways including: syringe, pressure-fed reservoirs or guns. Large or small amounts of paste can be dispensed this way. Single dots or strips may be controlled with timing devices or robotic applicators. Other methods of application include casting component contacts with solder paste before attachment to the PC board footprint. This is called the transfer method.

Screening is a widely used method of solder paste application. The paste is relatively easy to apply, and precision is controlled.

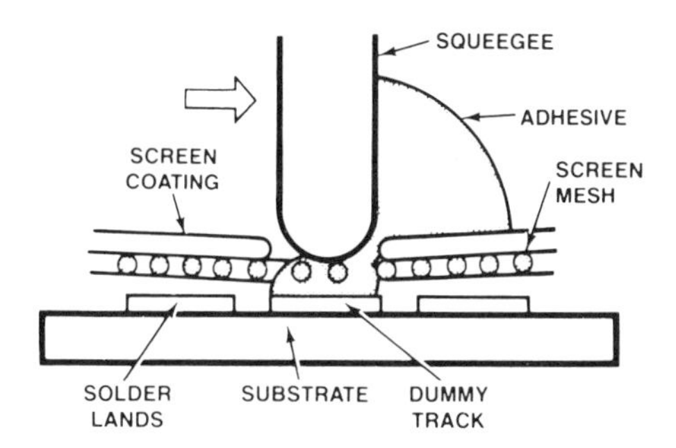

Fig. 4-2: Solder pastes can be accurately screened onto a work surface using screens ranging between 80-200 mesh.

Solder pastes can be accurately screened onto a surface, as shown in Figure 4-2, by a manual or automatic operation using screens ranging between 80-200 mesh. Mesh size depends on the definition required. Paste thickness is controlled by screen size, emulsion or mask thickness. A squeegee is normally employed to force paste through openings in the screen that contact print the areas where paste is to be applied. All other areas are sealed.

Curing Solder Paste

Because of its tackiness, wet solder paste serves as a fixing material. The retention of this tackiness must be considered in planning the production assembly. After this, the curing cycle prepares the board for the solder reflow. During the drying or curing process, the normal sequence of events is as follows:

1. The flux solvent starts to evaporate.
2. Viscosity starts to decrease (depending on solids in flux, percentage of metal, and thickness).
3. Solvent is completely volatilized.
4. Water-white rosin and/or heat-stabilized rosin/resin starts to melt and forms a protective envelope.

When curing is complete, components are held in place and the absorption of environmental moisture is eliminated.

An optimum drying system is one that slowly heats the assembly from the ambient to about 85°C. If some spread is acceptable, rapid drying using infrared heating is effective.

	COMPOSITION PERCENT													
Alloy Grade	Tin Desired	Lead Nominal	Antimony Min.	Antimony Desired	Antimony Max.	Silver Min.	Silver Desired	Silver Max	Bismuth Max	Copper Max	Iron Max	Aluminum Max	Zinc Max	Arsenic Max
70A	70	30	...	...	0.12	...	...	...						0.03
70B	70	30	0.20	...	0.50	...	...	...						0.03
63A	63	37	...	...	0.12	...	...	...						0.03
63B	63	37	0.20	...	0.50	...	...	...						0.03
60A	60	40	...	...	0.12	...	...	...						0.03
60B	60	40	0.20	...	0.50	...	...	...						0.03
50A	50	50	...	...	0.12	...	...	...						0.025
50B	50	50	0.20	...	0.50	...	...	...						0.025
45A	45	55	...	...	0.12	...	...	...						0.025
45B	45	55	0.20	...	0.50	...	...	...						0.025
40A	40	60	...	...	0.12	...	...	...						
40B	40	60	0.20	...	0.50	...	...	...						
40C	40	58	1.8	2.0	2.4	...	...	...						
35A	35	65	...	...	0.25	...	...	...						
35B	35	65	0.20	...	0.50	...	...	...						
35C	35	63.2	1.6	1.8	2.0	...	...	...						
30A	30	70	...	...	0.25	...	...	...	0.25	0.08	0.02	0.005	0.005	0.02
30B	30	70	0.20	...	0.50	...	...	...						
30C	30	68.4	1.4	1.6	1.8	...	...	...						
25A	25	75	...	...	0.25	...	...	...						
25B	25	75	0.20	...	0.50	...	...	...						
25C	25	73.7	1.1	1.3	1.5	...		...						
20B	20	80	0.20	...	0.50	...	...	...						
20C	20	79	0.8	1.0	1.2	...	...	...						
15B	15	85	0.20	...	0.50	...	...	...						
10B	10	90	0.20	...	0.50	...	...	...						
5A	5	95	...	...	0.12	...	...	...						
5B	5	95	0.20	...	0.50	...	...	...						
2A	2	98	...	...	0.12	...	...	...						
2B	2	98	0.20	...	0.50	...	...	...						
2.5S	0	97.5	...	...	0.40	2.3	2.5	2.7						
1.5S	1	97.5	...	...	0.40	1.3	1.5	1.7						
95TA	95	0.20 max.	4.5	5.0	5.5	...	...	...	0.15		0.04			0.05
95.5TS	96.5	0.20 max.	0.20	...	0.50	3.3	3.5	3.7	0.15		0.02			0.05

Table 4-2: Solder alloy composition chart.

The Reflow Process

There are several solder methods for reflow. These include: ovens, induction heating, infrared (IR), conveyorized hot plate, and hot air vapor phase.

A circulated-air oven or electrically heated convection furnace, with or without inert gas, can be used effectively to reflow solder paste. Both a batch-type oven and multi-zone conveyorized belt furnace can be used.

Infrared belt furnaces, with two or three heat zones and top and bottom heating, will provide rapid reflow. Absorption of IR energy varies with materials use. For example, the organic components of solder paste are excellent absorbers of IR, while gold or aluminum and reflowed solder are good reflectors. The use of focused IR is an efficient method of reflow soldering individual components.

An electric hot plate, or a series of hot plates, with a belt to transport parts, is an effective method of heating parts by conduction.

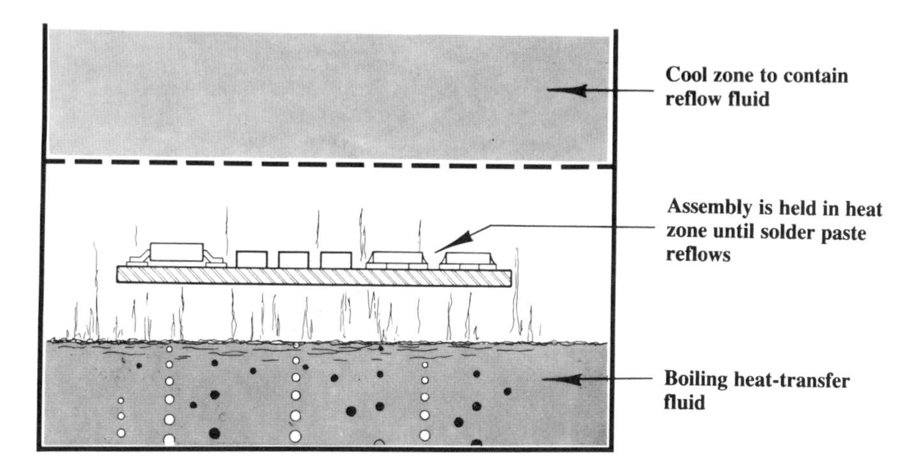

Fig. 4-3: Condensation soldering offers precise temperature control and is recommended where soldering of complex units is required.

Laser soldering with a microprocessor-controlled IR system is extremely fast. However, the paste must first be dried due to the rapid heating cycle of this method.

Vapor phase or condensation soldering offers precise temperature control. The entire assembly is heated to the boiling point of the heat-transfer fluid—and no higher. See Figure 4-3. By selecting solder alloys progressively from high to low melting points and by matching the heat-transfer fluids, it is possible to sequentially solder a complex assembly. Both batch and in-line vapor phase reflow systems are available.

Cleaning After Reflow Soldering

When cleaning to remove the flux residue, a solvent-type cleaner should be used for two reasons: 1) the flux, emulsifiers and thickeners in the solder paste do not readily dissolve in an aqueous solution; 2) the space under some of the surface mounted components is approximately .003 inch. The water molecules are not small enough to get under the components and flush out the contaminants. See Figure 4-4. Again, the equipment is dependent on volume and the company's specifications for assembly cleanliness.

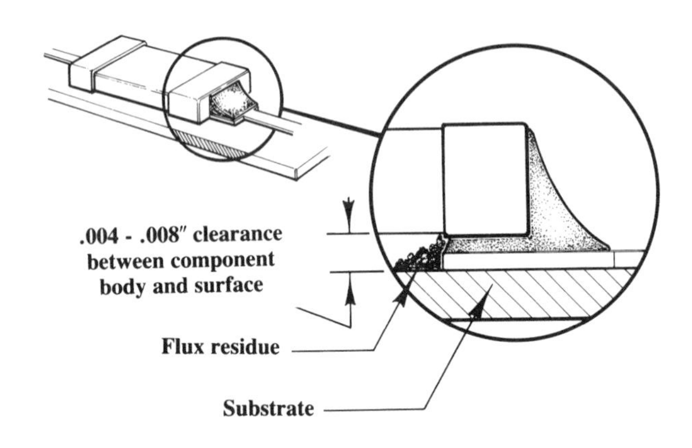

Fig. 4-4: Typical chip component after reflow soldering.

The cleaning method to remove the residues from rosin, resin, or related synthetic-solid-based solder paste must be matched to the components used in the soldered assembly. Fluorocarbon-type solvents with added polar solvent have excellent penetration characteristics and may be used in different cleaning cycles, including ultrasonic systems.

Bipolar chlorinated solvents are also effective, as is sparonification followed by water cleaning. Residues from water-soluble flux systems should be removed with agitated or boiling water, followed by de-ionized water rinses and forced air drying.

Assembly Methods

Placing the surface mounted devices onto their designated footprint or contact pattern is the operation referred to as pick and place. For very low-volume assembly or prototypes, hand placement with tweezers or vacuum pickup tools is adequate. As component count and density increases or assembly volume grows, the use of automation for this task becomes more practical. The speed, accuracy and component versatility will vary from one manufacturer to another. Many factors must be addressed if choosing equipment for your own factory, among them:

1. Maximum substrate or board size/placement area.
2. Volume projected for your product.
3. Placements per hour.
4. Accuracy of placement/repeatability.
5. Part mix, number of component stations.
6. Type of components/size limits.

After evaluating your product and component mix, it may be practical to have several different types of systems in line to efficiently process your assembly.

Assembly Options for SMDs

Type one: Type one is a one-or two-sided PC board, as shown in Figure 4-5. Surface mounted components are attached to contact or footprint patterns etched into the circuit. The assembly sequence would start with screen printing

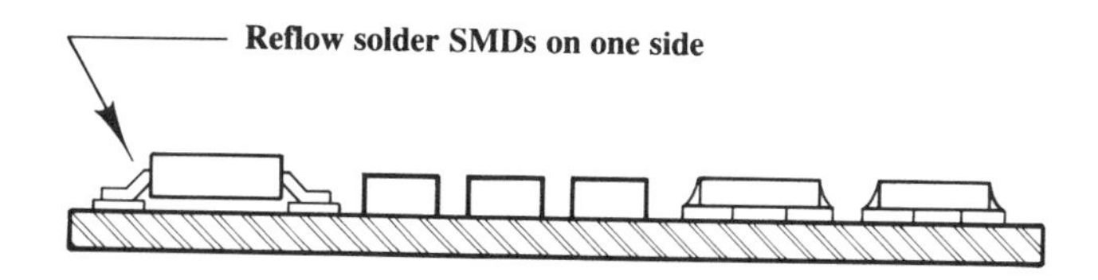

Fig. 4-5: A type one or two-sided PC board where SMD components are attached to contact or footprint patterns etched into the circuit.

solder paste—a mixture of tin/lead alloy powder, flux, solvent and binders—onto specific patterns on the board surface. The components are placed onto the PC board surface, contacts aligned with the footprint pattern and cured. The curing chamber raises the temperature of the assembly to extract any solvents or moisture still present from the paste or the board itself.

After curing, the assembly is placed into an even higher temperature chamber until the solder paste is converted into a liquid and reflows the solder, creating the electrical and mechanical bond from the component to PC board.

Type Two: Type two is a two-sided PC board with surface mounted components mounted on both sides, as shown Figure 4-6. The assembly sequence would first apply solder paste to the contact or footprint pattern on the PC board side one. The surface mounted devices are placed into the paste material and cured and reflowed in the same manner as explained for Type One assembly. Side two can be assembled in the same way or the components can be attached with adhesive epoxy and wave soldered. Solvent cleaning usually follows each assembly phase to prevent flux residue from hardening.

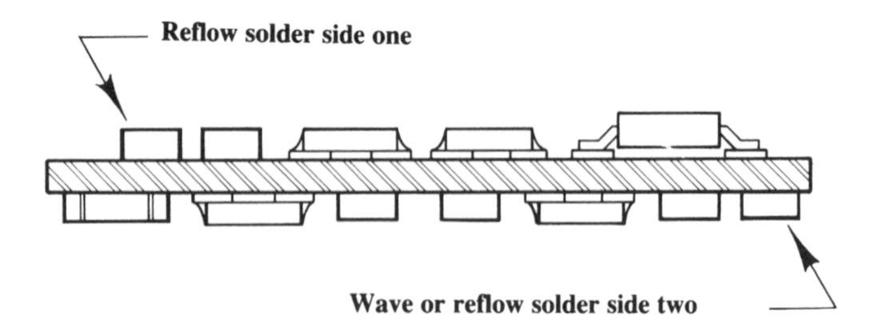

Fig. 4-6: Type two, two-sided PC board with SMD components mounted on both sides.

Type Three: Type three employs leaded or through-hole components mounted on side one and surface mounted parts attached to side two with a UV cured epoxy and wave soldered simultaneously as in Figure 4-7. The leaded components might include conventional DIP and SIP packages, axial leaded resistors, diodes, capacitors and jumpers or radial leaded devices.

On side two the surface mounted parts would incorporate chip resistor and capacitors, SOT and SO packages, diodes, etc. Tantalum capacitors in SMT packages may not always be suitable for side two mounting because of the higher part profile of the component. Wave soldering of the PCC (plastic chip carrier) is also not practical because of its high profile and lead configuration.

This type of assembly was the very first approach to adapting SMT to the high volume production of PC boards. The economic advantage is twofold:

1. Greater component density.
2. Utilization of common wave solder equipment.

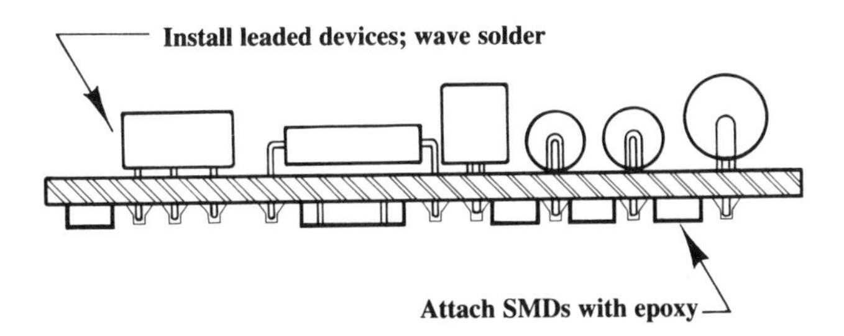

Fig. 4-7: Type three PC board untilizing a combination of through-hole and SMD components.

Type four: Type four attaches leaded and surface mounted devices on side one with additional surface mounted components attached to side two.

The goal of component specialists is to locate as many surface mounted alternatives to the leaded predecessor as possible. With few exceptions, most through-hole devices are furnished surface mounted packages. It will be necessary to continue to use leaded configurations until such time as the cost and availability of all SMDs become acceptable. For this reason, we have to contend with the "mixed technology" method of assembly, as shown in Figure 4-8.

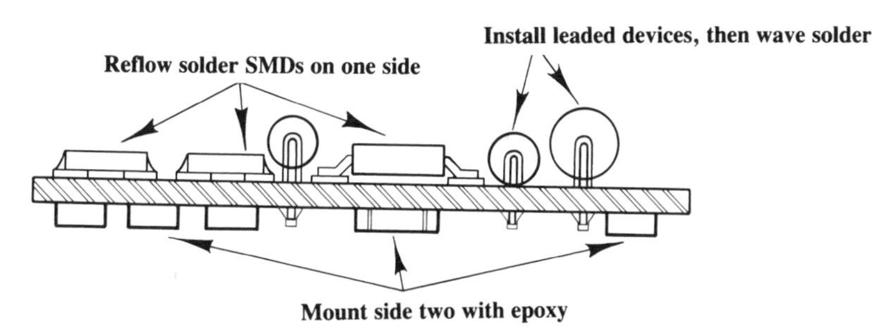

Fig. 4-8: This board utilizes leaded and surface-mounted devices on one side with additional SMD components on the opposite side.

The assembly sequence would start with:

1. Screen-printing solder paste to the surface mounted components' contact patterns.
2. Mounting the components in paste and curing.

3. The PC board passes through a high temperature chamber that melts or reflows the solder paste, leaving a finished solder connection on each contact of the surface mounted device. The assembly is then cleaned, ready for the next phase.

To mount the remaining surface mounted components on side two, an adhesive epoxy is applied to retain the device, as in Figure 4-9, and again cured to harden in position. With the parts now bonded to side two, the board is flipped back to side one and all remaining leaded components are inserted through plated holes in the board. Now the assembly is passed through the wave solder process and cleaned once again before inspection (and touch-up, if necessary.)

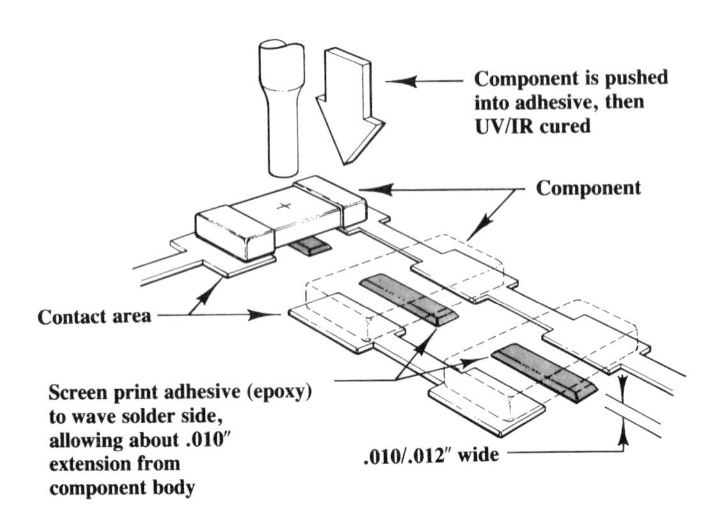

Fig. 4-9: SMDs are mounted with epoxy and then cured to harden in position.

In the case of smaller modules in the form of a dual in-line or single in-line pin configuration, several assemblies are usually attached together or panelized. The individual units are separated after assembly and leads are attached as a post operation, as shown in Figure 4-10.

Adhesive Applications and Curing

The use of adhesives in electronics is not new. Epoxies and silicones, for example, have been used for encapsulating and plotting components for years. But the application of an adhesive, either to component or substrate, as a means of holding the components in position prior to soldering, is new, and specific to surface mounted device (SMD) substrate assembly.

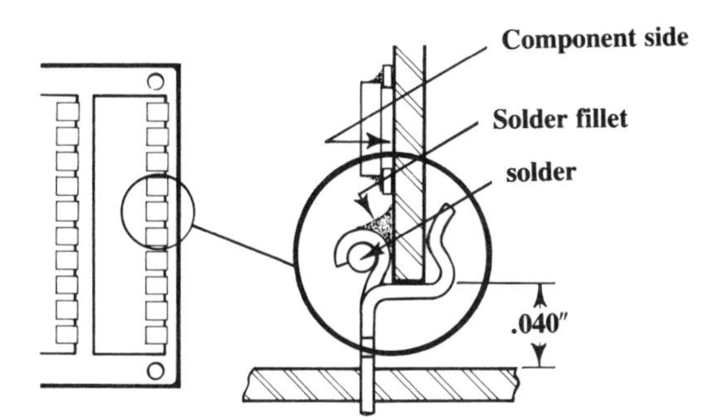

Fig. 4-10: When the smaller modules are manufactured, several assemblies are frequently attached together and then separated after assembly.

Unlike conventional through-hole (leaded) components which, once inserted into holes in a substrate, are held in place by clinching the leads, SMDs are simply placed onto the substrate surface with no inherent means of holding them in place as shown in Fig. 4-11. So in most cases, it is necessary to use an adhesive to form a bond until wave soldering can take place. Reflow-soldered SMD substrates are an exception to this as they use a solder paste applied to the substrate before the SMDs are placed. Here, the paste has sufficient adhesion to hold the SMDs in position until the solder is reflowed.

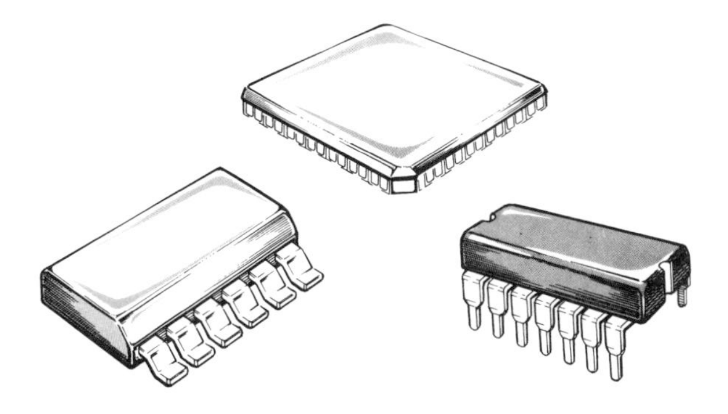

Fig. 4-11: SMDs, unlike through-hole components, have no convenient means of holding them in place prior to wave soldering.

However, reflow soldering is not suitable for all types of SMD assembly. For example, a Type II, mixed-print substrate (a combination of both SMD and through-hole components) must use the wave soldering technique, or combination wave/reflow soldring, so an adhesive is essential to hold the SMDs in place.

This adhesive must hold the device in the correct orientation upon placement, maintain it during the physical handling before final assembly, and withstand the adverse environments of fluxing plus the high temperatures of the solder wave.

After soldering, however, the adhesive's function is complete and it becomes a potential problem area. It may, for example, absorb moisture and become a conductive leakage path. Or it may degrade in an undesirable way. These considerations, plus methods of application and curing, govern the selection of an adhesive for SMD assembly.

To determine which is the best adhesive for SMD assembly, an outline specification or an "ideal" adhesive is used to evaluate the various adhesive types. This specification gives the characteristics of the adhesive itself, plus factors relating to the application and curing processes, and environmental elements such as toxicity and flammability.

Physical Characteristics

- Stable one-part (single component) system.
- Long shelf-life—without the need for refrigerated storage.
- Good void filling—fills gap between substrate and SMD.
- Thixotropic—good drop profile and no adhesive migration.
- Electrically non-conductive—no possibility of short-circuits.
- Non-corrosive—will not attack component or substrate.
- Sufficient pre-cure tackiness—to pull SMD from pick-and-place unit.
- Adequate post-cure bond strength—for reliable handling.

- Chemically stable—will not react with flux and cleaning solvents.
- Distinctive color—assists visual inspection.

Process Considerations

- Application method—pin-transfer, pressure-syringe, or screen-printing.

- Short cure time with low-energy cure—allows in-line curing.
- Resistance to high temperatures—encountered during wave soldering.
- Repair possibilities—removal of component after curing.

Environmental Considerations

- Non-flammable.
- Non-toxic.
- Odorless.
- Non-volatile.

Of the many adhesives available, none fits all these requirements exactly. There are, however, many that can bond ceramic or plastic SMDs to various substrate materials, each with its own advantages and disadvantages. Below is a brief description of four groups of adhesives, classified by chemical composition:

Thermosetting adhesives: Cured by chemical reaction to form a cross-linked polymer. The thermosets are strong, structural adhesives, and once cured, by heat or catalytic action, cannot be softened to establish new adhesive bonds. Examples include epoxy, acrylic, phenolic, polyester.

Thermoplastic adhesives: The thermoplastics do not change chemically when establishing a bond, and can be re-softened any number of times to form new adhesive bonds. Typically used in lightly stressed, non-structural, applications.

Elastomeric adhesives: A subset of thermoplastics, these adhesives have great elasticity, and are formulated from synthetic or naturally occurring polymers in solvents, latex cements, or dispersions. Particularly noted for their flexibility. Examples include rubber, silicone, neoprene, etc.

Toughened alloy adhesives: Formulated by blending rubbers and resins, both natural and synthetic, these adhesives are a special type of thermosetting adhesives. Developed for specific functions, these adhesives combine high strength with resistance to shock-loading. Examples include epoxy-nylon, phenolic-neoprene, etc.

Some of these adhesives have a proven track-record in the electronics industry. For example, the thermoset epoxies are probably the oldest and most-used engineering adhesives, and are available as one-part or two-part systems in the form of liquids, pastes, films, and powders. Cured by chemical reaction through heat, the epoxies are best suited to automated SMD placement processes.

Another popular group of adhesives, the cyanoacrylates, are not true one-part systems, but "one-part application" adhesives. This means that although they require two parts, only one need be applied as the second is supplied by the surfaces to be bonded in the form of absorbed moisture. Consequently, the polymerization is a surface initiated process, and cyanoacrylates function best with close fitting parts. This limits their application possibilities for SMD assembly, particularly when using MELF components (for example, leadless diodes with an SOD-80 encapsulation) which cannot use cyanoacrylates.

The alloys and elastomers, due to their specialized nature, are not generally used for SMD assembly, although an acrylic-rubber combination had been formulated as a pressure-sensitive adhesive for use with SMDs. Other adhesives to be considered for SMD applications are the acrylics and the anaerobics. Although they don't have an extensive history of use in the electronics industry, they do have certain characteristics that are suitable for use in SMD applications. Table 4-3 lists the advantages and disadvantages of these four types of adhesive in SMD assembly.

Comparing the characteristics of the adhesives given in Table 4-3 with those for the "ideal" adhesive for SMD assembly illustrates that some trade-off has to be made. In many cases, the choice of adhesive limits the choice of application method. For example, the cyanoacrylates may only be applied by syringe.

Ultimately, the user should first determine the application process best suited to the type of substrate to be loaded, then choose a machine capable of this process (either a stand-alone adhesive application station or one integrated into a pick-and-place machine), and finally, select the ideal adhesive for the machine. Such an adhesive will quite often be the responsibility of the machine manufacture, as they will have designed the machine with a specific adhesive type in mind.

So the selection process is:

1. Process
2. Machine
3. Adhesive

Experience has shown that a thermoset epoxy adhesive is best suited to pin-transfer application or pick-and-place machines. It has excellent precured properties (simple, trouble-free dispensing, good fill characteristics and good dot formation); effective curing chemistry (no undesirable gas release or adhesive migration); and ideal post-cured properties (resistance to damage during handling and to the high temperature of the soldering operation). It also has shown no long term influence on device or substrate.

Adhesive Application

There are three methods to be considered for automatically applying adhesives in surface mounted device assembly. They are:

Pin-transfer—in which a pin picks up a droplet of adhesive from a reservoir and transfers it to the surface of the substrate or component; surface tension causes a portion of the droplet on the pin to form a dot on the substrate (or component). See Fig. 4-12.

Screen-printing—in which a fine mesh screen, coated except for the areas where adhesive is required, is placed over the substrate. A squeegee passes across the screen and forces the adhesive through the uncoated areas of the mesh and onto the substrate. See Fig. 4-13.

Pressure-syringe—a software-controlled syringe dispenses the adhesive from an enclosed reservoir by applying air pressure for a fixed time. The adhesive dot size is determined by the size of the syringe nozzle, and by the duration and magnitude of the applied pressure. See Fig. 4-14.

Manual application of adhesive can only be considered for prototyping, development work or repair, with an operator using a hand-held dispenser. The advantages and disadvantages of each automated system are summarized in the table in Fig. 4-15.

Pin-Transfer

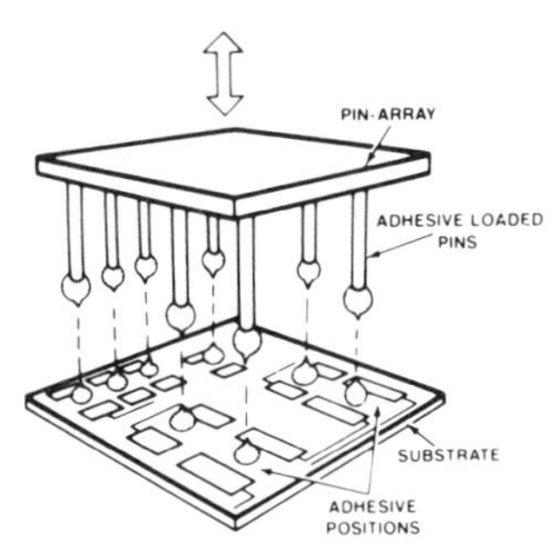

Fig. 4-16: Pin-transfer by pin-array directly onto the substrate.

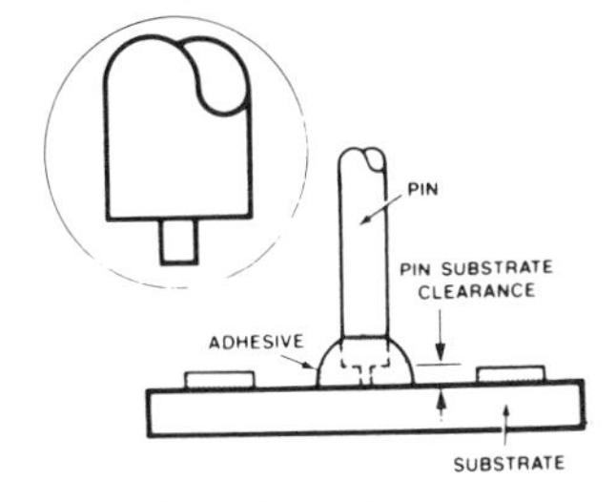

Fig. 4-17: The pin mustn't touch the substrate surface.

Pin-transfer is the simplest mass application method for automated high-volume production machines. With this method you can either apply the adhesive to the substrate, or directly onto the component.

Application of adhesive to the substrate requires a fabricated pin-array that exactly corresponds to the adhesive positions on the substrate (See Fig. 4-16), and the adhesive application stage must be incorporated in the substrate assembly line immediately before the SMDs are placed. Curing of the adhesive must take place shortly after SMD placement. The pin diameter and shape, the depth to which the pin is dipped into the reservoir and the viscosity of the adhesive determines the dot size.

During the transfer of the adhesive dot from pin to substrate, the pin must not touch the substrate surface as this will distort the resulting dot profile. Therefore, the substrate must be relatively flat and free from distortion. However, to ensure that the pin does not distort the adhesive dot, the tip of the pin is modified as shown in Fig. 4-17. This smaller diameter pin tip prevents the main bulk of the pin touching the substrate, but is of too small a diameter to have any significant effect on the resulting dot profile.

The nature of the pin-array makes it possible to apply the adhesive to a substrate that has already been loaded with through-hole (leaded) components on the other side (a mixed-print substrate) since the pins fit between the clinched leads of components as shown in Fig. 4-18.

Once placed, the adhesive dot must accept the SMD from the pick-and-place unit, and these may be of two types: a vacuum pipette that holds the SMD by vacuum alone; or a jawed chuck that holds the SMD incorporating a vacuum pipette to sense the presence of a component. For this latter system, the adhesive must have sufficient uncured tackiness to pull the SMD from the vacuum sensing pipette.

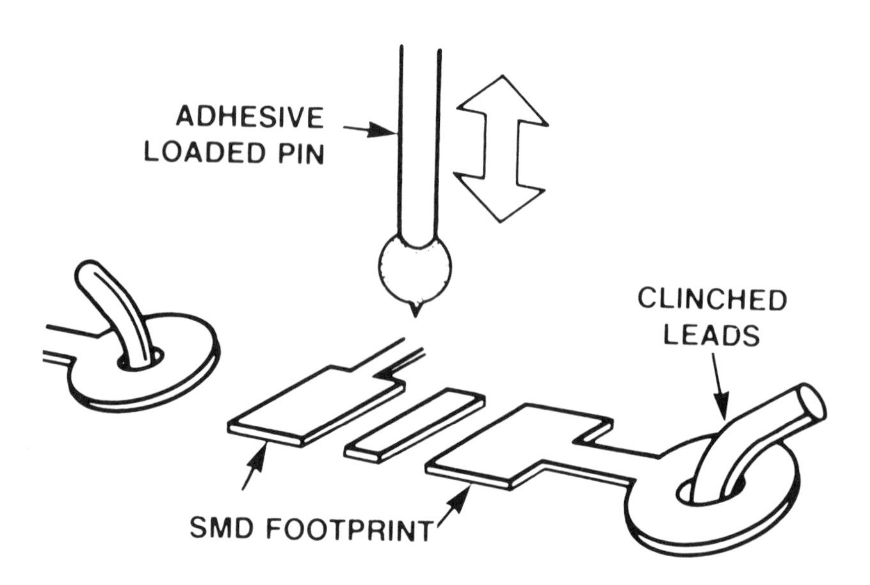

Fig. 4-18: The pin can apply adhesive between the clinched leads of through-hole components.

The alternative pin-transfer technique, where adhesive is applied directly to the SMD, is shown in Fig. 4-19. Here, the pins apply the adhesive to the component from below, immediately after they have been taken from the tape reels by the pick-and-place units. This system has the advantage of not requiring a purpose-made pin-array and is therefore suited to smaller production runs where several slightly different versions of a design may be required. The tooling costs are lower as the number of pins employed by the machine is software programmable.

Screen-Printing

Like pin-transfer, the screen-printing technique places all adhesive dots simultaneously. Also like pin-transfer, it requires the substrate to be flat and distortion-free, but unlike pin-transfer, there can be no protrusions or obstructions on the surface as this will prevent the screen making contact with the substrate. Therefore, screen printing cannot be used for adhesive application of pre-loaded mixed-print substrates.

The adhesive dot size is determined by the size of the exposed area in the coating on the screen-mesh, the thickness of this coating (shown in Fig. 4-20), the mess density, and the viscosity of the adhesive. Although essentially a very simple application method, screen-printing does have certain drawbacks regarding the high level of maintenance required. For example, the screen must be kept scrupulously clean as a blocked mesh inhibits dot formation.

A similar method of adhesive application is that of stenciling. Here, a rigid stencil without a mesh is used in place of the screen, and the adhesive is sprayed through the apertures onto the substrate.

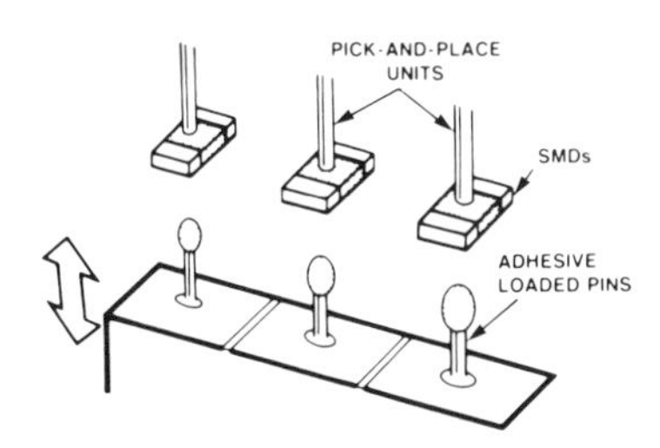

Fig. 4-19: Pin-transfer of adhesive directly to the SMDs.

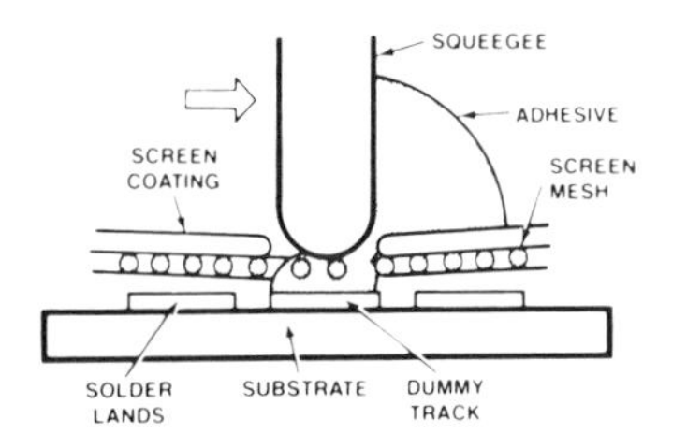

Fig. 4-20: Section showing the principle of screen printing adhesive.

Pressure Syringe

The main advantage of the pressure-syringe method is that it allows different amounts of adhesive to be applied to different positions. Moreover, by using an adhesive with a specific rheological (flow) characteristic, relatively high dots can be realized as shown in Fig. 4-21.

This method, however, is difficult and slower than the others as fewer dots can be placed simultaneously, thus limiting its high-volume production potential. But for smaller production runs, and substrates with fewer SMDs, the flexibility provided by the software programmability of this system is certainly advantageous.

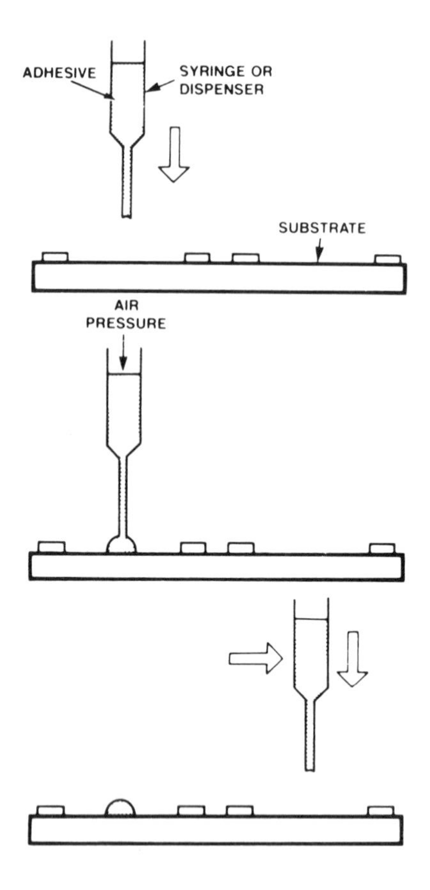

Fig. 4-22: Pressure-syringe adhesive application.

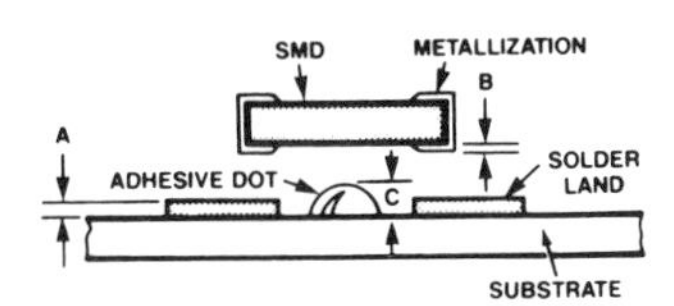

Fig. 4-23: Variation in substrate metallization height.

Adhesive Dot Height Criteria

The amount of adhesive applied to each position is critical for two reasons: first, the adhesive dot must be high enough to reach the SMD, and second, there must not be too much adhesive which may foul the solder land and prevent the formation of a good solder joint.

The two dimensions (substrate metallization height, A) and SMD metallization height, (B) that determine the minimum height of the adhesive dot (C) are shown in Fig. 4-23. This diagram illustrates that C must be greater than (A + B) for the dot to reach the SMD when placed. In practice, C should be at least twice (A + B) for the formation of a good, strong bond.

Taking these parameters in turn, the substrate metallization height (A), can range from about 35 μm for a normal print-and etch PCB, to 135 μm for a plated-through-hole board (see Fig. 4-24). And the component metallization height (B), for example of R/C1206 passive devices, varies between 10 and 50 μm.

The variation in the lead standoff height (corresponding to dimension B) for SOT-23 and SOT-143 devices of SO ICs is even greater. The standoff distance (B) is 178 ±76 μm for the SO and 203 ±101 μm for the SOL.

These packages were primarily intended for reflow soldered substrates since a standoff distance much greater than 100 μm is not advisable for use with adhesives. Clearly, this poses a problem for the dot height criteria, for (A + B) can vary considerably on any one substrate. But, as it is advantageous to apply the same amount of adhesive to all placement positions, some compromise is necessary.

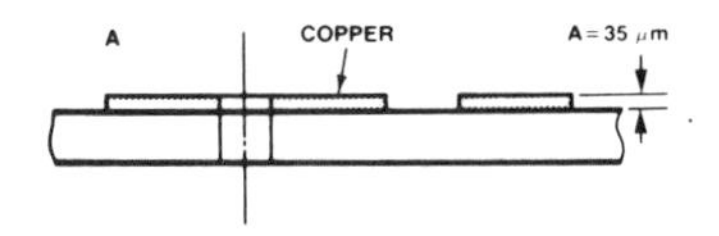

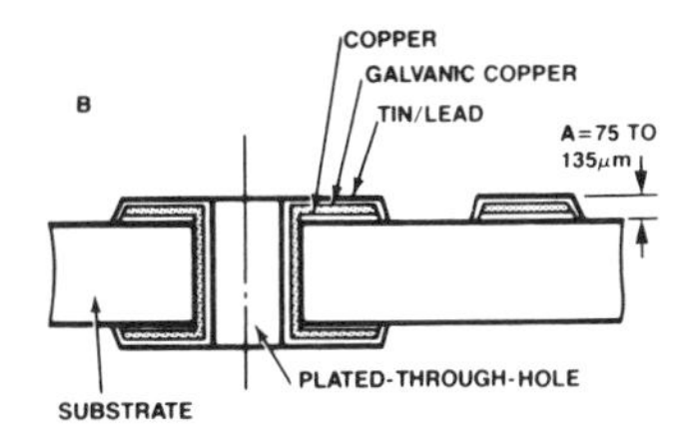

Fig. 4-24: Variation in SMD metallization height/lead standoff height for: R/C 1206 (top), and SOT-23 or SOT-143 (bottom).

Dummy-Tracks

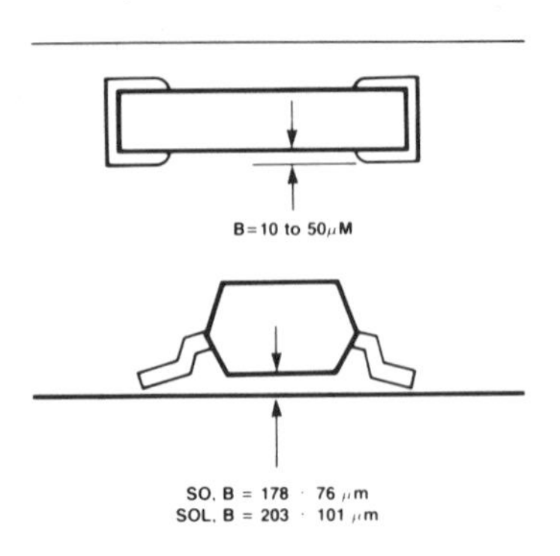

Fig. 4-24: Variation in SMD metallization standoff height for: R/C1206 (top), and SOT-23 or SOT-143 (bottom).

A solution to this problem is to put a track under the device, as shown in Fig. 4-25. This will help eliminate the substrate metallization height (A), from the adhesive dot-height criteria (although, it must be borne in mind that a track 0.3-0.5mm [.012-.020″] will not completely support a drop with a diameter of 1.0-1.2 mm [.040-.048″]). Consequently, (C) can be a virtual constant. Quite often, the high component density of SMD substrates makes it essential to route tracks between solder lands and where it doesn't, a short dummy-track should be introduced.

In practice, it is possible to run two tracks under the R/C12106 size case and most other SMDs, and these are often covered with a solder-resist, as shown in Fig. 4-26. The resist further reduces the amount of adhesive required and the double tracks provide a broader base on which the dot can be applied.

One dot of adhesive is sufficient for bonding discrete SMDs and small-outline transistors and diodes, and for SO ICs up to 16 pins. The larger SO IC packages require two dots of adhesive, so the through-tracks (or dummy-tracks) must be positioned beneath the IC accordingly.

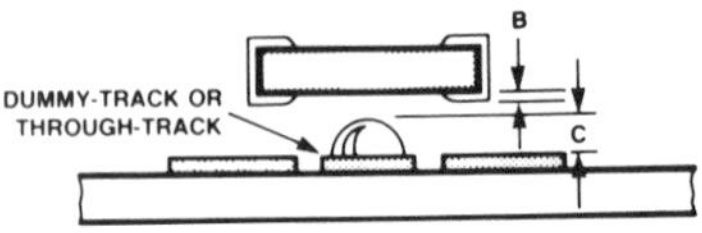

Fig. 4-25: Modifying adhesive dot-height criteria with a dummy-track.

Solder Land Contamination

Contamination of the solder land with even a very thin layer of adhesive has very serious consequences for the soldering process. Many commercial adhesives are transparent when spread into thin layers, and if such a layer spreads onto the solder land, the solder will not form a good joint. For this reason, the adhesive viscosity must be correct to prevent "bleeding" and the adhesive placement must be very accurate.

MELF Component Placement

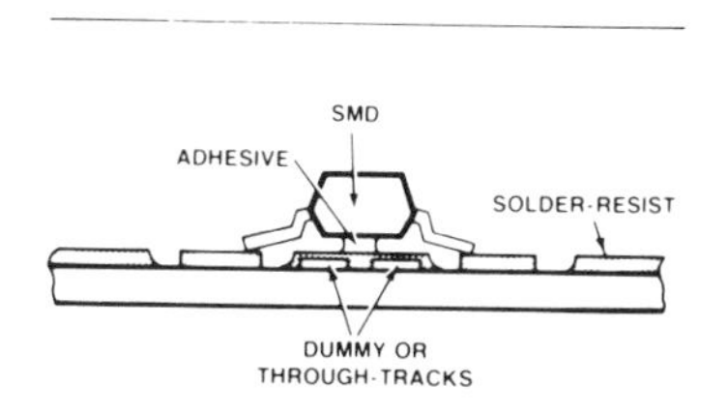

Fig. 4-26: Double tracks with solder resist.

MELF component (cylindrical SMDs, for example the SOD-80 diode encapsulation, see Fig. 4-27) placement can present a problem if the component body dos not hit the adhesive dot right in the middle. For when a cylindrical body is brought into contact with the hemispherical adhesive dot, the area of contact is very small indeed. Any deviation to either side of the center of the dot and the integrity of the adhesive bond may be seriously affected.

Take for example, the SOD-80 package. Along the center of this type of MELF, the lead stand-off height (dimension B in the dot height criteria) is about 0.1 mm (.040″). A deviation of 0.5 mm. (.020″) either side of center results in dimension (B) increasing to nearly 0.4 mm (.016″) unacceptable for the formation of a good bond.

Another problem with MELF components can arise if a substrate loaded with them is stored in a non-horizontal position (in an inclined rack, for example) prior to curing the adhesive contact area and the cylindrical shape, MELFs have a tendency to break free and cool down the substrate. Therefore, MELF loaded substrates should be kept in the horizontal position until the adhesive is fully cured.

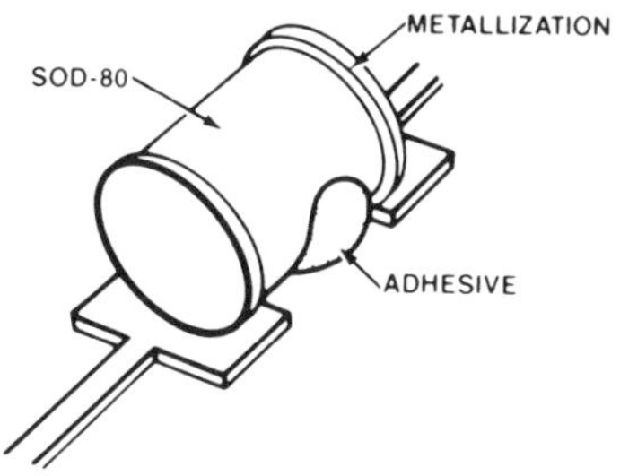

Fig. 4-27: Typical MELF component; the SOD-80 package.

Adhesive Curing

The method of curing an adhesive depends largely on the type of adhesive used. For automated placement machine applications, adhesives will generally use one or more of the following methods:

- Heat/Time—conventional oven or infra-red radiation.
- Catalytic action—mixing with a hardener.
- Ultra-violet (UV) radiation.
- Anaerobic—the absence of oxygen.

Curing By Heat/Time Plus Catalyst

Two-part epoxy systems consist of a resin and a hardener. When both constituents are mixed together, curing will proceed at room temperature, or be accelerated at elevated temperatures. These two parts can be combined by pre-mixing, or mixing at the application site, this latter method requiring a complicated metering and delivery system as accurate mixing in the correct proportions is necessary for rapid and complete polymerization.

A pre-mixed two-part epoxy has the hardener in suspension with the resin, only to be released when subjected to elevated temperature. This simplifies metering and application but limits the shelf-life.

A one-part epoxy has a limited shelf life but also requires refrigerated storage. In general, polymerization of one-part systems require moderately high temperatures for extended periods (although the actual requirements depend on the chemistry of the epoxy). A typical curing characteristic of a two-part thermoset epoxy is shown in Fig. 4-28 (warm-up time is excluded from the time axis).

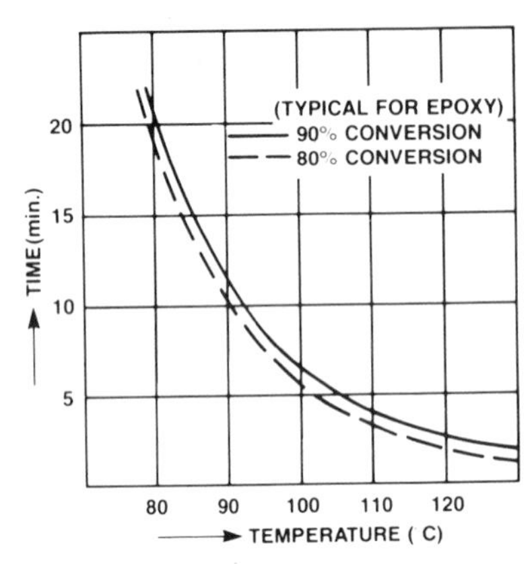

Fig. 4-28: Typical curing characteristic of a thermoset epoxy adhesive.

The heat for curing is provided by a conventional batch oven or a infra-red radiation, and be either an on-line or an off-line (remote) process. Off-line curing places greater demands on the adhesive, since in the uncured state, it must have sufficient adhesion to withstand the mechanical handling involved in transferring substrates from the placement station to the curing station.

The manufacturing process also places certain demands on the adhesive. The two main processes for mixed print substrates are outlined in the flow charts of Figures 4-29 and 4-30. For the first process (through-hole components inserted before SMDs), the cured adhesive requirements are some minimal bonding and resistance to the solder wave. For the second process (SMDs placed before through-hole components), the demands on the cured adhesive are greater; it must now also withstand the mechanical shock, brushing and flexing incurred during the through-hole insertion process.

The maximum cure temperature is also influenced by these production techniques. For example, in the mixed-print design where the through-hole components are inserted before SMD placement and adhesive curing, the maximum cure temperature is determined by what the components (both SMD and through-hole) can withstand. If, however, the SMDs are placed and the adhesive cured before the through-hole components go in, it is only the SMDs and substrate material limitations that will influence the maximum cure temperature. The user must always check the components' specifications before subjecting them to elevated cure temperatures.

Table 4-1: Advantages and disadvantages of various adhesive types.

Adhesive	Advantages	Disadvantages
Epoxy (one and two part systems)	• Proven history in electronics • High temperature use • Excellent solvent resistance • Excellent moisture resistance • Good void filling characteristics • UV cure systems available	• Limited shelf life • Longer cure time (one-part) • Higher cure temperature (one-part) • Complex application system (two-part) • Single application method (two-part) • Refrigerated storage (one-part)
Cyano-acrylate	• Very fast bonding • One-part system • Long shelf-life • Room temperature storage	• Single application method • Hazardous application • Bad void filling • Fair moisture resistance
Acrylic	• Moderate cure time • Good moisture resistance • Good solvent resistance	• Complex application system
Anaerobic	• One-part system • Unlimited shelf-life • Simple, inexpensive cure • High temperature resistance • UV cure systems available • Room temperature storage • Good solvent resistance	• Incompleteness of cure • Chemical activity • Low bond strength

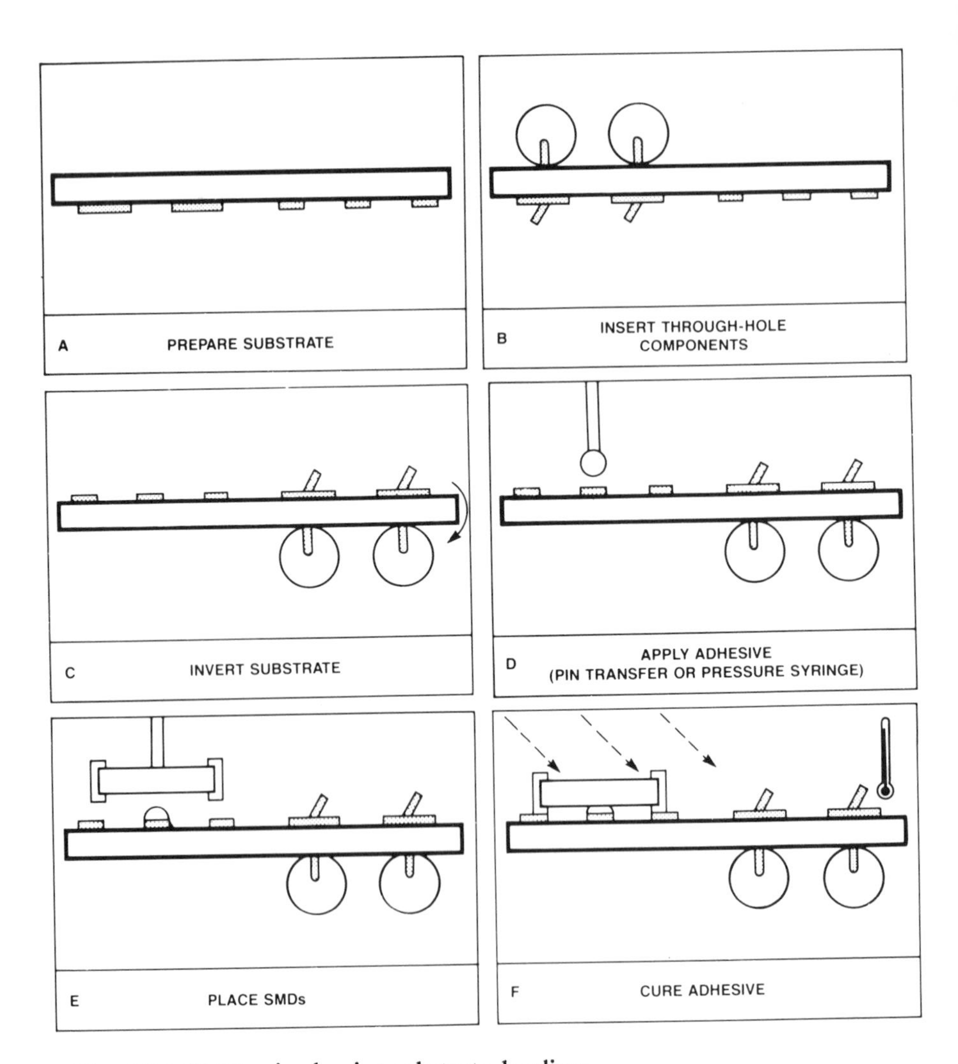

Fig. 4-29: SMD mixed print substrate loading: through- hole components inserted first.

Curing By Catalytic Action Alone

Reactive acrylics are two-part systems (adhesive plus catalyst) that cannot be cured by any means other than the mixing of the two parts. When the two parts are mixed, polymerization starts and proceeds very rapidly. Consequently, they cannot be pre-mixed; the adhesive must be applied to one surface, the catalyst to the mating surface and curing takes place on contact. This places considerable limitations on reactive acrylics for SMD assembly as it involves a complex application system—one part to the substrate and the other part to the SMD.

Another adhesive group that can only be cured by catalytic action is the cyanoacrylates. But unlike the acrylics, they are "one-part application" systems as the catalyst is water, obtained as minute amounts of moisture occurring naturally on the surfaces to be bonded. Strong handling bonds are formed in 5 to 10 seconds and full cure strength is achieved in less than one hour.

Due to this rapid curing, they can only be dispensed from a closed system (pressure-syringe). Another drawback for their use in SMD assembly is their bad fill characteristics (they function best with close fitting parts—any gap greater than 0.25 mm (.010″) requires an additional primer).

Additionally, because they are almost universal adhesives which will bond skin to skin, machine part to machine part (and any combination thereof), care is needed when working with them.

Anaerobic And UV Assisted Curing

For anaerobic and UV curing material, the position of the adhesive dot with respect to the SMD is critical. Anaerobic adhesives, for example, which cure only in the absence of oxygen, must be situated so that the SMD when placed cuts out as much air as possible. This will assure the most rapid and complete curing.

On the other hand, with adhesives that cure by exposure to UV you must make sure that a portion of the adhesive is directly exposed to the UV source. This poses obvious problems with SMD assemblies, particularly with small-outline ICs, as the adhesive dot will be shielded from the UV by the component body. Polymerization may still occur, but at a much slower rate which means that curing may not be complete, and chemically-active material may remain under the device.

The alternative approach is to combine the properties of both groups and initiate the polymerization of anaerobic adhesives with long-wavelength UV radiation.

Pressure-sensitive acrylic alloy adhesives are one such example. In general, these adhesives are formulated from acrylates and rubber "tackifiers" (added to give the adhesive an initial tackiness). Curing is accomplished by exposing the

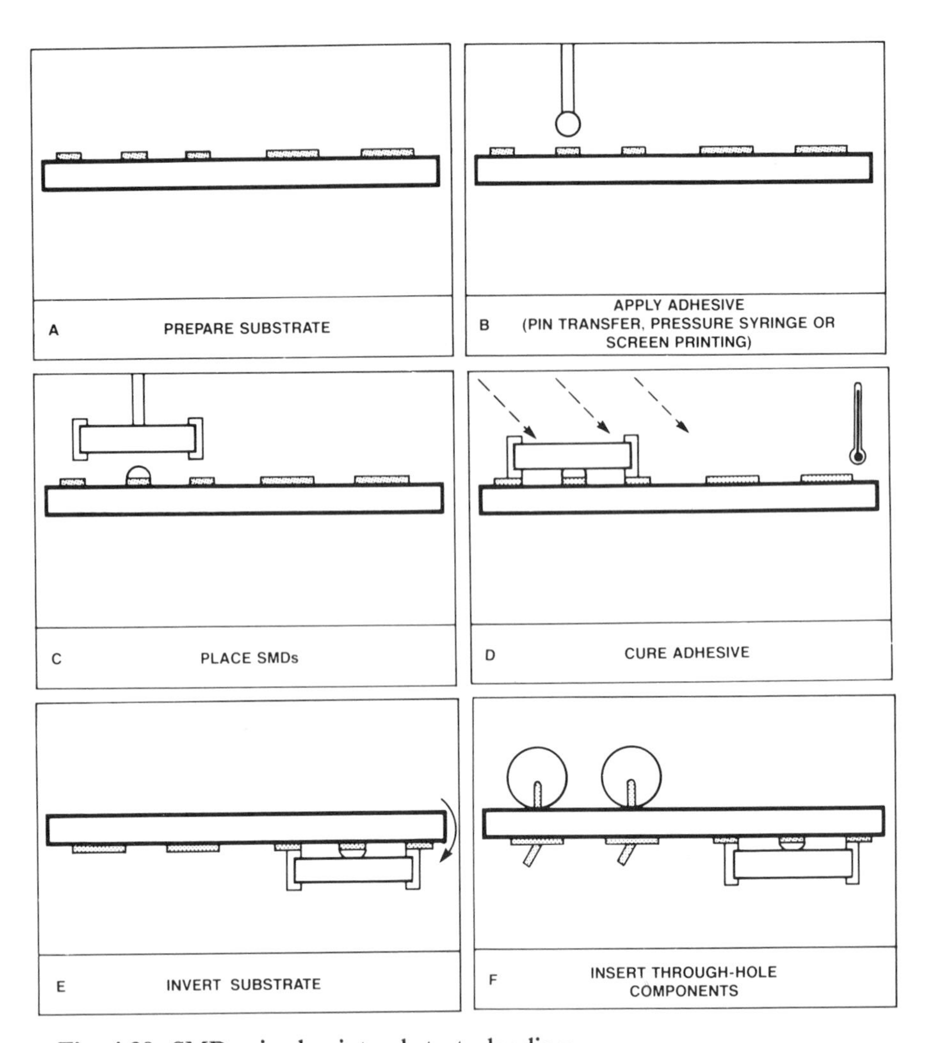

Fig. 4-30: SMD mixed print substrate loading: SMDs placed first.

deposits of adhesive to UV before the SMD had been placed. Exposure time is in the order of one to 10 seconds, and after exposure, the adhesive is ready to accept the device without further curing.

Fluxing and Cleaning

The adoption of mass soldering techniques by the electronics industry was prompted not only be economics, and a requirement for high throughput levels, but also by the need for a consistent standard of quality and reliability in the finished product simply unattainable with manual methods. With surface mounted device (SMD) assembly, this need is even greater.

The quality of the end-product depends on the measures taken during the design, and with correct choice of components and substrate configuration. It is, however, at the manufacturing stage where the greatest number of variables, both with respect to materials and techniques, have to be optimized to produce high quality soldering, a prerequisite for reliability.

Of the two most commonly used soldering techniques, wave and reflow, wave soldering is by far the most widely used and understood. Many factors influence the outcome of the soldering operation, some relating to the soldering process itself, and others to the condition of components and substrate to which they are to be attached. These must e collectively assessed to ensure high quality soldering.

And one of the most important, most neglected and least understood of these processes is the choice and application of flux. This publication outlines the fluxing options available, and discusses the various cleaning techniques that may be required, for SMD substrate assembly.

Fluxes

Populating a substrate involves the soldering of a variety of terminations simultaneously. In one operation, a mixture of tinned copper, tin/lead or gold plated nickel-iron, palladium-silver, tin/lead plated nickel-barrier, and even materials like Kovar, each possessing varying degrees of solderability, must be attached to a common substrate using a single solder alloy.

It is for this reason that the choice of the flux is so important. The correct flux will remove surface oxides, prevent reoxidation, help to transfer heat from source to joint area, and leave non-corrosive, or easily removable corrosive residues on the substrate. It will also improve wettability of the solder joint surfaces.

The wettability of a metal surface is its ability to promote the formation of an alloy at its interface with the solder to ensure a strong, low-resistance joint.

However, the use of flux does not eliminate the need for adequate surface preparation. This is very important in the soldering of SMD substrates, where any temptation to use a highly-active flux in order to promote rapid wetting of ill-prepared surfaces should be avoided because it can cause serious problems later when the corrosive flux residues have to be removed. Consequently, optimum solderability is an essential factor for SMD substrate assembly (this is covered in another publication titled Component and Substrate Solderability).

Flux is applied before the wave soldering process, and during the reflow soldering process (where flux and solder are combined in a solder cream). By coating both bare metal and solder, flux retards atmospheric oxidization which would otherwise be intensified at soldering temperature. In the areas where the oxide film has been removed, a direct metal-to-metal contact is established with one, low energy interface. It is from this point of contact that the solder will flow.

Types of Flux

There are two main characteristics of flux. The first is efficacy—its ability to promote wetting of surfaces by solder within a specified time. Closely related to this is the activity of the flux, that is, its ability to chemically clean the surfaces.

The second is the corrosivity of the flux, or rather the corrosivity of its residues remaining on the substrate after soldering. This is again linked to the activity; the more active the flux, the more corrosive are its residues.

Although there are many different fluxes available, and many more being developed, they fall into two basic categories: Those with residues soluble in organic liquids, and those with residues soluble in water.

Organic Soluble Fluxes

Most of the fluxes soluble in organic liquids are based on colophony or rosin, (a natural product obtained from pine sap that has been distilled to remove the turpentine content). Solid colophony is difficult to apply to a substrate during machine soldering, so it is dissolved in a thinning agent, usually an alcohol. It has a very low efficacy, and hence limited cleaning power, so activators are added in varying quantities to increase it. These take the form of either organic acids, or organic salts that are chemically active at soldering temperatures. It is therefore convenient to classify the colophony-based fluxes by their activator content.

Non-Activated Rosin (R) Flux

These fluxes are formed from pure colophony in a suitable solvent, usually isopropanol or ethyl alcohol. Efficacy is low and cleaning action is weal. Their uses in electronic soldering are limited to easily wettable materials with a high level of solderability. They are used mainly on circuits where no risk of corrosion can be tolerated, even after prolonged use, (implanted cardiac pacemaker for example). Their flux residues are noncorrosive and can remain on the substrate, where they will provide good insulation.

Rosin, Mildly Activated (RMS) Flux

These fluxes are also composed of colophony in a solvent, but with the addition of activators, either in the form of di-basic organic acids (such as succinic acid), or organic salts (such as dimethylammonium chloride or diethylammonium chloride). It is customary to express the amount of added activator as mass percent of the chlorine ion on the colophony content, as the activator-to-colophony ratio determines the activity, and hence the corrosivity. In the case of RMA activated with organic salts, this is only some tenths of one percent.

When organic acids are used,m a higher percentage of activator must be added to produce the same efficacy as organic salts, so frequently, both salts and acids are added. The cleaning action of RMA fluxes is stronger than that of the R type, although the corrosivity of the residues is usually acceptable. These residues may be left on the substrate as they form a useful insulating layer on the metal surfaces. This layer can, however, impede the penetration of test probes at a later stage.

Rosin, Activated(RA) Flux

The RA fluxes are similar to the RMA fluxes but contain a higher proportion of activators. They are used mainly when component or substrate solderability is poor and corrosion-risk requirements are less stringent. However, as good solderability is considered essential for SMD assembly, highly activated rosin fluxes should not be necessary. The removal of flux residues is optional; usually dependent upon the working environment of the finished product and the customers requirements.

Water Soluble Fluxes

The water soluble fluxes are generally used to provide high fluxing activity. Their residues are more corrosive and more conductive than the rosin-based fluxes, and consequently must always be removed from the finished substrate. Although termed water soluble, this does not necessarily imply that they contain water; they may also contain alcohols or glycols. It is the flux residues that are water soluble. The usual composition of a water soluble flux is shown below.

Although these substances can be dissolved in water, other solvents are generally used, as water has a tendency to spatter during soldering. Solvents with higher boiling points, such as ethylene glycol or polyethylene glycol are preferred.

1. A chemically active component for cleaning the surfaces.
2. A wetting agent to promote the spreading of flux constituents.
3. A solvent to provide even distribution.
4. Substances such as glycols or water soluble polymers to keep the activator in close contact with the metal surfaces.

Water Soluble Fluxes with Inorganic Salts

These are based on inorganic salts such as zinc chloride or ammonium chloride or inorganic acids such as hydrochloric. Those with zinc or ammonium chloride must be followed by very stringent cleaning procedures as any halide salts remaining on the substrate will cause severe corrosion. These fluxes are generally used for non-electrical soldering. Although the hydrazide halides are among the best active fluxing agents known, they are highly suspect from a health point of view and are therefore no longer used by flux manufacturers.

Water Soluble Fluxes with Organic Salts

These fluxes are based on organic hydrohalides such as dimethylammonium chloride, cyclo hexalamine hydrochloride and aniline hydrochloride, and also on the hydrohalides of organic acids. Fluxes with organic halides usually contain vehicles such as glycerol or polyethylene glycol, and non-ionic surface-active agents such as nonylphenol polyoxyethylene.

Some of the vehicles, such as the polyethylene glycols, can degrade the insulation resistance of epoxy substrate material and, by rendering the substrate hydrophilic, make it susceptible to electrical leakage in high humidity environments.

Water Soluble Fluxes with Organic Acids

Based on acids such as lactic, melonic or citric, these fluxes are used when the presence of any halide is prohibited. However, their fluxing action is weal and high acid concentrations have to be used. On the other hand, they have the advantage that the flux residues can be left on the substrate for some time before washing without the risk of severe corrosion.

Solder Creams

For reflow soldering, both the solder and the flux are applied to the substrate before soldering and can be in the form of solder creams (or pastes), preforms, elector-deposit, or a layer of solder applied to the conductors by dipping. For SMD reflow soldering, solder cream is generally used.

Solder cream is a suspension of solder particles in flux to which special compounds have been added to improve the rheological properties. The shape of the particles is important and normally spherical particles are now being added, particularly in very fine-line soldering.

In principle, the same fluxes are used in solder creams as for wave soldering. However, due to the relatively large surface area of the solder particles (which can oxidize), more effective fluxing is required and in general solder creams contain a higher percentage of activators than the liquid fluxes. The drying of the solder paste during pre-heating (after component placement) is an important stage as it refuses any tendency for components to become displaced during soldering.

Flux Selection

Choosing an appropriate flux is of prime importance to the soldering system for the production of high quality, reliable joints. When solderability is good, a mildly activated flux will be adequate, but when solderability is poorer, a more effective, more active flux will be required, The choice of flux, moreover, will be influenced by the cleaning facilities available, and if in fact, cleaning is even feasible.

With water-soluble fluxes, aqueous cleaning of the substrate after soldering is mandatory. If through cleaning is not carried out, severe problems may arise in the field, due to corrosion or short circuits caused by too low a surface resistance of the conductive residues.

For rosin based fluxes, the need for cleaning will depend on the activity of the flux. Mildly activated rosin residues can, in most cases, remain on the substrate where they will afford protection and insulation. In practice, for the great majority of electronic circuits, the choice will be between an RA or an RMA rosin based flux.

Application of Flux

Three basic factors determine the method of applying flux: the soldering process (wave or reflow), the type of substrate being processed (all-SMD or mixed print), and the type of flux.

For wave soldering the flux must be applied in liquid form before soldering. While it is possible to apply the flux at a separate fluxing station, with the high through-put rates demanded to maximize the benefits of SMD technology, today's wave soldering machines incorporate an integral fluxing station prior to the pre-heat stage to be used to dry the flux as well as preheat the substrate to minimize thermal shock.

Foam Fluxing

Foam flux is generated by forcing low pressure clean air through an aerator immersed in liquid flux (see Fig. 4-31). The fine bubbles produced by the aerator are guided to the surface by a chimney-shaped nozzle. The substrates are passed across the top of the nozzle so that the solder side comes in contact with the foam and an even layer of flux is applied. As the bubbles burst, flux penetrates any plated-through holes in the substrate.

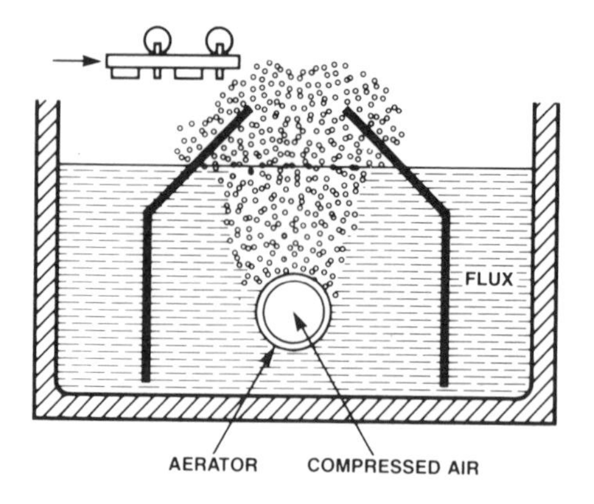

Fig. 4-31: Cross-section of a foam fluxer.

Wave Fluxing

A double-sided wave can also be used to apply flux, where the washing action of the wave deposits a layer of flux on the solder side of the substrate (see Fig. 4-32). Wave-height control is essential and a soft, wipe-off brush should be incorporated on the exit side of the fluxing station to remove excess flux from the substrate.

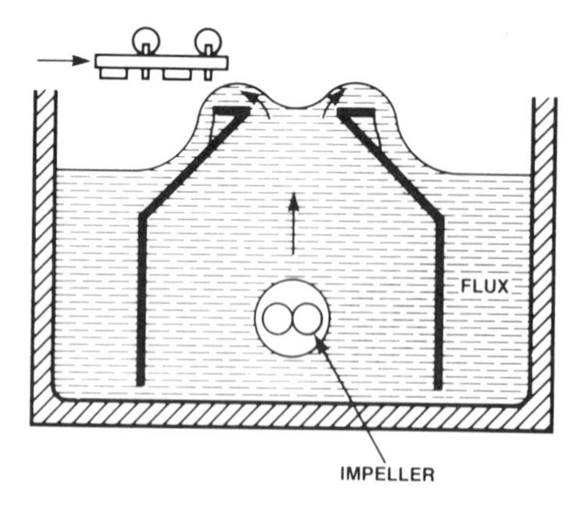

Fig. 4-32: Cross-section of a wave fluxer.

Spray Fluxing

Several methods of spray fluxing exist; the most common involves a mesh drum rotation in liquid flux. Air is blown into the drum which, when passing through the fine mesh, directs a spray of flux onto the underside of the substrate (see Fig. 4-33). Four parameters affect the amount of flux deposited: conveyor speed, drum rotation, air pressure and flux density. The thickness of the flux layer can be controlled using these parameters, and can vary between 1 and 10um.

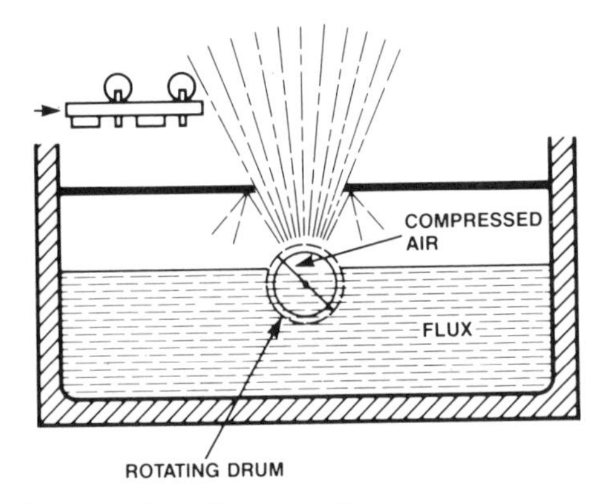

Fig. 4-33: Cross-section of a spray fluxer.

The advantages and disadvantages of these three flux application techniques are outlined in the Table in Fig. 4-34.

Flux Density

One of the main control factors for fluxes used in machine soldering is the flux density. This provides an indication of the solids content of the flux, and is dependent on the nature of the solvents used. Automatic control systems, which monitor flux density and inject more solvent as required, are commercially available and it is relatively simple t6o incorporate them into the fluxing system.

Pre-Heating

Pre-heating the substrate before soldering serves several purposes. It dries the flux to evaporate most of the solvent, thus increasing the viscosity. If the viscosity is too low, the flux may be prematurely expelled from the substrate by the molten solder. This can result in poor wetting of the surfaces, and solder spatter.

ADVANTAGES AND DISADVANTAGES OF FLUX APPLICATION METHODS

Method	Advantages	Disadvantages
Foam fluxing	• Compatible with continuous soldering process • Foam crest height not critical • Suitable for mixed print substrates	• Not all fluxes have good foaming capabilities • Losses through evaporation may be appreciable • Prolonged pre-heating because of high boiling point of solvents
Wave fluxing	• Can be used with any liquid flux • Compatible with continuous soldering process • Suitable for densely populated mixed print	• Wave crest height is critical to ensure good contact with bottom of substrate without contaminating the top
Spray fluxing	• Can be used with most liquid fluxes • Short pre-heat time if appropriate alcohol solvents are used • Layer thickness is controllable	• High flux losses due to non-recoverable spray • System requires frequent cleaning

Drying the flux also accelerates the chemical action of the flux on the surfaces, and so speeds up the soldering process. During the pre-heating stage, substrate and components are heated to between 80 degrees and 90 degrees C (solvent based fluxes) or to between 100 degrees and 110 degrees C (water based systems). This reduces the thermal chock when the substrate make contact with the molten solder, and minimizes any likelihood of the substrate warping.

The most common methods of preheating are convection heating with forced air, radiation heating using coils, infra-red quartz lamps or heated panels, or a combination of both convection and radiation. The use of forced air has the added advantage of being more effective for the removal of evaporated solvent. Optimum preheat temperature and duration will depend on the nature and design of the substrate and the composition of the flux.

Figure 4-35 shows a typical method of pre-heat temperature control. The desired temperature is set on the control panel and the microprocessor regulates pre-heater No. 1 to provide approximately 60 percent of the required heat. The IR detector scans the substrate immediately following No. 1 heater and reads the surface temperature. By taking into account the surface temperature, conveyor speed, and the thermal characteristics of the substrate; it then calculates the amount of additional heat required by heater No. 2 to attain the pre-set temperature. In this way, each substrate will have the same surface temperature on reaching the solder bath.

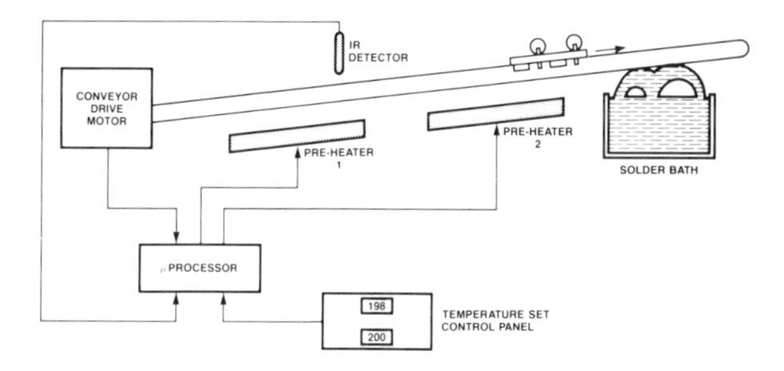

Fig. 4-35: Schematic diagram of a typical controlled pre-heat system.

Post-Soldering Cleaning

Now that world-wide efforts in both commercial and industrial electronics are converting old designs from conventional assembly to surface mounting, or a combination of both, it can also be expected that high volume cleaning systems will convert from in-line aqueous cleaners to in-line solvent cleaners or in-line saponification systems (a technique that uses an alkaline material in water to react with the rosin so that it becomes water soluble). These systems may, however, become subject to environmental objections, and new governmental restrictions on the use of halogenated hydrocarbons.

The major reason for this is that the water soluble flux residues, containing a high concentration of activators, or showing hygroscopic behavior, are much more difficult to remove from SMD populated substrates than rosin based flux residues. This is primarily because the higher surface tension of water, compared to solvents, makes it difficult for the cleaning agents to penetrate beneath SMDs, especially the larger ones, with their greatly reduced off-contact distance (the distance between component and substrate).

Post-soldering cleaning removes any contamination, such as surface deposits, inclusions, occlusions or absorbed matter which may degrade to an unacceptable level, the chemical, physical or electrical properties of the assembly. The types of contaminant on substrates that can produce either electrical or mechanical failure over short or prolonged periods are shown in the table in Fig. 4-36. The contaminants listed in this table include organic and inorganic compounds (soluble and insoluble), organo-metallic compounds, and particle matter.

All of these contaminants, irrespective of their origin, fall into one of two groups: *polar* and *non-polar*.

SUBSTRATE CONTAMINANTS

Contaminant	Origin
Organic compounds	Fluxes, solder mask
Inorganic insoluble compounds	Photo-resists, substrate processing
Organo-metallic compounds	Fluxes, substrate processing
Inorganic soluble compounds	Fluxes
Particle matter	Dust, fingerprints

Polar Contaminants

These are compounds that dissociate into free ions which are very good conductors in water, quite capable of causing circuit failures. They are also very reactive with metals and produce corrosive reactions. It is essential that polar contaminants are removed from the substrates.

Non-Polar Contaminants

These compounds do not dissociate into free ions or carry an electrical current and are generally good insulators. Rosin is a typical example of a non-polar contaminant. In most cases, non-polar contamination does not contribute to corrosion or electrical failure and may be left on the substrate. They may however, impede functional testing by probes and prevent good conformal coat adhesion.

Solvents

The solvents currently used for the post-soldering cleaning of substrates are normally organic based and are covered by three classifications: hydrophobic, hydrophilic, and azeotropes of hydrophobic/hydrophilic blends.

Azeotropic solvents are mixtures of two or more different solvents which behave like a single liquid in so much that the vapor produced by evaporation has the same composition as the liquid, which has a constant boiling point between the boiling points of the two solvents that form the azeotrope. The basic ingredients of the azeotropic solvents are combined with alcohols and stabilizers. These stabilizers, such as nitromethane, are included to prevent corrosive reaction between the metallization of the substrate and the basic solvents.

Hydrophobic solvents do not mix with water at concentrations exceeding 0.2% and consequently have little effect on ionic contamination. They can be used to remove non-polar contaminants such as rosin, oils and greases.

Hydrophilic solvents do mix with water and can dissolve both polar and non-polar contamination, but at different rates. To overcome these differences, azeotropes of the various solvents are formulated to maximize the dissolving action for all types of contamination.

Solvent Cleaning

Two types of solvent cleaning systems are in use today, batch and conveyorized systems; either can be used for high-volume production. In both systems, the contaminated substrates are immersed in the boiling solvents, and ultrasonic baths or brushes may also be used to further improve the cleaning capabilities.

The washing of rosin based fluxes offers advantages and disadvantages. Washed substrates can usually be inserted into racks easier, as there will be no residues on their edges; test probes can make better contact without a rosin layer on the test points, and the removal of the residues makes it easier to visually examine the soldered joints. On the other hand, washing equipment is expensive, and so are the solvents, and some solvents present a health or environmental hazard if not correctly dealt with.

Aqueous Cleaning

For high-volume production, special machines have been developed in which the substrates are conveyor fed through the various stages of spraying, washing, rinsing and drying. The final rinse water is blown from the substrates to prevent any deposits from the water being left on the substrate.

Where water soluble fluxes have been used in the soldering process, substrate cleaning is mandatory. For the rosin based fluxes it is optional and is often at the discretion of the customer.

Conformal Coatings

A conformal, or protective coating on the substrate, applied at the end of processing prevents or minimizes the effects of humidity and protects the substrate from contamination by airborne dust-particles. Substrates that are to be provided with a conformal coating (dependent on the environmental conditions to which the substrate will be subjected) must first be washed.

Environmental And Ecological Aspects Of Fluxes And Solvents

Fumes and vapors produces during soldering processes, or during cleaning, will not, under normal circumstances, present a health hazard, providing relevant health and safety regulations are observed.

Fumes originating from colophony can cause respiratory problems, so an efficient fume-extraction system is essential. The extraction system must cover the fluxing, pre-heating and soldering stations, remain operational for at least one hour after machine shutdown, and conform to local regulations. Today, the problem of noxious fumes is unlikely to concern the cleaning station as all commercial systems are equipped to condense the vapors back into the system. In the future, however, it can be expected that a much lower degree of escape of noxious fumes from any system will be allowed, and all systems may have to be reviewed.

Certain fluxes, particularly some water soluble ones, contain highly aggressive substances, and must not be allowed to come into contact with the skin or eyes. Any contamination should immediately be removed with plenty of clean, fresh water. Deionized water should also be readily available as an eye-wash. Should contamination occur, a qualified medical practitioner should be consulted. Protective clothing should be worn during cleaning or maintenance of the fluxing station.

Conclusions

SMD technology imposes tougher restraints on fluxing and cleaning of substrate assemblies. Traditionally, rosin based fluxes have been used in electronic soldering where residues were considered "safe" and could be left on the board. However, increased SMD packing density, fine-line tracks and more rigid specifications have resulted in changes to this basic philosophy.

There is now a demand for surfaces free from residues; test probes are more efficient when they do not have to penetrate rosin flux residues, and conformal coating and board inspection benefit from the absence of such residues.

Cleaning also poses problems for SMD substrates. The close proximity of component and substrate means that solvents cannot effectively clean beneath devices. Components must also be compatible with the cleaning process. They must, for example, be resistant to the solvents used and to the temperatures of the cleaning process. They must also be sealed to prevent cleaning fluids from entering the devices and degrading performance.

So eliminating the need for cleaning is better than poor or incomplete cleaning. And in a well balanced system, mildly activated rosin based fluxes, leaving only non-corrosive residues, can be successfully used for SMD substrate soldering without subsequent cleaning.

Much research is at present being carried out into fluxes and solder creams; for example, the production of synthetic resin, with superior qualities to colophony at a lower cost. Another area of research is that of solder creams with non-melting additives that increase the distance between component and substrate—making it easier for cleaning fluids to penetrate beneath the component. It also increases the joint's ability to withstand thermal cycling.

Rosin-free and halide-free fluxes are also being developed with similar activities to conventional rosin-based fluxes. These new types will combine the "safety" of rosin fluxes with easier removal in conventional solvents. Using non-polar materials, ionizable or corrosive residues are eliminated, and the need for cleaning immediately after soldering is avoided.

Solder Joint Criteria

Although the advent of high-speed, accurate SMD placement machines and improved mass-soldering methods has greatly reduced the human involvement in the assembly process, one area remains where skill, training and judgment are the most important factors: inspection of the solder joints.

The increasing component density and decreasing component size of SMD assemblies is stimulating research into solder-joint inspection methods for substrates either totally or partially populated with SMDs. Although systems are being developed that use pattern recognition to check a substrate for missing or misaligned components, visual examination by a trained inspector still remains the most practical method for verifying the quality and reliability of the soldered joint.

Inspection

Several forms of inspection methods are available: the naked eye, microscope, and video are examples. Of these, the most widely used is the eye, with or without the aid of a magnifier having 2 x or 3 x magnification and integral lighting. A magnifier is recommended for the smaller, more densely populated substrates, since, apart from making inspection easier, it also reduces operator fatigue.

Typically, a general inspection is first made of the soldered side of the substrate. Overall appearance should be pleasing to the eye, all joints having a similar appearance, the length of the through-hole component terminations should be within the prescribed limits and the substrate should be clean. This will be followed by an appraisal of the upper side to check for solder flow in plated through-holes, displaced or missing components and general cleanliness. Next, the soldered side should be re-examined in detail for such obvious defects as missing SMDs, solder bridges, residues and the individual soldered joints assessed for quality and reliability.

Unlike leaded component joints, where the lead also provides added mechanical strength, the SMD must rely on the quality of the soldering for both electrical and mechanical integrity. It is therefore necessary for the inspector to be trained to be able to make a visual assessment with a view to long term reliability.

Defects can occur in either the substrate or the individual joint. Substrate, or circuit defects, such as misaligned components, bridges, and solder balls are inclined to be the more obvious ones. Joint defects, however, are much more a matter of judgment and interpretation of the visual aspects, such as the degree of wetting, volume of solder or the shape of the joint.

Defect Classification

Defective solder joints are classified under three headings, as follows:
Major defects—are likely to result in the failure of the circuit to perform its intended function. These must always be rectified. The application of Statistical Quality Control (SQC) to frequently recurring defects can be very helpful in identifying the cause and indicating the appropriate action.
Minor defects—are unlikely to result in circuit failure. They form departure from established standards that has little bearing on the effective use or operation of the unit. The need for rectification depends upon the level of reliability required. Minor defects occurring in circuits intended for use in life support systems, or in the aerospace industry would be rectified, while those destined for the consumer market may well be ignored.
Cosmetic defects—those that have no effect on the function or reliability of the circuit, and therefore need not be rectified. However, their presence should invite an investigation into their cause.

Soldering Defects

Substrate, or circuit related defects will generally fall into the following categories: solder bridges, solder icicles and solder residues. Another defect, specific to reflow soldering of leadless components, is that known as "draw-bridging." In wave soldering, adhesive contamination of contact areas can also introduce defects.

Solder bridges: Bridges are unwanted solder connections between metal parts of the circuit and can occur between adjacent conductors or terminations. When such a bridge causes a short-circuit between points of differing potential, or when it completes an unwanted ground-loop, it becomes a major defect. When it connects points at the same potential, it may be regarded as cosmetic, provided of course, that it does not form a ground-loop.

Bridging between passive SMDs placed end to end and connected by a conductor constitutes a minor defect. This is because, even though the two terminations are at the same potential, the extra solder in the bridge produces a degree of rigidity that may well be unacceptable. Solder bridges occur much more often with wave soldering than with reflow methods.

Possible causes of solder bridges:

Design Faults:

- Insufficient spacing of conductors, solder lands or components.
- Incorrect orientation of components.

Process Faults:

- Excess solder; solder wave too high.
- Warped substrate resulting in flooding of substrate upper surface.
- Insufficient flux.
- Damage or misregistration of solder resist.
- Contamination of the solder bath.
- Insufficient oil intermix.

Solder icicles, or spikes: Apart from solder bridges, excess solder on the underside of the substrate can also result in the formation of icicles or spikes. These are formed when a filament of solder is drawn off the component terminations as the substrate breaks contact with the solder wave.

Icicles are usually regarded as cosmetic defects, but when they significantly reduce the air insulation distance, especially on high-voltage circuits, they constitute a major defect. Like bridges, icicles are a product of wave soldering.

Possible causes of solder icicles:

- Contamination of the solder bath.
- Insufficient oil intermix.
- Uneven distribution of flux.

Solder residues: Solder residues can be identified as webs, skins, spatters or balls.

A web is a fine oxide thread, sometimes with solder particles attached, connected to the substrate or terminations. Being very thin, it is often difficult to see, particularly when occurring above the substrate surface. Webs can cause short-circuits and are always major defects.

A skin is a thin film of solder attached to the substrate and conductors. Like webs, skins can also cause short-circuits and are, therefore, major defects.

Solder spatters (or splashes) may be in the form of either small isolated solder skins attached to the substrate, or solder balls. Apart from causing short-circuits immediately after soldering, they can also cause failure in the field if they become detached during use. Their classification as major or minor defects will depend largely on their size and location.

Webs, skins and spatters are usually the result of wave soldering. Solder balls occur with reflow soldering.

Possible causes of solder residues:

Wave Soldering:

- Incorrect choice or application of flux.
- Solder contamination.
- Incorrect pre-heat time or temperature.

Reflow Soldering:

- Incorrect pre-heat time or temperature.
- Incorrect choice of solder paste.

Drawbridging

Drawbridging is the standing up of leadless SMDs on one of their end faces, particularly during vapor phase reflow soldering.

This phenomenon occurs when the solder cream reflows at one end of leadless SMD before the other, and the forces exerted by the surface tension of now molten solder "pulls" the SMD away from the still solid solder cream at the other end. These forces can be strong enough to up-end the SMD.
Drawbridging is always regarded as a major defect.

Possible causes of drawbridging:

- Insufficient width of metallization under component.
- Wrong size of solder land.
- Incorrect measure of solder cream.
- Wrong type of solder cream.

Adhesive Contamination

In wave soldering, the prior application of adhesive is a critical step (see Adhesive Application And Curing). As the tolerances are very tight, there is a great risk of the leads or solder lands becoming contaminated with adhesive.The result is that a joint will not be formed at all, or that only the extremity of the lead or metallization is soldered. This is a major defect.

Possible causes of adhesive contamination:

- Misplacing of adhesive on solder land or component.
- Adhesive with the wrong viscosity.
- Movement of SMD after placement on adhesive (particularly MELF).

Blow Holes

During soldering, gas or flux can become entrapped in the solder and cause a void. These holes in solder joints vary in both size and type. As they are completely enclosed in solder, they cannot be detected visually, but they are usually present in most soldered joints.

If, however, a void ruptures during solidification, then a blow hole results. Blow holes are irregular holes in the surface of the solder whose defect classification will depend on their size and location. True holes extend right through the joint and may well be large enough to significantly reduce the amount of solder in the joint, thus producing a major fault.

General Solder Joint Criteria

A good, reliable solder joint will perform both its electrical and mechanical function, without failure, during the lifetime of its associated assembly. In assessing the quality and reliability of the joint, irrespective of whether it is a through-hole component joint or an SMD joint, and irrespective of whether it was formed by wave or reflow methods, the inspector has three basic criteria upon which to base his judgment. These are:

- Good wetting of the surfaces.
- A sound and smooth joint surface.
- The correct amount of solder.

Good Wetting

Good wetting is indicated by an even flow of solder over the surfaces of the solder land and component termination, and thinning toward the edges of the joint. A metallic interaction will have taken place, resulting in a smooth, unbroken and adherent layer of solder on the joint.

Non-wetting is a failure of the surface metal to be coated by molten solder, so that the original surface remains partially or totally visible. This may be due to contamination of the joint surfaces, insufficient fluxing, or too short a preheat time.

If, after wave soldering, any joints exhibit total non-wetting, whereby the joint has failed to come into contact with the molten solder due to the "shadow effect" (shielding of the joint by the component body or other adjacent components).

De-wetting on the other hand, is the withdrawal of molten solder from the surface after initial wetting. It is characterized by dispersed and irregular solder droplets on the joint surface, often with a thin film of solder between them. De-wetting becomes very evident at high soldering temperatures and long dwell-times.

This is usually indicative of contamination of the solder or joint surfaces, poor solderability or dissolution of the SMD metallization. Severe de-wetting often results in insufficient volume of solder at the joint.

Sound Smooth Surface

The surface of the solder should be smooth, shiny and uninterrupted. Eutectic tin/lead solder is uniformly shiny, except perhaps for the center of the shrinkage dimple sometimes present. However, the surface of 60/40 tin/lead alloy may have a rough surface after slow cooling. This is cosmetic and does not detract from the quality of the joint.

Correct Amount of Solder

A good solder joint should have neither too much nor too little solder. For mixed-print substrates with through-hole components, the shape of the leads should be recognizable within the contour of this joint. This applies particularly to clinched leads on the soldered side of a substrate without through-hole plating.

The solder should be seen to have penetrated completely through the plated holes in substrates loaded with components on both sides. In the case of single-sided substrates without plated holes, the solder should wet the entire solder land and should increase uniformly from the edge of the land up to the termination, forming a neat fillet.

SMD Joint Assessment

Although the criteria for the assessment of an SMD joint are basically the same as those for through-hole components, that is, good wetting, the right amount of solder, and a sound, smooth surface, there is an additional factor to be considered that is not applicable to leaded components. This is the misalignment of the component on the solder lands.

Some misalignment is usually inevitable, due to the tolerances on the component, the solder land, and the placement machine. Provided at least half the termination is within the boundaries of the solder land, then misalignment is acceptable (see Fig. 4-37). It may not, however, be cosmetically acceptable.

If the width of the solder land is greater than the width of the metallized connecting area of the component, then a projection smaller than half the width of the connecting area is a minor defect.

If, on the other hand, the width of the solder land is smaller than the width of the metallized connecting area of the component, then a projection is unavoidable. This is acceptable provided that the width of the solder land is completely covered. If the width of the solder land is not covered completely, but the projection is still smaller than one-half of the width of the connecting area, the defect is a minor one.

When the degree of misalignment extends beyond these limitations, the fault is a major one. The resulting reduced volume of solder in the joint greatly affects the mechanical strength and hence the reliability of the joint. Over and above this, excessive misalignment may interfere with adjacent components. A detailed account of these tolerances can be found in Substrate Design Guidelines.

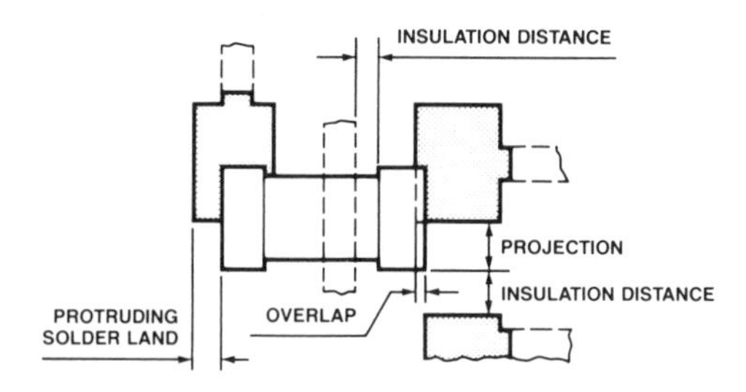

Fig. 4-37: Misalignment of leadless SMDs.

The part of the solder land protruding in the longitudinal direction of the component should be at least equal to the required height of the meniscus.

Leadless SMDs

When a surface mounted leadless device is accurately placed on the solder land, and hence does not protrude from it, it should have solder meniscus along the whole periphery. If the device is misplaced (but conforms to the criteria discussed above), then the meniscus should still extend over the complete width of the connecting area situated over the solder land. If it does not, it is a major defect.

The metallization should show no signs of dissolution and should have a solder meniscus along its full length. The height of the meniscus at its centerline must be at least one-third of the component metallization height for components less than 1mm (.040") high, or 0.4mm (.016') high for larger components. The shape of the solder fillet may be either concave or convex. THe surface of the solder should be smooth and shiny, with little or no evidence of de-wetting or holes.

SMDs with Few Short Leads

Surface mounted devices with a few short leads, such as the SOT-23, SOT-89 and SOT-143, introduce new criteria. The foot and heel of the lead should be positioned within the boundaries of the solder land. A projection of half the width of the lead over the edge of the solder land constitutes a minor defect, while more than half is a major defect.

The solder should be smooth and shiny, and the sides of the lead should be fully wetted. In the case of these SMDs with a few short leads, wave soldering usually leaves a large amount of solder, surrounding the entire foot. However, the maximum amount of solder is not indicated, and the only criteria is that the joint does constitute a major defect.

SO IC Packages

For surface mounted IC packages which have a larger lead count, for example, small-outline (SO) ICs, a projection of more than half the width of the lead is again a major defect. A projection smaller than half the width of the lead is a minor defect. Shifting in the longitudinal direction of the lead is not a problem, as long as the whole foot of the lead is situated on the solder land.

Inspection can be confined to those areas of the lead that form the actual joint, that is, the sides and heel of the foot, and the space between the heel and the solder land. The top surface of the foot should show evidence of wetting, although an homogeneous coat is not necessary. The cut end of the foot need not be wetted, but a meniscus is usually present.

For the optimum joint, the space between the heel and the solder land should be filled with solder with a meniscus height equal to the thickness of the lead. The solder fillets on the sides of the foot should also be this height. In the case of projection of the leads beyond the solder lands, this requirement is valid only for the side of the lead situated on the solder land. A meniscus height, of less than half the lead thickness is unacceptable.

VSO IC Packages

The criteria for assessing a good soldered joint on very small outline (VSO) packages are largely the same as for the SO with regard to the meniscus height and degree of wetting. However, at minimum joint, the foot of the lead should be secured over at least three quarters of its length, and the sides of the leads are wetted over the secured length to a height equal to half the thickness of the lead.

PLCC with J-leads

For ICs in plastic leaded chip carriers (PLCCs), part of the lead and its associated solder fillet will be hidden from vies beneath the component. It is necessary, therefore, to assess the quality of the joint from the quantity of solder and the appearance of the fillet between the outside bend of the lead and the solder land.

In the ideal joint, the sides of the lead should be wetted and the area between the outside bend and the solder land should be filled with solder to a height equal to the thickness of the lead (see Fig. 4-38). A meniscus extending to a height equal to half the thickness of the lead is the acceptable minimum. In both cases, the solder should wet the entire solder land.

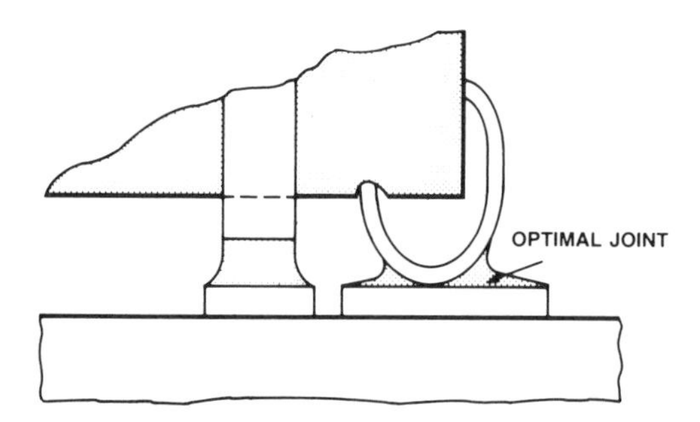

Fig. 4-38: Acceptable joint on a PLCC J-lead.

Chip Carriers with Metallized Castellations

As with SOTs and SOs, the metallization of leadless ceramic chip carriers (LCCCs) should be aligned with the solder lands. Up to half a width overlap is a minor defect, more than half a width is a major defect. The whole of the metallization and the solder lands should be wetted. The solder fillet should extend to the full height of the metallization and to the edges of the land. A concave surface is indicative of good wetting beneath the component by capillary action. Figure 4-39 illustrates the unacceptable and acceptable volumes of solder for LCCC packages with castellated terminations.

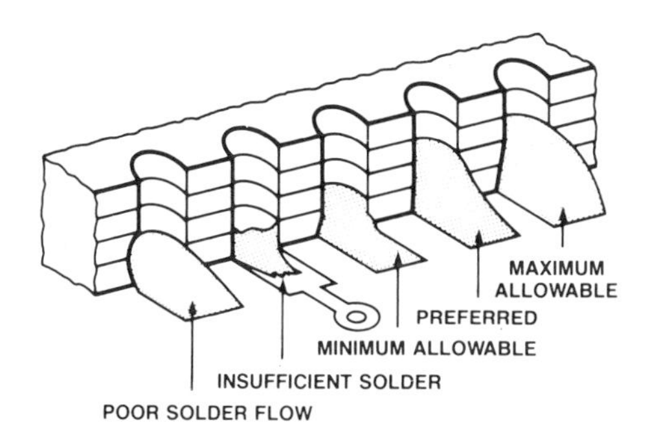

Fig. 4-39: Joint criteria for LCCCs with castellated connecting areas.

Inspection Systems

Although visual inspection with the aid of a magnifier is the most cost-effective method, and adequate for most consumer products, the choice of systems may well be influenced by criteria other than cost, for example, high throughput speed or ultra-high reliability.

It is necessary, therefore, to consider the following factors when evaluating an inspection system:

Speed of inspection: the higher the throughput, the more cost effective the system.

Operator fatigue: results in lack of concentration; more defects get through.

Judgment: consistency in operator judgment is difficult to maintain.

Resolution: high resolution can often be at the expense of other factors.

Parts manipulation: the more automation the better. Excessive handling increases the probability of mechanical or electro-static damage.

Depth of field: for optical systems only. Evaluation is easier if the whole joint can be viewed at once.

Field of view: should be as large as possible to allow for a maximum number of joints to be viewed with a minimum of manipulation.

ADVANTAGES AND DISADVANTAGES OF INSPECTION SYSTEMS

SYSTEM TYPE	ADVANTAGES	DISADVANTAGES
Binocular microscope	• Easy manipulation • Low cost • Good depth of field • Good field of view	• High operator fatigue • Good judgment required • Low inspection speed
Video systems	• High inspection speed • Lower operator fatigue • Reasonable resolution • Reasonable field of view	• Good judgment required • Shallow depth of field • Poor parts manipulation
X-ray inspection	• High resolution • Good judgment not needed • Good field of view • Can see hidden defects	• Low inspection speed • High cost of system • Poor parts manipulation • High level of operator training
Laser inspection	• High resolution • Good judgment not needed • Can see some hidden defects	• Very low inspection speed • High cost of system • Requires programming

Component and Substrate Solderability

In theory, practically every metal can be soldered given an active enough flux, a high enough temperature, and a long enough dwell-time. In practice, however, as far as electronic circuits are concerned, the flux must be only mildly activated, soldering temperatures must be relatively low, and dwell-times kept as short as possible. So good, solderable components must be offered to meet this criteria.

A mildly activated flux will make post-soldering cleaning easier, or even unnecessary, and minimize corrosion problems. Relatively low soldering temperatures and short dwell-times will prevent damage to both substrates and SMDs. It is, therefore, essential that substrate and components have optimum solderability before assembly.

Aspects of Solderability

The solderability of components and substrates can be defined by their suitability for an industrial soldering process. It is determined by:

Thermal demand—the thermal aspects of component or substrate must allow for heating of the joint area to the desired soldering temperature within the time available without adversely affecting component or substrate.

Wettability—the nature of the component metallization, or conductor, must be such that the surface is wetted with molten solder within the specified time available for soldering, without subsequent de-wetting.

Resistance to dissolution of metallization—the component and substrate metallization must be able to withstand soldering temperatures without dissolving.

During the wave soldering process, for example, the joints must be formed within a few seconds. Dwell-times of two to four seconds after preheating are common, and within these short times, the component terminations and substrate solder lands must reach a sufficiently high temperature to cause the solder to flow evenly.

SMD substrates, however, can be soldered by several different processes, and each of these places specific thermal demands on the loaded substrate. These thermal demands must be evaluated during the design stage, and component/substrate suitability established by consulting relevant specifications before starting the process of populating the substrate.

Wettability of the surfaces is affected by aging, and by the conditions of storage, transit, and handling, and therefore, more under the control of the user. Communication between manufacturer and user via standards and specifications, with regard to protection, storage, and packaging, is an important factor in ensuring that components and substrates are at a sufficient level of wettability at the start of the soldering process.

The dissolution of the component metallization during the soldering process can cause problems. SMDs in particular, must be highly resistant to dissolution, for if the metallization is lost into the solder, the integrity of the joint will be seriously affected. It is the metal used for plating the component metallization that determines the rate of dissolution.

The smaller dimensions of SMDs, together with the trend toward blister-tape packaging, makes component solderability testing by the user a difficult problem, one that has yet to be standardized. The simple globule test, as used for leaded components cannot be applied to SMDs, and quantitative analysis using the wetting balance method is impractical, owing to the very small areas of metallization.

Should the user wish to carry out solderability tests, he must, at present, rely on dipping the component vertically into molten solder, followed by a visual inspection of the terminations using a stereoscopic microscope with between x 10 and x 20 magnification. Good solderability usually requires that 95 percent of the metal surfaces are wetted by the molten solder during a dwell-time of approximately two seconds at 235 Degrees C.

For leadless SMDs with metallized electrodes, the dipped surface should be covered with a bright solder coating with no more than small amounts of scattered imperfections (such as pinholes, non-wetted or de-wetted areas). Attention should be given to quality of the coating (appearance of the surface), and attack of the metallization (any decrease in area).

Surface mounted components with short leads, such as SO ICs with gull-wing leads, have the leads divided into three regions, each with different requirements.

Protective Coatings

The types of coating applied to substrates to ensure continued wettability are generally either fusible or soluble. Fusible coatings, deposited wither by electroplating with tin/lead, or coating with molten solder during the soldering process.

Soluble coatings may be either of electro-deposited noble metal, that is, gold, silver, platinum or palladium, which will dissolve into the molten solder, or of an organic nature which will be dissolved by, or will mix with, the flux.

Fusible Coatings

Electro-Plated Tin/Lead

This method is used mainly in the production of substrates by the pattern plating process. A plating mask is applied over the unwanted areas of copper. The exposed pattern is electro-plated with copper, and then tin/lead. When the mask is removed, the tin/lead deposit serves as a resist during the etching process. The plating may then be reflowed using hot liquid, hot air or IR radiation, to restore its wettability.

Reflowing the electro-plated tin/lead layer fuses it and forms an intermetalic zone of a copper/tin alloy at the boundary, the thickness of which is a function of temperature and time. As this copper/tin intermetallic layer has poor solderability, it is desirable to keep its thickness to a minimum, so reflowing the electro-plated tin/lead is best avoided if possible.

The electro-plated tin/lead coating, if correctly applied to a clean substrate, can provide good wettability without flowing, if storage and transit conditions are not too humid. Substrates coated in this way give a good indication of future wettability, and if showing no de-wetted or non-wetted areas, will retain it for an appreciable time.

However, with this practice, bridging problems can arise during wave soldering when the circuit resolution calls for very fine lines and spacing. Although a solder resist can help to a certain extent, the tin/lead layer beneath the resist can remelt during soldering, causing it to wrinkle and blister. Not only does this spoil the appearance of the finished substrate, but particles of the resist may fall off and contaminate the solder bath.

Flux may also become entrapped in cracks in the resist, leading to cleaning or corrosion problems. To overcome these problems, it is possible to strip off the tin/lead layer after etching and apply a solder resist directly onto the bare copper. This is followed by solder-coating the lands, and levelling the solder, usually with hot air or liquid.

Solder Coating

This process involves applying a solder resist, and then dipping the substrate, usually vertically, into molten eutectic solder for a pre-determined dwell-time. During rapid withdrawal, both sides of the substrate are exposed to either high pressure hot air or liquid. This removes excess solder and clears any plated-through holes. The result is a bright, uniform solder layer on the conductors and on the walls of plated-through holes.

The thickness of the coat is determined by the operating parameters of the machine, such as speed of withdrawal, air or liquid temperature, and the flow rate. In general, coated and levelled substrates maintain good wettability over long periods.

Solder coating and levelling also produces the copper/tin intermetallic layer, and the levelling process results in a convex meniscus of solder in the land. When plated-through holes are present (as in mixed print), this results in a very thin layer of solder at the edges of the holes, exposing the intermetallic layer, and wettability becomes inadequate soon after fusion. During subsequent wave soldering, this lack of wettability can prevent solder flowing past the upper edges of the hole, resulting in non-wetting of the upper solder lands.

The meniscus effect has other drawbacks, particularly in SMD assembly. It can, for example, raise the SMD above the adhesive dot, resulting in a poor bond. This problem is compounded as small solder lands develop a higher solder meniscus than larger lands, further complicating the adhesive dot height criteria. And as a convex meniscus provides a very small point of contact, the device may slip off-center of the solder land.

Levelling also has other limitations. Hot air levelling is an inherently dirty process, since air under pressure will inevitably blow flux and dross around. This, together with the high heat levels developed within these systems, means that the operator must spend a significant amount of time cleaning the machine to ensure that the solder-coat thickness is uniform and consistent.

Liquid levellers on the other hand, at least those that use hot oil, can contaminate the substrate material by oil impregnation. To overcome this, machines are available employing fluorocarbon liquids or water soluble fluids. Levelling processes, whether using air or liquid, must always be followed by cleaning, because of the higher than normal activity of the flux used.

Solder Land Contamination

Whichever method is used to provide the protective layer, great care must be taken during the application of the solder resist. The slightest contamination by the solder resist of the areas to be soldered, either by misregistration of the solder resist pattern or "bleeding" of the resist, renders these areas irreversibly unsolderable, and usually results in the scrapping of the substrate.

Another problem which seriously affects the solderability of substrates is that of the adhesive used for bonding the SMDs bleeding onto the solder land. Many commercial adhesives are almost transparent when forming very thin layers, and if such a layer spreads onto the solder land, it renders it unsolderable. Therefore, adhesives of the correct viscosity must be used and placement must be very accurate.

chapter 5

CONTACT (FOOTPRINT) DESIGN

Contact patterns for surface mounted devices will vary due to the package shape, lead spacing and contact type.

An effort to standardize package configuration has been made by the electronics industry through various organizations.

In this chapter, we illustrate process-proven footprint patterns for each of the component types presently used, as well as guidelines for creating a suitable pattern on future products. The footprint geometry and spacing follow the recommendations and proposals furnished by the Surface Mount Technology Association (SMTA), The Institute for Interconnecting and Packaging Electronic Circuits (IPC), Component Manufacturers and NuGrafix Group, Inc.

Through the years, there have been subtle changes in contact shapes for SMDs as a result of improved processes and refined component quality. The contact area shapes shown are primarily for the reflow solder process. However, variations for wave solder applications are also discussed.

Note: The terms footprint patterns, pad geometry, contact pattern are all interchangeable and refer to the area which is the contact for the surface mounted component.

Designing for Producibility

Pad geometry plays a prominent role in the SMT assembly. It is the most significant key to successfully controlling the reflow solder process. Designers who are new to SMT must avoid the temptation to take short cuts in pad geometry. The following examples will illustrate a few of the DO's and Don'ts of pad geometry.

The width of the pad for chip components is the first consideration. A footprint pattern that is too long will cause the chip component to float to one side. If the pattern is too wide, the chip component may rotate, as illustrated in Figure 5-1. The ideal pattern will cool evenly, centering the component equally on both pads.

Fig. 5-1: The ideal pattern will cool evenly, centering the component equally on both pads.

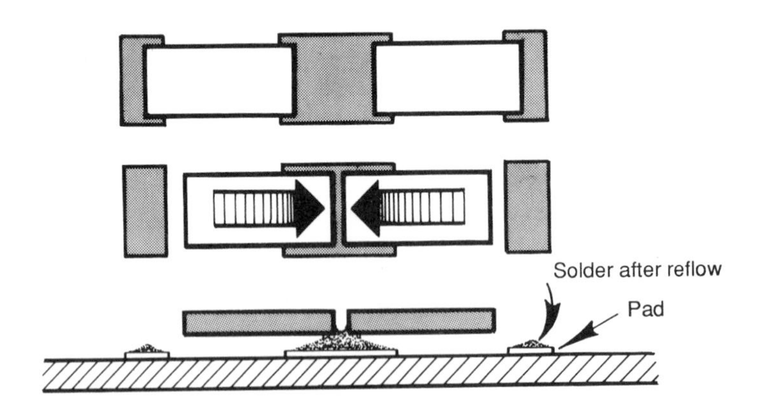

Fig. 5-2: Using a single large pad between two components will overwhelm the outer pads—drawing the components to the higher deposit of solder.

There is also a tendency to use a single large pad between two components, as shown in Figure 5-2. The solder on the larger pad will overwhelm the outer pads, drawing the components to the higher deposit of solder.

Separation of the footprint patterns of two chip components will insure containment of the solder paste. When it is important to increase conductor width between components, the best results will be achieved with two narrow traces, rather than one wide one. The containment of the solder within the footprint patterns is the key to controlling the solder process. See Figure 5-3.

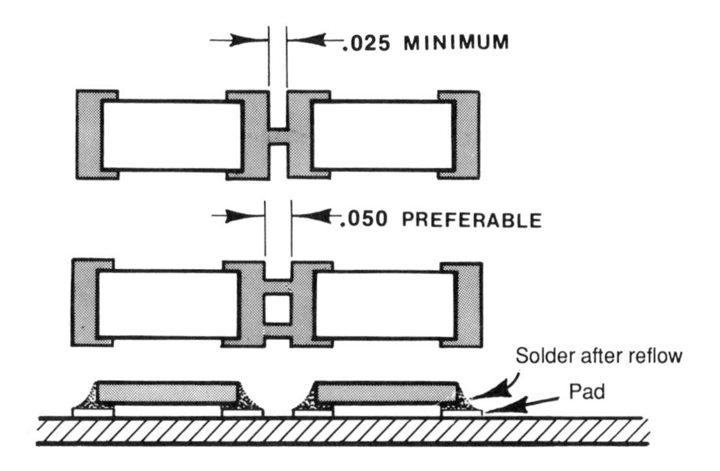

Fig. 5-3: The containment of the solder within the footprint patterns is the key to controlling the solder process.

Too large a footprint pattern for chip components will encourage an excess of solder build-up. To reduce unwanted reworking, match the component to the proper size pad geometry. When wave soldering discrete components, a narrow pad geometry can be adapted to limit the amount of solder on the component connection. Be aware of the component body when positioning the component connection or the footprint pads. If the narrow pattern is used, it will appear that the designer has adequate clearance, but when a component is placed the potential for solder bridging increases. Tolerances on component size, placement accuracy, and PC board itself must be taken into account, as these could add up to .010″ or .020″. Specific details of pad geometry are shown in Figure 5-4.

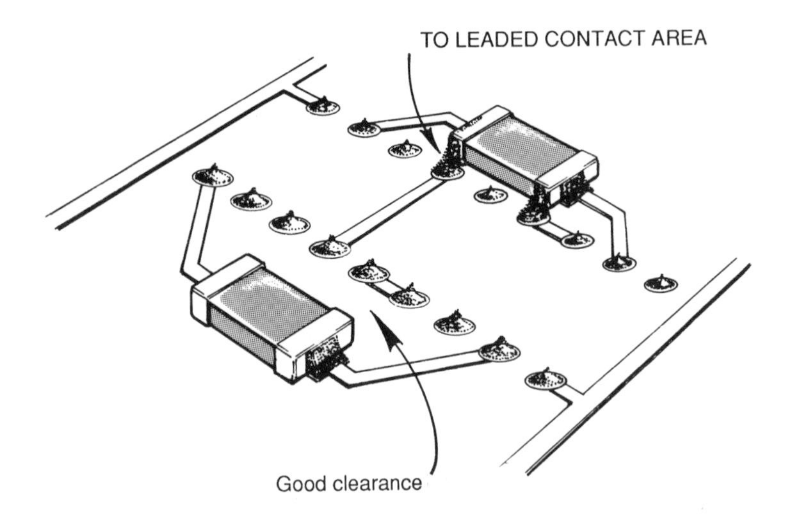

Fig. 5-4: Specific details of pad geometry.

Component Spacing

During the reflow soldering process, the solder is in a liquid form, and for a time, the components will float on the high point of the pad. Problems will occur when a component is too closely spaced to the next. One component may dray toward the other or slide off the center onto the adjacent pad. This problem is generally eliminated with adequate clearance between footprint patterns, as shown in Figure 5-5.

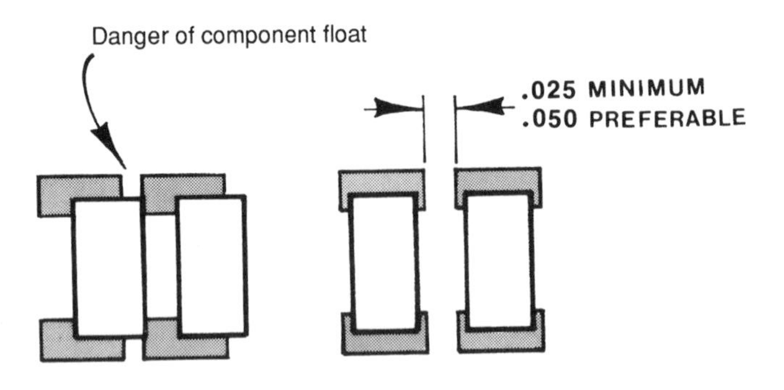

Fig. 5-5: Adequate clearance between footprint patterns will help eliminate shifting.

To compensate for less than ideal pad geometry, it will be necessary to use component mounting epoxy that is normally reserved for wave solder attachment. Adding epoxy to the reflow solder process, as a band-aid for poor design, will increase the cost of the assembly.

The disadvantages of using epoxy in the reflow procedure are:

1. Particles from the solder or flux can be trapped in the epoxy material.
2. Cleaning under the parts is difficult and may allow metallic bridging to occur.

The footprint pattern shown in Figure 5-6 will give the most satisfactory results in the reflow process. This pattern also works well in wave solder, when the placement of the component is accurate. Ideally, a finished solder connection will encase the end-cap area of the component as shown in Figure 5-7.

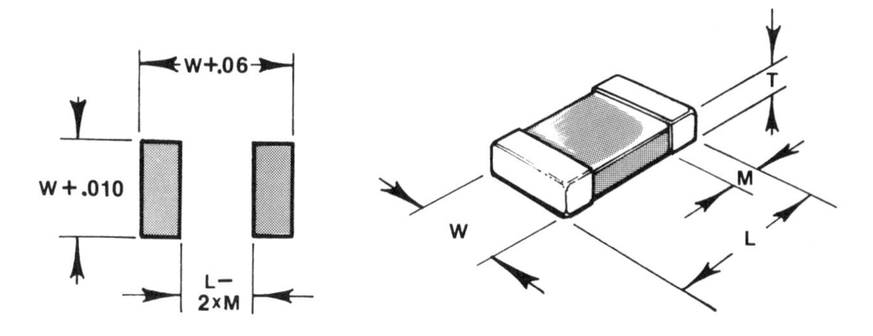

Fig. 5-6: Footprint patterns that will give the most satisfactory results in the reflow soldering process.

Wave soldering chip components on the opposite side of leaded devices is very popular, especially for PC boards with mixed technology. Often a company's first application of SMT will employ this technique because capital equipment cost is not as significant and existing conventional wave solder machines can be used. The full contact pattern can work well in wave solder. However, some companies will choose a narrow contact area to control solder build-up. See Figure 5-8.

Clearances between lead pads of a component must be maintained. When component bodies are too close to an adjacent leaded contact, bridging may occur as shown in Figure 5-9.

Clearance between chip components must accommodate inspection and rework tools. With adequate space, the danger of solder bridging and voids are eliminated. Figure 5-10 is an international guideline for component spacing of these devices; it includes the contact pattern and component body.

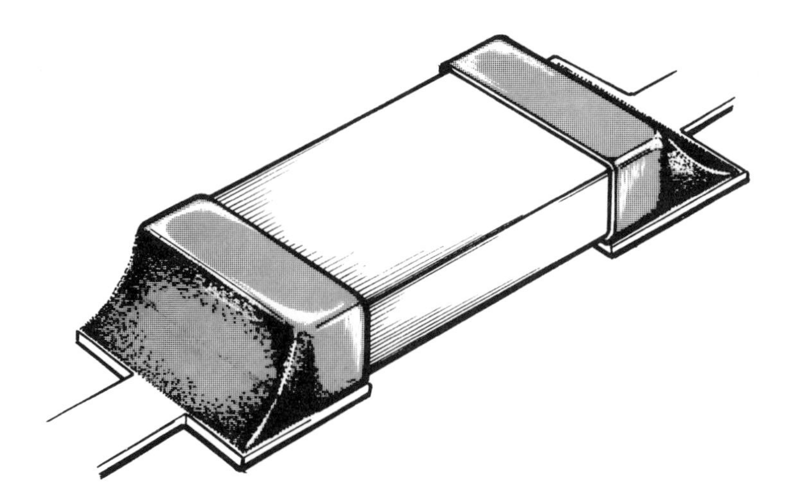

Fig. 5-7: The ideal solder connection should encase the end-cap area of the SMD.

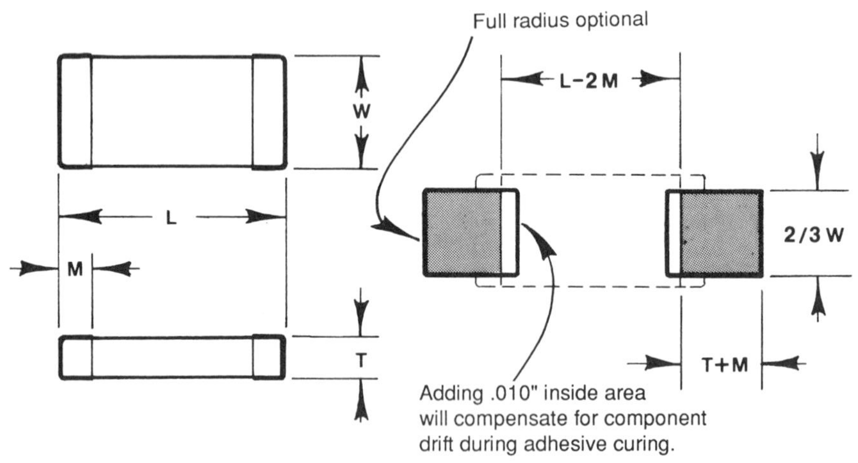

Fig. 5-8: Some designers choose a narrow contact area to control solder build-up.

Discrete Component Contact Design

While resistors and capacitors are available in many sizes, a uniform size should be established for general use. The standard footprint pattern gives the assembly personnel greater control over the equipment and processes. The most acceptable resistor and capacitor, as we mentioned earlier, has been the 1206 size, which is used for ⅛- and ¼-watt applications. Most mid-range capacitors are available from several sources in the 1206 configuration to further standardize pad geometry and uniformity of the assembly.

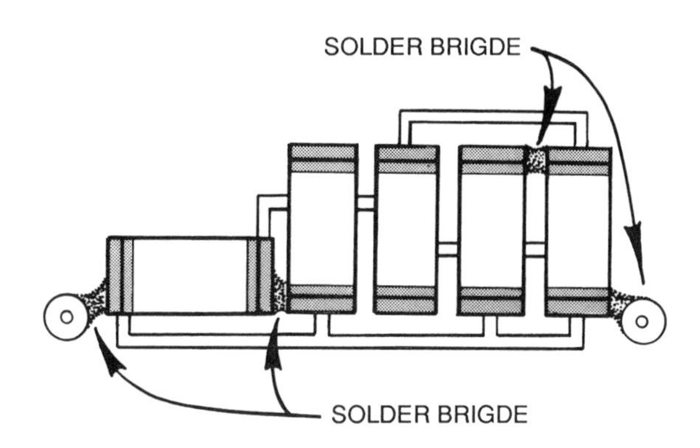

Fig. 5-9: Bridging often occurs when component bodies are too close to an adjacent lead contact.

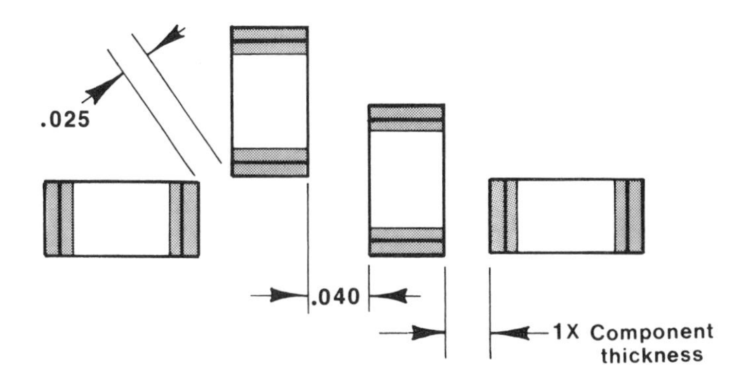

Fig. 5-10: International spacing guidelines for chip components.

Assembly equipment manufacturers suggest the 1206 or the 1210 size capacitor as the preferred choice because they are larger and easier to handle than the 0805. For values under 330ρF, the 0805 is widely available. For values between 100ρF and 330ρF, the designer has a choice of either the 0805 or 1206 size. Capacitor values between 300ρF and .18μF, are widely available from several sources in the 1206 or 1210 size capacitor. The selection of size should be made early in the design because the footprint patterns of the 0805 and 1206 or 1210 are very different. See Figure 5-11 for details.

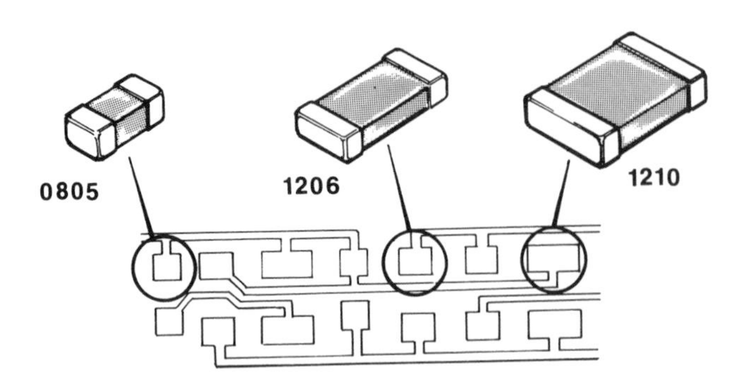

Fig. 5-11: The selection of component size should be selected early in the design stage, since there are wide variations in dimensions.

When placing plated through-hole pads on your artwork, do not allow the hole pad area to make direct contact with the component contact area. The clearance between the footprint pattern and feedthrough pads will allow a desirable barrier of solder mask coating that will stop the migration of liquid solder during the reflow process. The conductor trace connecting the contact area with the feedthrough pad should be approximately .010″ to .015″ wide. See Figure 5-12.

Feedthrough pads, against or within the footprint pattern, will cause migration of the liquid solder away from the contact area during reflow. Feedthrough pads under chip components cannot be seen and may cause failure during additional processes. Unlike the reflow solder process, feedthrough pads and heavy traces adjoining the contact area will not have significant negative results in the wave solder process.

A feedthrough pad under a chip device (Figure 5-13) is not recommended. Solder or adhesives may migrate during the secondary assembly operation—thereby causing additional rework.

The migration of solder will cause one end of the component to pull away from the contact area or flow into unwanted places during secondary wave solder procedures. Figure 5-14 illustrates the migration of liquid solder through a closely connected feedthrough hole.

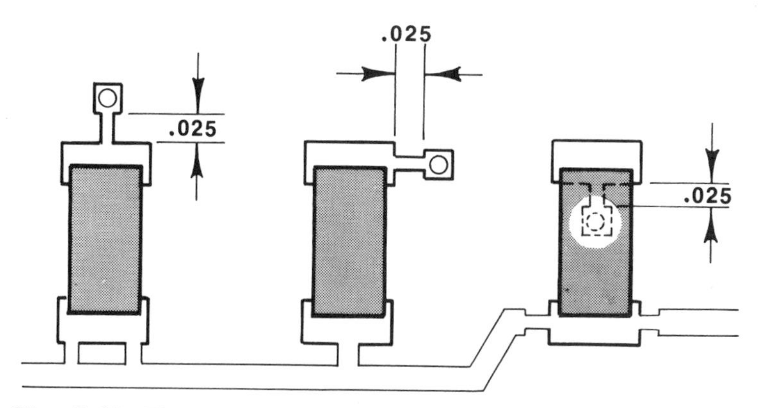

Fig. 5-12: The conductor trace connecting the contact area with the feedthrough pad should be approximately .015″ to .015″ wide.

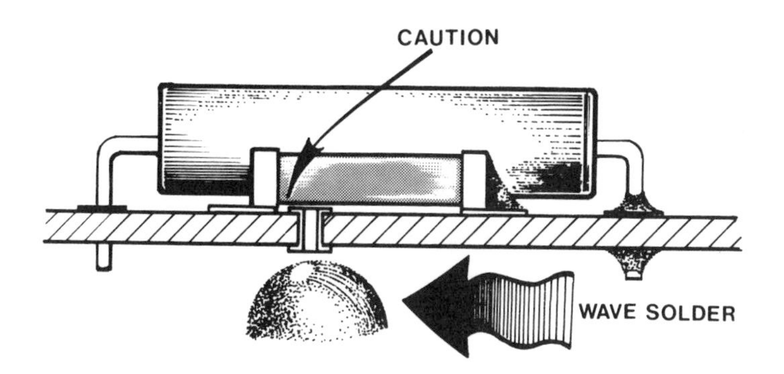

Fig. 5-13: Feedthrough pads, under a chip device, is not recommended in most situations.

The footprint geometry for chip components using reflow solder will be described later in this chapter.

Many of the same clearance rules used for chip components apply to the SOT components. When arranging the SOT components on the PC board the designer must provide enough clearance to allow for the placement accuracy of the assembly equipment used. The guidelines for contact area to feedthrough pad distance, which are used for chip components, are also valid for SOT devices, as shown in Figure 5-15.

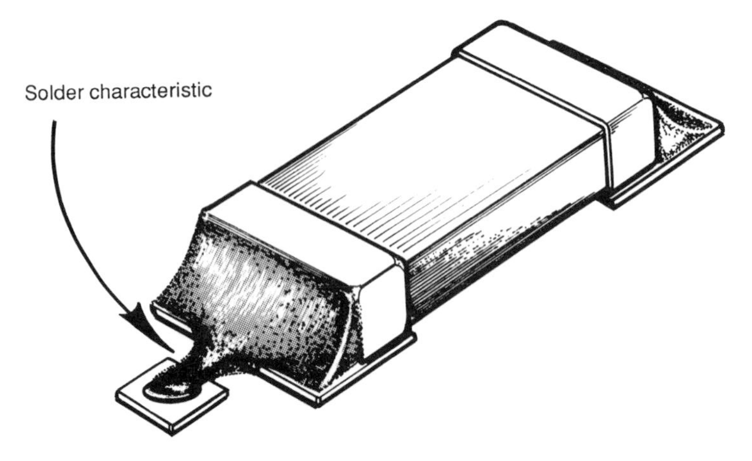

Fig. 5-14: Migration of liquid solder through a closely connected feedthrough hole.

SOT devices are furnished in tape and reel and/or tubes, with either left or right orientation. If the assembly equipment will not rotate each part before placement, cycle time may be slowed. Plan the layout carefully and maintain a consistent orientation when possible, as in Figure 5-16. This, in turn, will contribute to the reduction of manufacturing costs.

Preferred Component Orientation

The direction of chip components and SOT devices should be in one direction to take advantage of high speed robotics.

Reflow soldering requires a space between pad area contacts as compared in Figure 5-17 "A" and 5-17 "B". The distance specified in "A" furnishes an adequate solder-mask area to prevent solder migration away from the component lead. Maintaining distance between contacts of the SOT device will also reduce excess solder build-up, bridging and voids when the wave solder process is applied.

Commercial IC Footprint Planning
SOIC, PCC, and QUAD Lead Packages

The Small Outline IC (SOIC) is assembled internally in the same way as the familiar DIP package. The same die is attached to a smaller lead frame. The wire is bonded from the chip to the lead frame, then a plastic body is molded

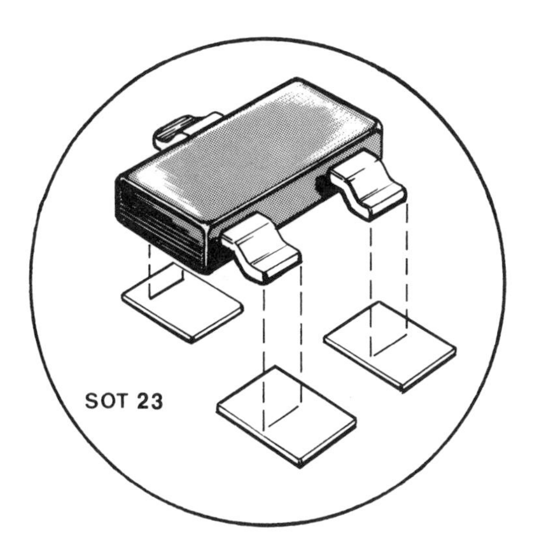

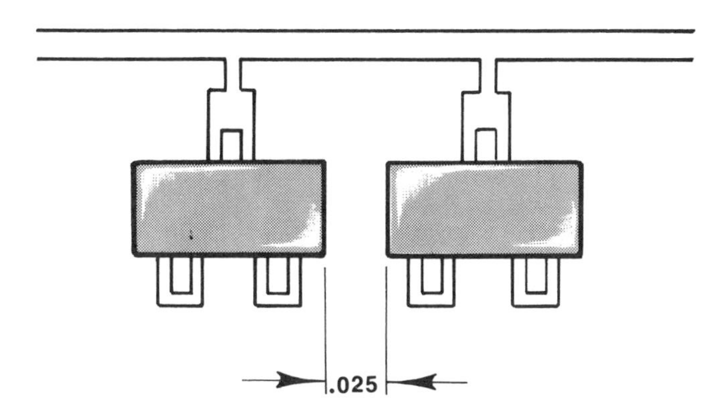

Fig. 5-15: The guidelines for chip component contact area to feedthrough pad distance are also valid for SOT devices.

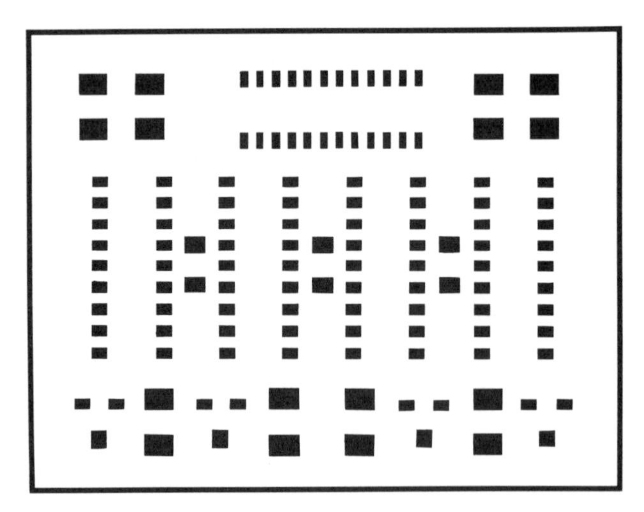

Fig. 5-16: A consistent PCB layout is always best.
Plan the layout carefully to achieve this result.

over the assembled lead frame. The major differences is that the SOIC will be mounted on a footprint instead of through-holes and 50% less surface area will be required.

The SO-8, SO-14, and SO-16 use a smaller die size. The pinout is usually the same as the larger DIP package. However, it is advisable to check the specifications carefully. Some SO-16 devices require a larger lead frame to accommodate the dies size. This wide format extends to the SO-20, SO-24 and SO-28 as well. See Figure 5-18.

The SO package shown in Figure 5-19 is offered by several IC manufacturers. The Signetics, National, T.I. and Motorola configurations meet the JEDEC registered specifications. Always check the manufacturer's specifications before beginning your design.

The dimensions shown in Figure 5-18 reflect the JEDEC (American) standard. Other American and Japanese manufacturers may vary from the JEDEC package. Even though the .050″ contact spacing is the same, the body width and the distance between contact rows may be inconsistent.

Pad geometry can be adjusted to allow substitution of non-JEDEC SOIC's. By lengthening the contact area, a wider than standard part can be mounted to the PC board with minimal impact on the assembly process. Figure 5-20 illustrates a typical solder connection to the PC board.

The Plastic Chip Carrier (PCC) is a commercial high-volume package that resembles the early ceramic Leadless Chip Carrier (LCC). The PCC footprint is smaller than the LCC, but the superior lead design is more suitable to the current production equipment, reflow solder processes and the larger selection of substrate material.

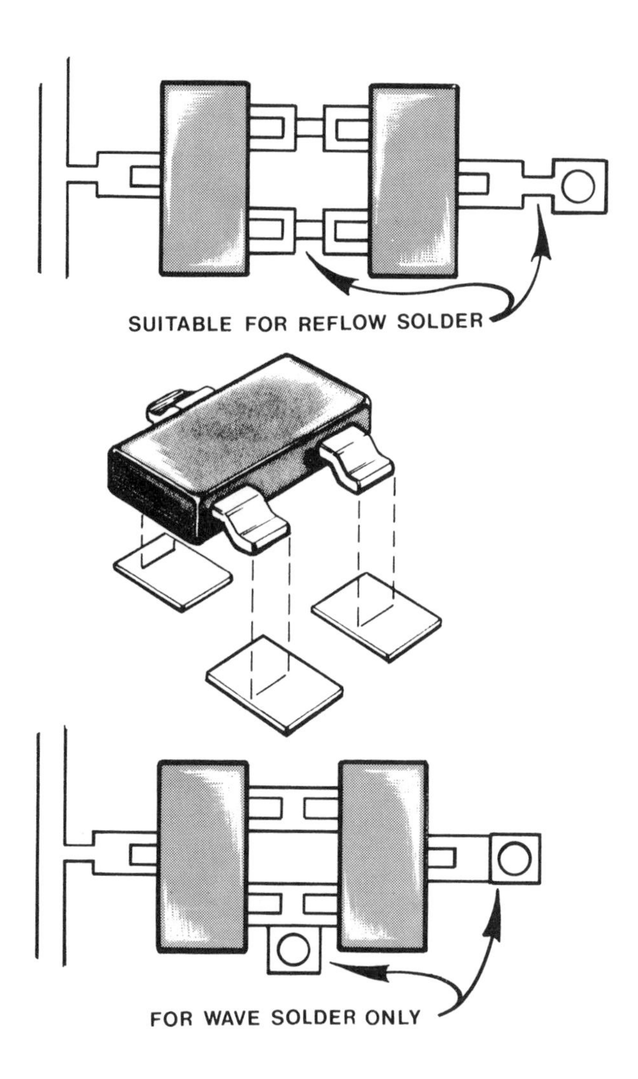

Fig. 5-17: Reflow soldering requires a space between pad area contacts.

The even-number quad pattern provides consistent wire-bonding length and lead frame design. The quad lead packages are available in 20, 28, 44, 52, 68 and greater lead patterns with contacts on preferred .050″ grid spacing.

Figure 5-21 illustrates the subtle contact pattern offset that will increase producibility and provide for inspection or rework.

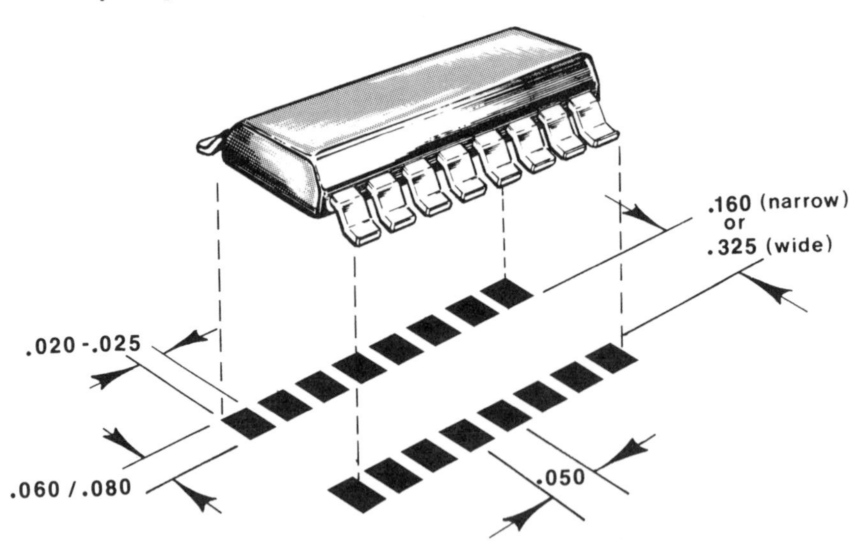

Fig. 5-18: Some SO-16 SMDs require a larger lead fame to accommodate the die size.

The ideal footprint will allow a good solder fillet toward the outside of the IC package. Extending the footprint inward only increases the chance of component drift during the reflow process. Extending the footprint's length to the outside is acceptable if test-probe access must be provided. However, the longer contact areas will reduce usable board area and component density. See Figure 5-22.

The distance between contact groups is usually determined by the complexity of the circuit. When bussing, a greater number of contacts of ICs can be interconnected and higher density of the components may be possible. In these cases, a minimum of .025″ between rows is possible, but for access and inspection, allow .150″ or more space between the PCC. See Figure 5-23.

All footprint patterns for the PCC-ICs will allow the designer the option of routing a .008″ trace between contacts while maintaining a .008″ air gap on both sides. The patterns shown in Figure 5-24 will accommodate the most common JEDEC PCC devices. Components with greater lead counts are also available. Use the manufacturer's specifications to generate appropriate footprint patterns.

The footprint family for the PCC-IC is described in detail later in this chapter. As with the passive device, the feedthrough hole and pad will be separated from component contact area by a narrow connecting trace.

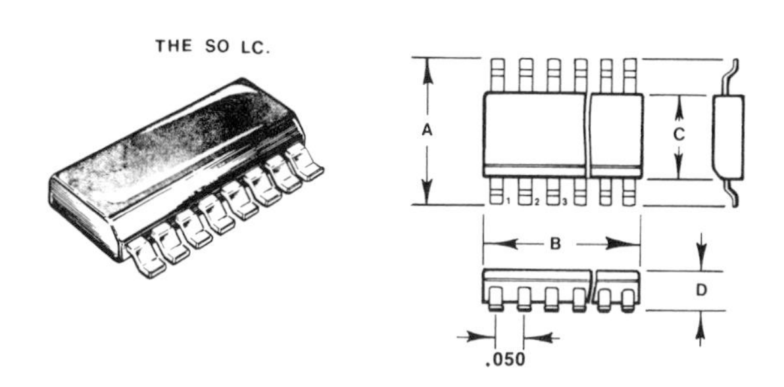

	SO-8	SO-14	SO-16	SO-16L	SO-20	SO-24	SO-28
A	.240	.240	.240	.415	.415	.415	.415
B	.195	.340	.390	.410	.510	.610	.710
C	.155	.155	.155	.295	.295	.295	.295
D	.070	.070	.070	.103	.103	.103	.103

Fig. 5-19: This SO package is now offered by several different manufacturers.

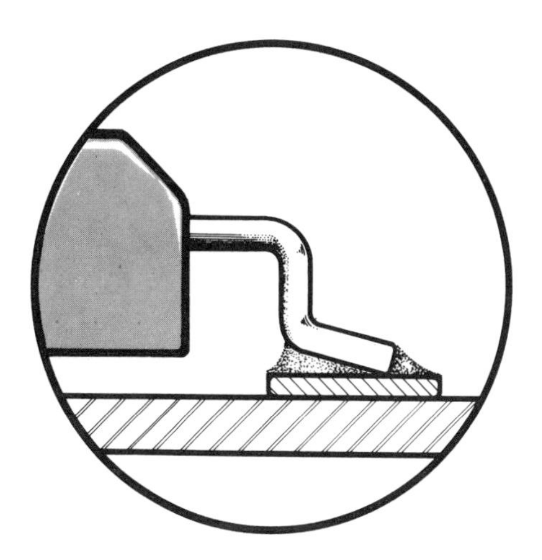

Fig. 5-20: Typical solder connection to a PCB.

Figure 5-25 is typical of the contact and feedthrough pad interconnect on the PC board surface. The shaded area represents the solder mask coating over all conductor traces.

Use a pattern similar to the one shown in Figure 5-26 as a solution to the second source problem when the same function is available in the narrow package from Vendor "A" while Vendor "B" only supplies the wide package.

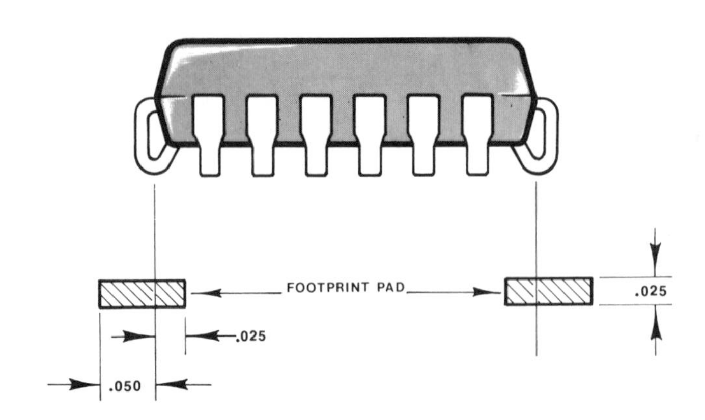

Fig. 5-21: Contact pattern offset that will increase producibility and enhance inspection or rework.

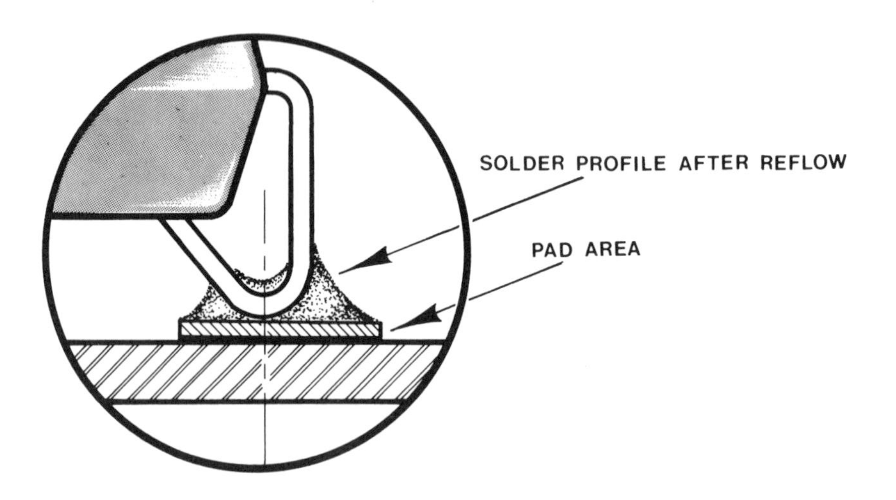

Fig. 5-22: Longer contact areas will reduce usable board area and component density.

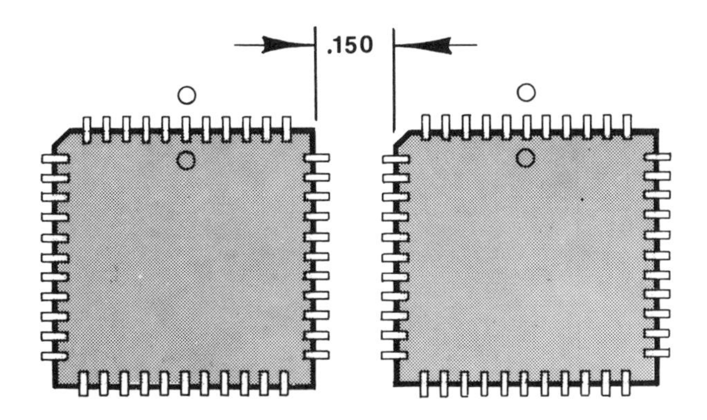

Fig. 5-23: For access and inspection try to use .150″ or greater spacing between the PCCs.

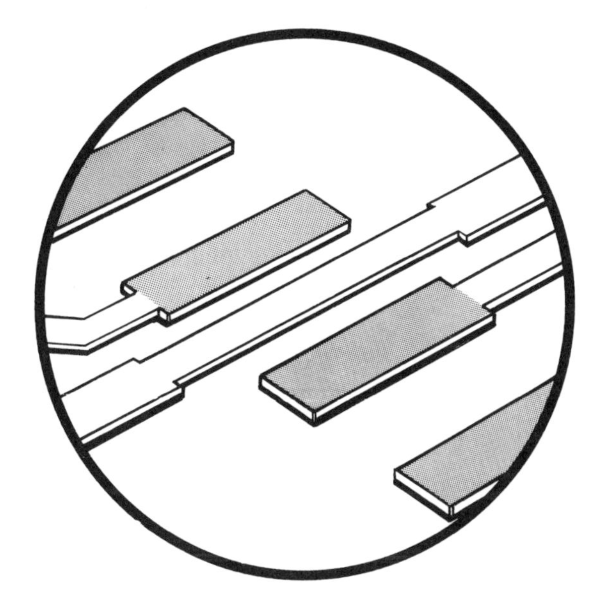

Fig. 5-24: This pattern will accommodate the most common JEDEC PCC devices.

Quad lead flat packs are one of the popular surface mounted packages for custom and semi-custom ICs. Contact spacing will vary from one manufacturer to another and could range anywhere from .050″, .040″, 1mm, 0.8mm or as close as .025″ and .020″.

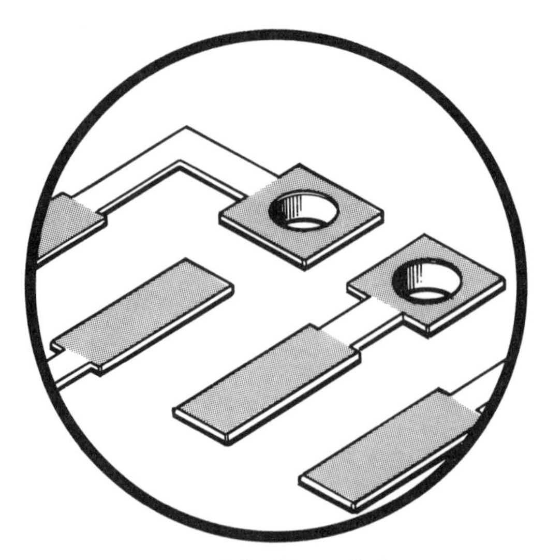

Fig. 5-25: Typical contact and feedthrough interconnect on a PCB.

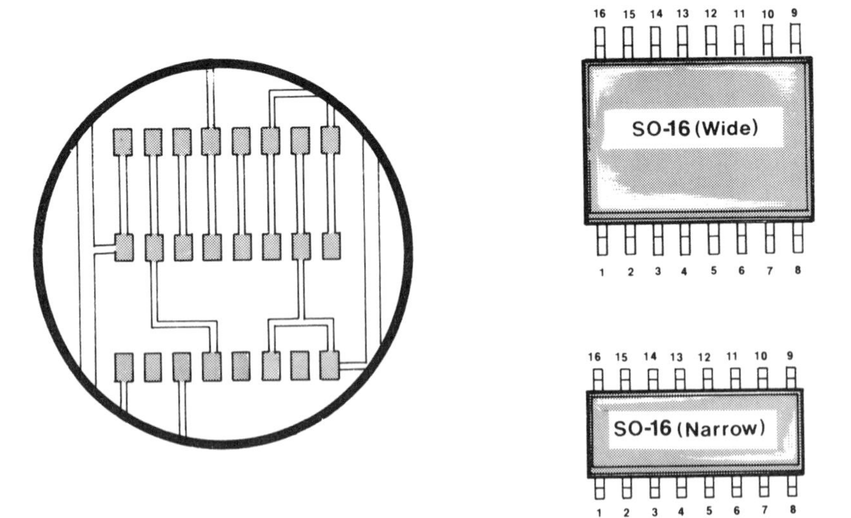

Fig. 5-26: Pattern showing solution to second sourcing problem.

Note: The pattern in Fig. 5-26 will take more space on the board and is not a practical solution for all situations.

The Quad Flat Pack IC in Figure 5-27 is more difficult to handle in a volume production environment. Closer lead spacing is very fragile and reflow techniques could require hand alignment prior to processing.

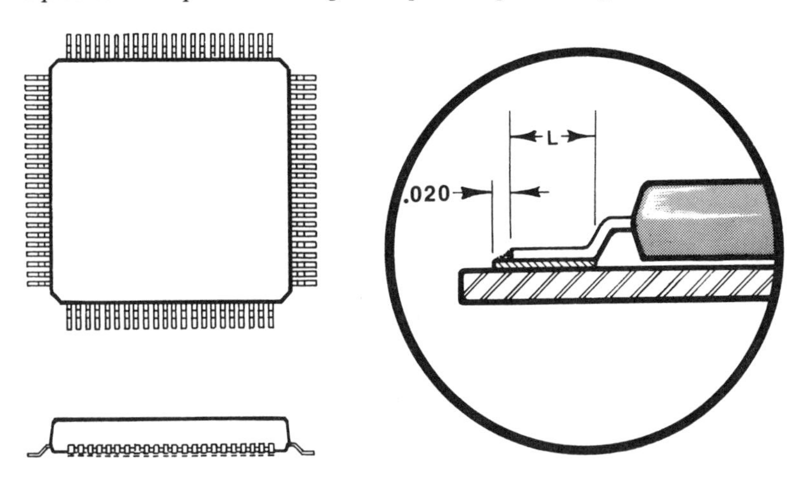

Fig. 5-27: The Quad Flat Pack IC is more difficult to handle in a volume production environment.

The footprint pattern will vary depending on lead spacing and length. The guidelines, shown in Figure 5-28, are generally accepted as reflow solder compatible. Components are typically supplied in tray carriers with an area for each device. Not all equipment will adapt to this system and special fixtures must be developed. Components should be furnished with the leads pre-formed to mount directly onto the surface assembly. If the leads are not pre-formed, your company will want to specify bend and length necessary to your needs.

Ceramic IC Packages

For military or other extreme environments, it is necessary to specify ICs in a ceramic package. As with the commercial products, there are several configuration options:

1. Leadless ceramic chip carrier (LCC)
2. Ceramic quad with J-bend leads (LDCC)
3. Ceramic quad flat pack (QFP)

Ceramic packages, as shown in Figure 5-29, are packaged for lower volume applications, but can be handled by many of the assembly systems available.

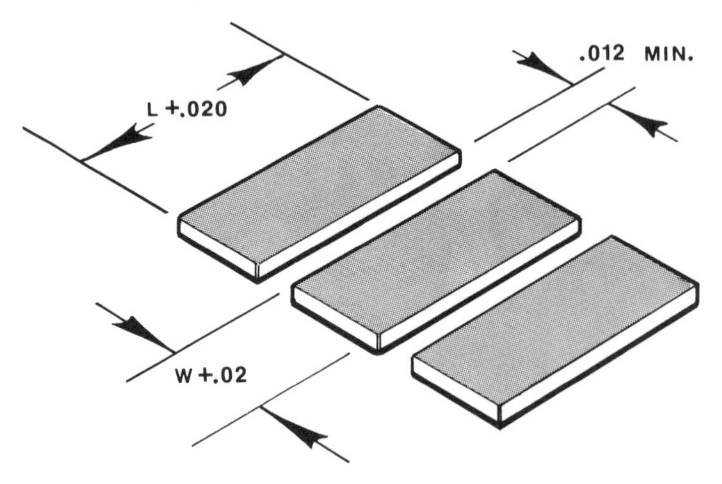

Fig. 5-28: These guidelines are generally accepted as reflow solder compatible.

The pad geometry for ceramic packages will be explained later in this chapter, but first, a few considerations should be noted. The leadless devices must be mounted to a more stable material than the commercial substrate equivalent. The reason for this is the difference in thermal coefficient of expansion(TCE). This will cause failure of the solder connection with repeated temperature cycling of unlike materials.

Ceramic components (FP) with leads extending straight from the side will usually be modified before attaching to the substrate. See Figure 5-30.

The advantage of using the leaded ceramic products is compatibility with a large variety of substrate types. The leads will flex sufficiently during the temperature variables and act as a shock absorber for the solder bond, eliminating the TCE problems.

Contact Design for Other SMT Products
DIP and SIP Module Design

The Dual Inline Package (DIP) module is a method of taking advantage of high-speed robotic assembly technology on a smaller, controlled scale. See Figure 5-31. The size of the DIP module could conform to the limits of standard ICs or take an outline adapted to individual needs. It is a common practice in SMT, as well as in hybrid assemblies, to use both surfaces to increase the density of the module.

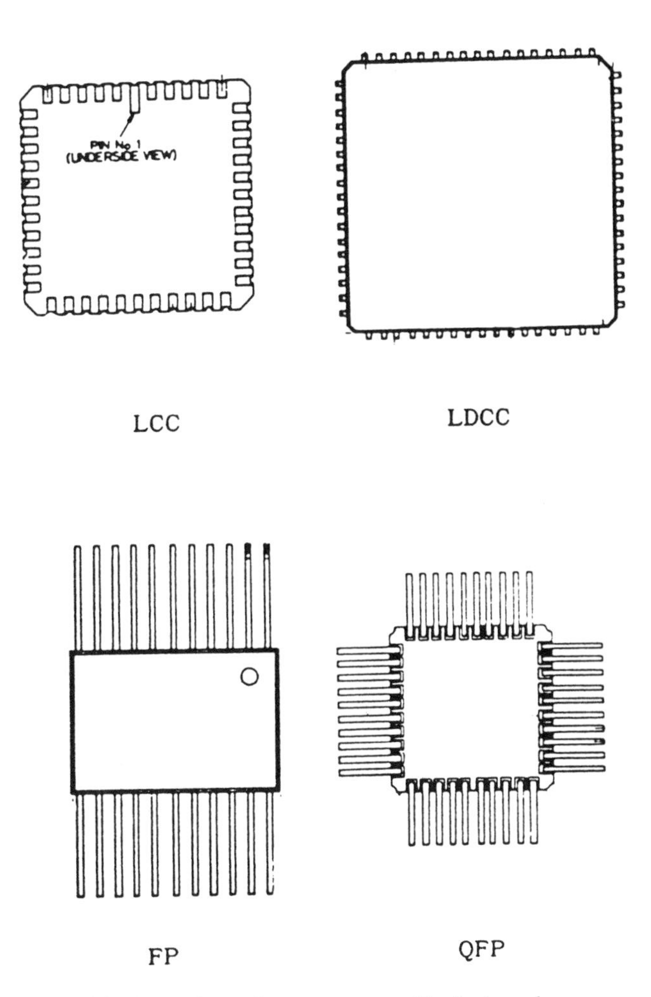

Fig. 5-29: Ceramic packages are actually designed for lower volume applaications, but can be handled by many of the assembly systems currently in use.

Dual inline contact-strips are the most economical method of terminating this module. The contact strips work well when the module is mounted into assemblies that will be wave soldered. The footprint pattern shown in Figure 5-32 is for solder reflowing the contacts to the module. Pin and socket strips or headers should be used for more durable contact requirements. Typically, pin and socket connectors are used when a module needs to be added or replaced into a socket without special tools.

The edge-mount contacts are furnished in a continuous strip with a common breakaway bar to help alignment. The contacts can be supplied with solder preforms built into the contact to insure an even flow and a strong connection. The contact strips are best suited for modules requiring a low profile.

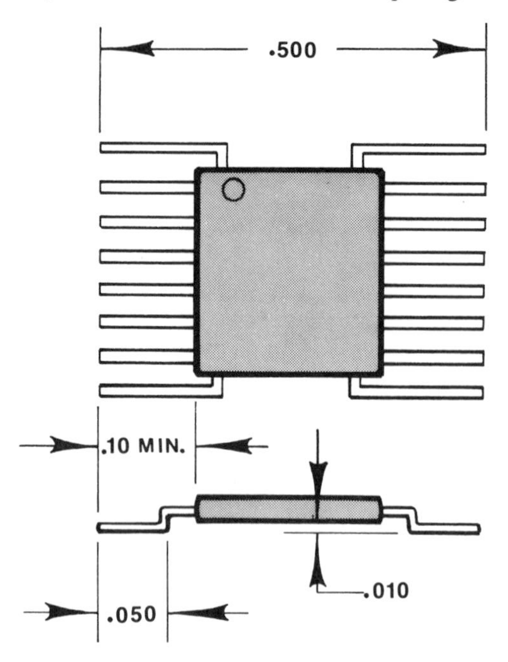

Fig. 5-30: Ceramic components (FP) with leads extending straight from the side will usually be modified before attaching to the substrate.

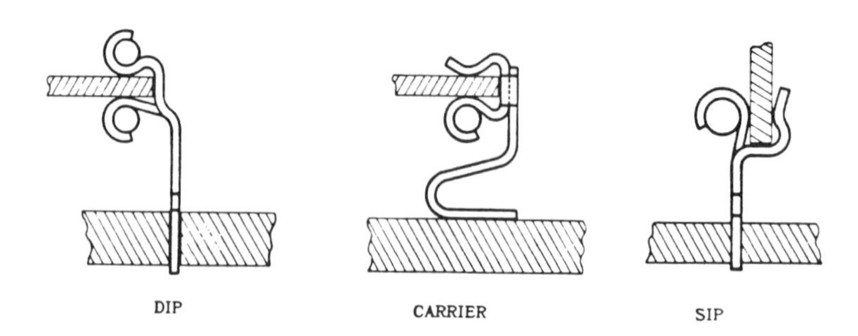

Fig. 5-31: Optional edge clip contact designs for DIP modules.

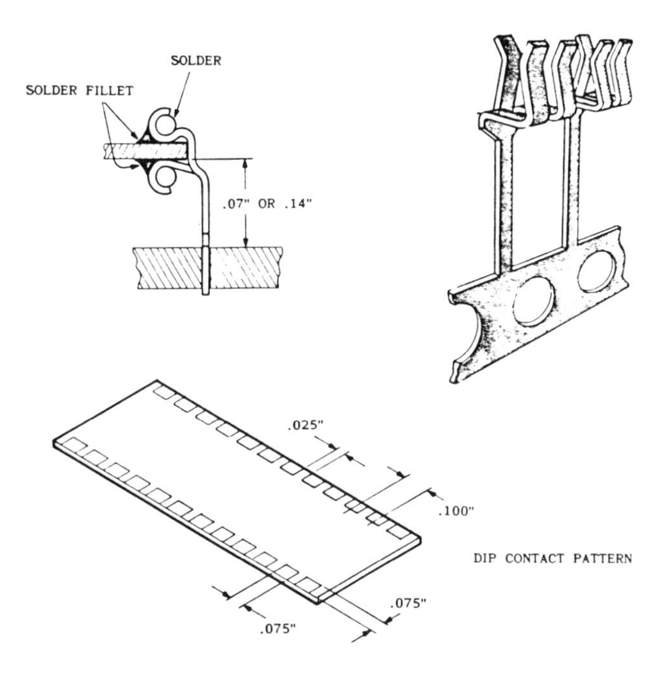

Fig. 5-32: Footprint pattern for solder reflowing the contacts to the module.

The Single Inline Pin (SIP) module is a popular form factor for SMT. Many expensive ceramic hybrid modules can be converted to surface mounted assemblies, thereby reducing substrate costs by 50% to 60%. It is common to partition the electronic functions into a module that can be used in many products. See Figure 5-33. The SIP module is handled as a component, tested and easily soldered into larger PC board assemblies.

Contacts for SIP modules are also supplied plain or with solder to help the uniformity of the solder connection. SIP contacts can be mounted to the module prior to the solder reflow process or added as a post assembly procedure.

The footprint pattern shown in Figure 5-34 will provide excellent electrical and mechanical characteristics after reflow solder. When allowing for overall height of your SIP module, don't overlook the stand-off height built into the contact. This height is optional and must be specified by the user. It is important to study the manufacturer's specifications closely. Contact pins are designed to mount into hole patterns and sizes generally used for DIP ICs or other leaded devices.

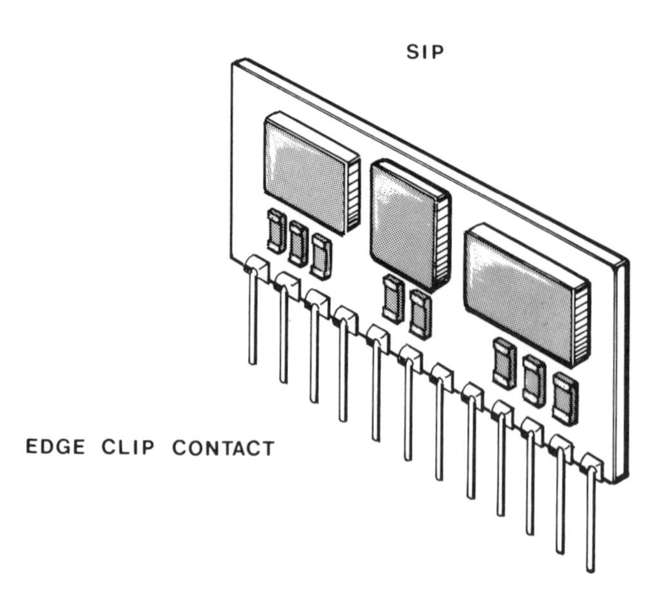

Fig. 5-33: It is common practice to partition the electronic function into a module that can be used in many products.

Chip Carrier Design

Using the .050″ space contact leads the designer can develop an SO and quad lead module or adapt a ceramic LCC to a conventional fiberglass PC board. Use the carrier contact when possible and design the footprint pattern around one of the JEDEC standard arrangements. If the IC is then converted to a plastic molded device, the main PC board will not require modification.

Carrier contacts, shown in Figure 5-35 are supplied on a .050″ spacing and ideally suited for miniature applications. The contact design allows the mating of substrate materials, which absorbs the stress caused by different rates of expansion during thermal cycles.

Further complex interfacing between a module and the main PC board is possible with closer lead spacing. The chip carrier design is often used to provide a working model of a proposed custom gate array or for a product that is not presently available from the component manufacturer. Using this carrier module concept may provide for the strategic, early entry of your product into a highly competitive marketplace.

Building Contact (Footprint) Libraries for SMT

As mentioned previously, resistors and capacitors are available in many sizes. We will illustrate the footprint pattern for each of the five sizes recommended as standard. The contact design and spacing is typical of the formula shown in Figure 5-36. This formula is adaptable to any chip style device, even though it is not specifically covered in the following pages.

The geometry illustrated in Figure 5-36 provides a good electrical and mechanical interface, as well as promoting a self-centering, reflow-solder process.

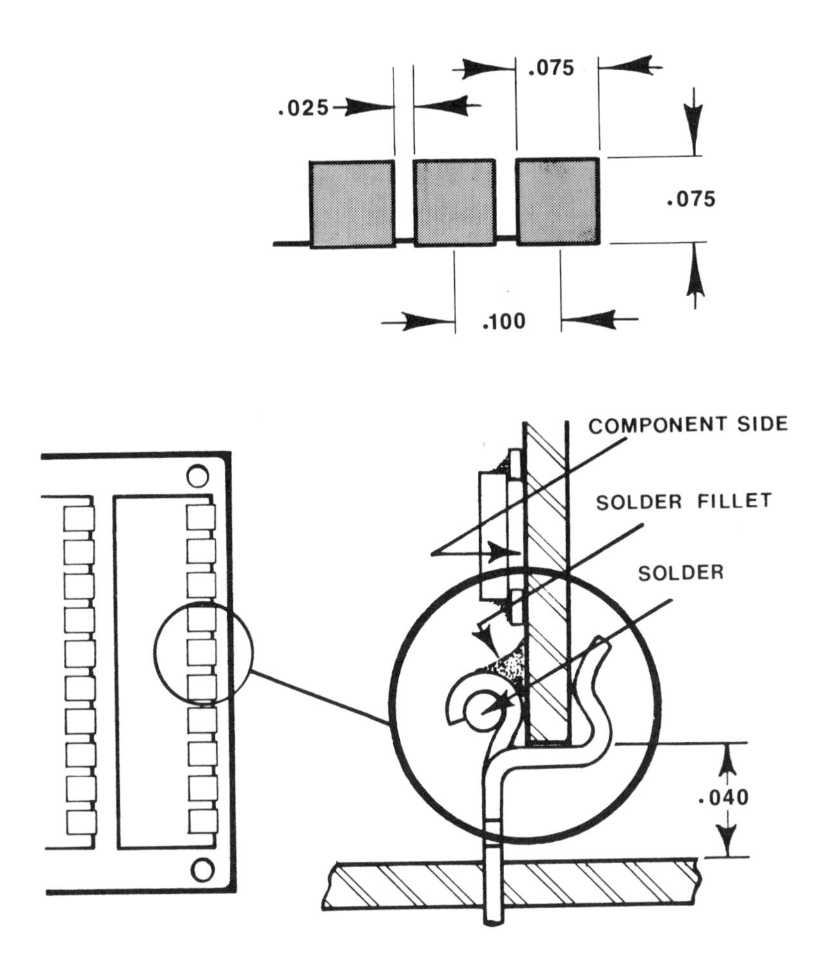

Fig. 5-34: This footprint patterns provides excellent electrical and mechanical bonding characteristics after reflow soldering.

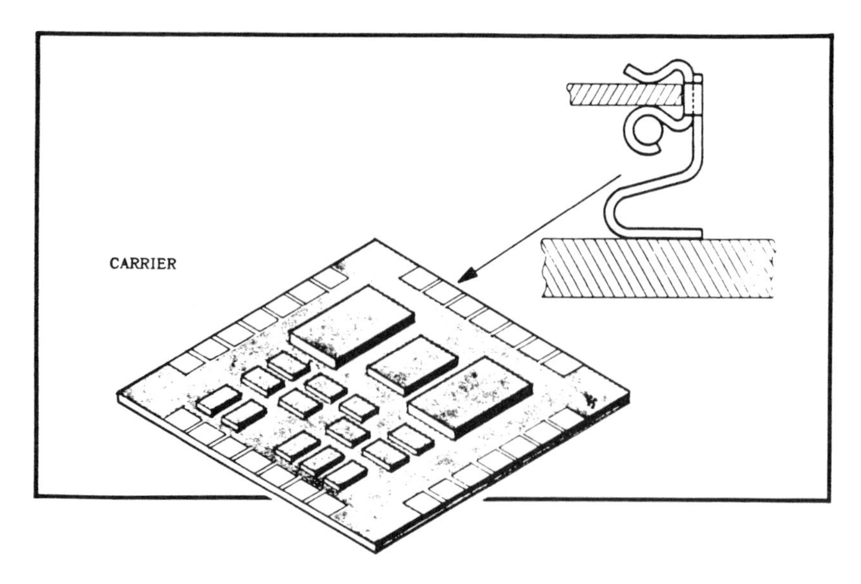

Fig. 5-35: Carrier contacts are supplied on a .050″ spacing and are ideally suited for miniature applications.

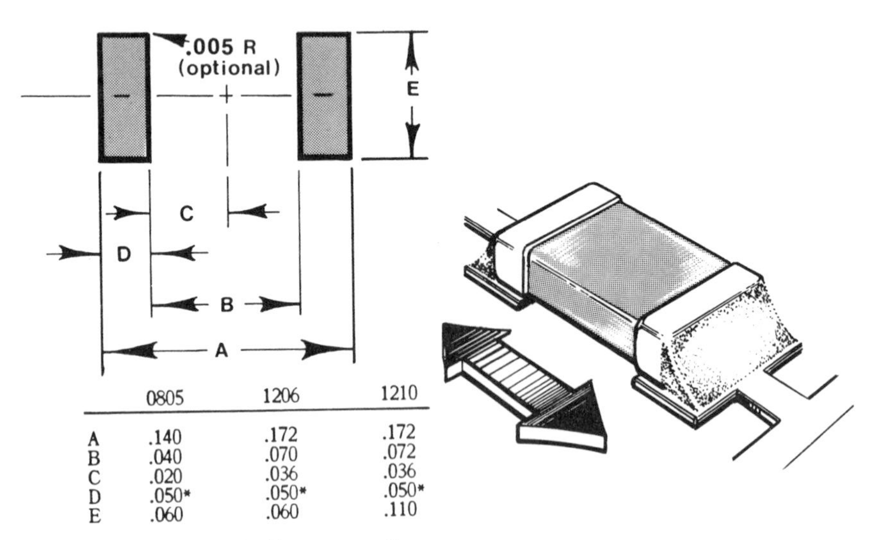

	0805	1206	1210
A	.140	.172	.172
B	.040	.070	.072
C	.020	.036	.036
D	.050*	.050*	.050*
E	.060	.060	.110

* .060 Max. is acceptable if room permits

Fig. 5-36: Typical contact design and spacing of resistors and capacitors.

For those using wave solder to terminate chip components, it is acceptable to reduce the contact width pad geometry. Although the pattern shown previously works well in the wave solder application, the narrow pattern will minimize excess solder on the miniature devices.

Figure 5-37 illustrates how pad geometry is designed to maintain a reliable solder connection while allowing for:

1. Placement accuracy of assembly equipment
2. Component shift during the epoxy cure.

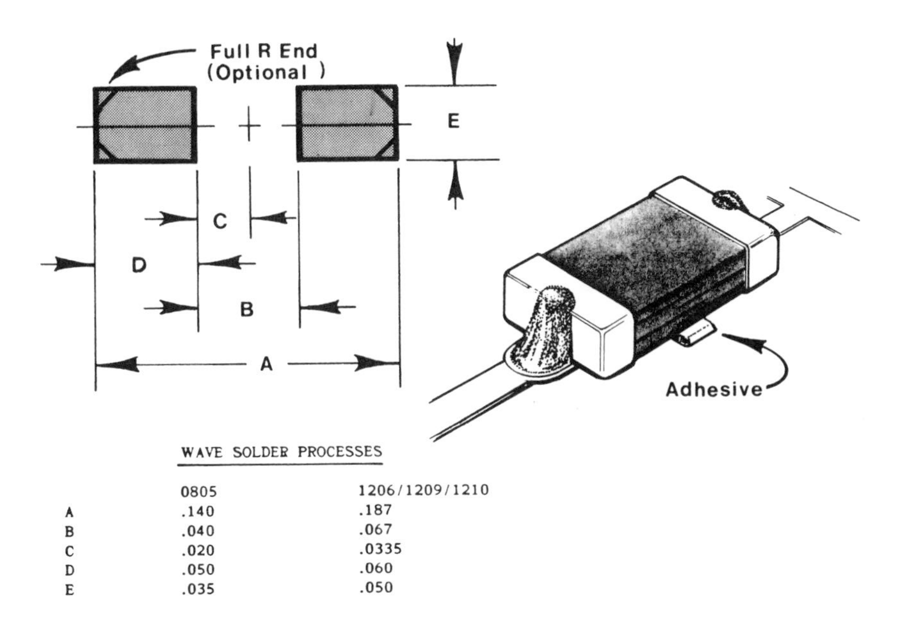

WAVE SOLDER PROCESSES

	0805	1206/1209/1210
A	.140	.187
B	.040	.067
C	.020	.0335
D	.050	.060
E	.035	.050

Fig. 5-37: Pad geometry designed to maintain a reliable solder connection.

Contact Geometry for Chip Components

The footprint details shown on this page are the full-width general-purpose style of pad geometry which is suitable for reflow or wave solder process. See details in Figure 5-38.

For those that require a visible side castilation of solder on capacitors, .010″ can be added to the “B” or width dimension.

The pattern shown in Figure 5-38 will accommodate .025″ grid positioning, which should be compatible with most CAD systems.

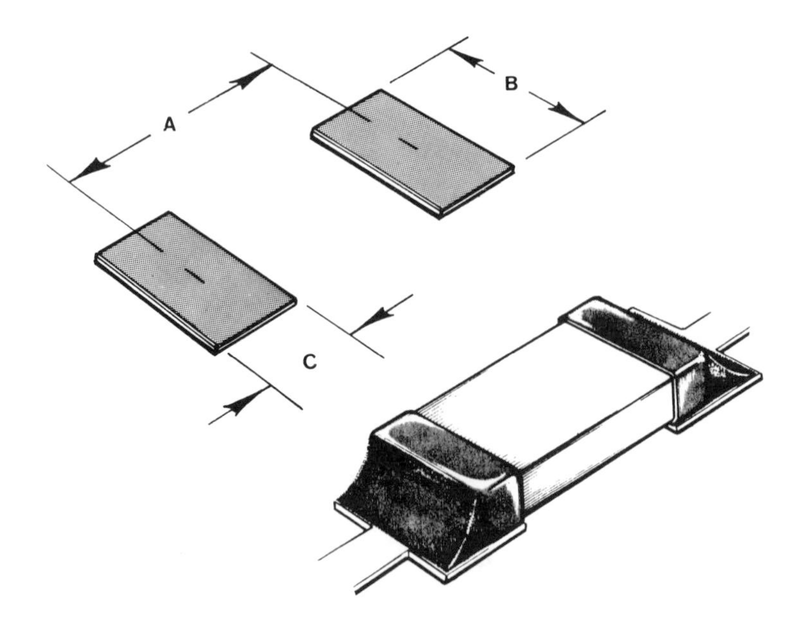

Fig. 5-38: Full-width, general-purpose style of pad geometry that is suitable for reflow or wave solder processes.

Optional Wave Solder Contact Geometry for Chip Components

An optional contact pattern for chip devices attached to the PC board with epoxy adhesive for wave solder is described in Figure 5-39.

Caution: When using the narrow footprint, compensate for the component body and contact area. When spacing the device, allow for an additional .040" clearance to an adjacent component or feedthrough hole.

Tantalum Capacitor Contact Geometry

Molded devices such as tantalum capacitors and inductors will be difficult to find in multiple sources. Several companies can supply a given value, but most devices are not mechanically the same.

A careful selection of a reliable supplier for tantalum capacitors is the safest route to follow. Compare your choice with the proposed standard tantalum configuration and value/voltage range in manufacturer's specifications.

The details illustrated in Figure 5-40 will work well with the proposed A,B,C, and D case sizes.

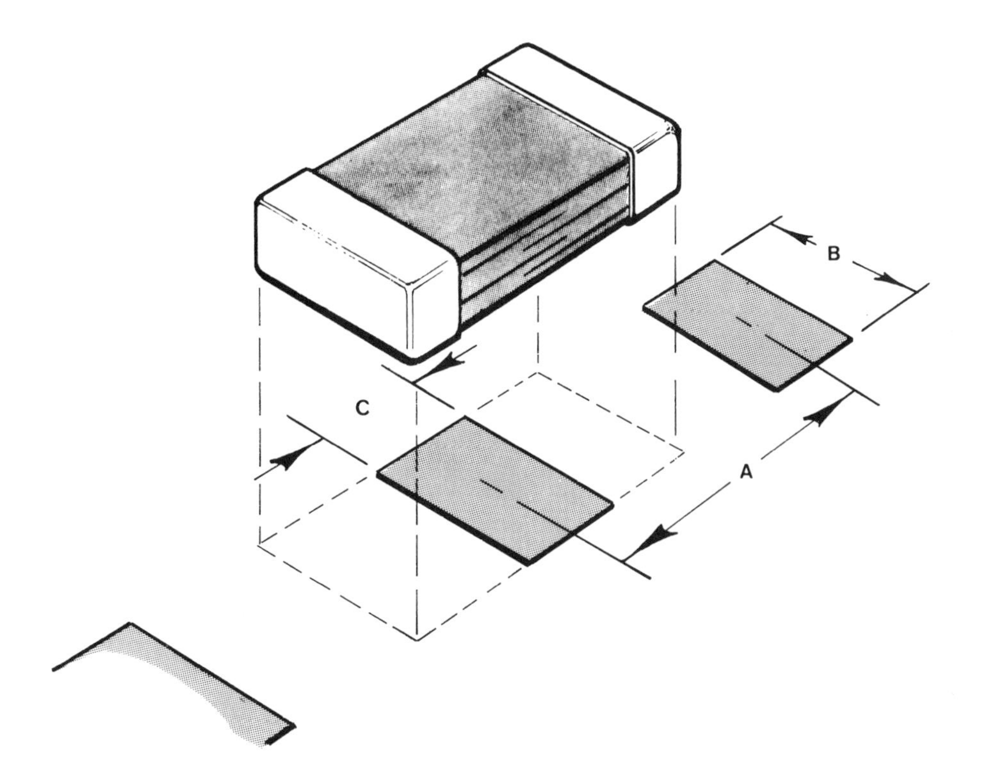

Fig. 5-39: Optional contact patter for chip devices for attachment to PCBs with epoxy adhesive for wave soldering.

MELF Component Contact Geometry

The MELF, a round shaped component, is dimensionally similar to the chip device and will adapt to the standard patterns previously detailed.

When reflow-solder processes are used, the MELF device will occasionally roll off the center line. For that reason, adhesive epoxy—as used in wave solder—can be applied to retain the component on the component side. A modification to the standard rectangular footprint will also reduce the tendency to roll. The notch pattern on adjacent pads, as shown in Figure 5-41, forms a nesting characteristic with the solder paste. If the paste is the proper consistency and thickness, the epoxies can be eliminated when the MELF is reflow soldered.

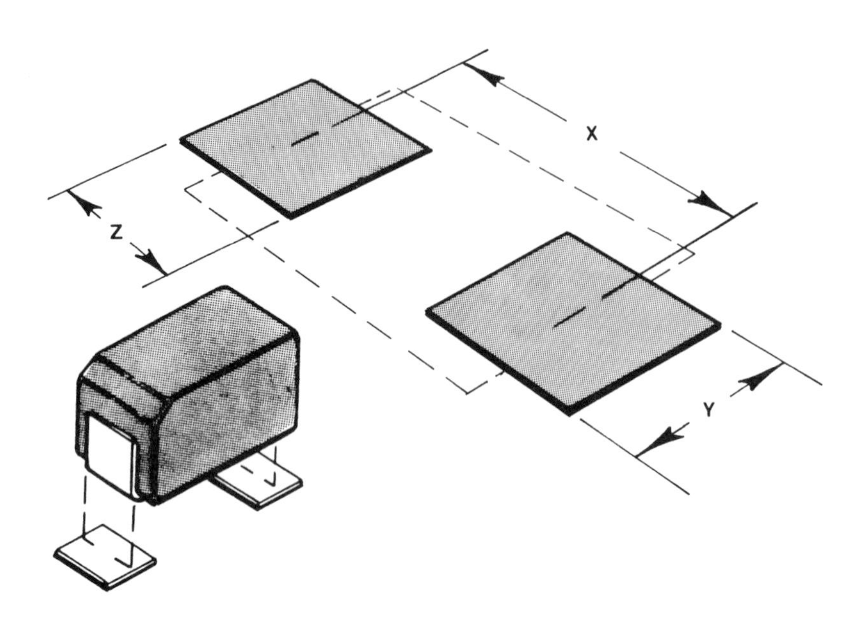

Fig. 5-40: Spacings that work well with A, B, C, and D case sizes.

End-cap termination on MELF components are not always consistent from one manufacturer to another. Careful selection of your supplier will help reduce the necessity of artwork changes or process variation.

The most popular packaging for robotic assembly of MELF components is the tape and reel. When using the MELF diode, keep in mind that the orientation of the component is supplied in a standard direction.

Figure 5-42 illustrates the standard diode direction with the cathode end toward the index holes of the type carrier. By maintaining a consistent direction on diodes, placement time per component can be minimal and inspection by QA is easier. Of course, one direction is not always feasible but, when possible, will contribute to a more efficient assembly.

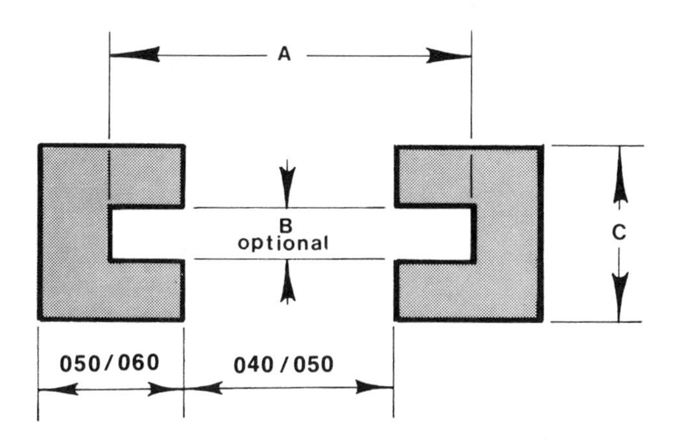

Dimension	
A	Component length + .005" / -.000"
B	Notch width = .012" / .015"
C	Component diameter +.010" / -.000"

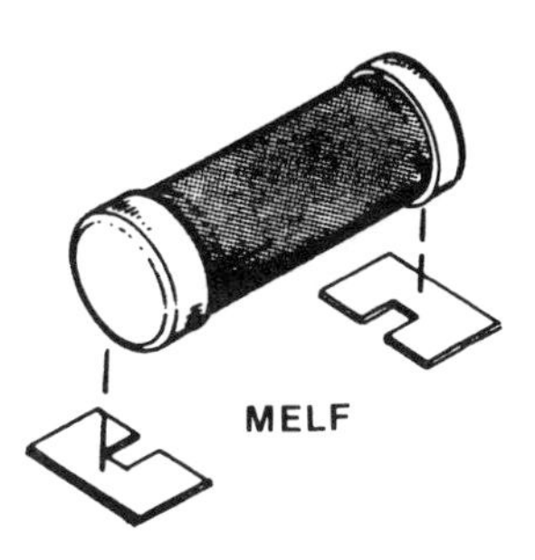

Fig. 5-41: Notch patterns on adjacent pads form a characteristic with the solder paste.

SOT-23 Contact ACT Geometry

The footprint pattern used for SOT-23 is a pad for each of the three legs of the device. The SOT-23 pattern shown in Figure 5-43 "A" is recommended by the manufacturers and works best in wave solder applications. The patterns in Figures 5-43 "B" & "C" have an extra long footprint pattern on one side, which compensates for the imbalance of solder area during the reflow process. The solder does not cool equally on all three pads and the side with two contact areas will dominate. Without this compensation, the component tends to lift away from the board surface on one side.

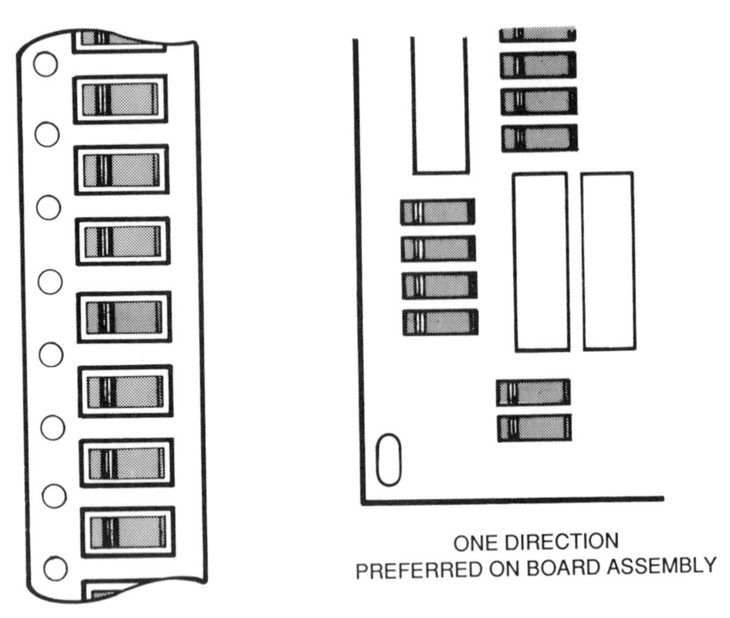

Fig. 5-42: A standard diode direction with the cathode end toward the index holes of the type carrier.

Figure 5-43 "C" is a pattern for a MMD/MMT or micromin device but can be used as a universal pattern which will also accommodate the SOT-23. Another feature that should be noted is the grid position of the contacts.

Four leaded devices are also supplied in the SOT-23 size body. FET's and multifunctional diodes sometimes require this fourth position at the same time retaining a very small package. Check the specifications for proper lead spacing and adjust width of pads when necessary.

SOT-89 Contact Geometry

The SOT-89 package is used to accommodate the larger die sizes and high power operation of some devices. The large area of the center tab will aid heat dissipation away from the component body to the surface of the board.

The dimensions shown in Figure 5-44 do not lend themselves to grid position, but they are recommended by the component manufacturer.

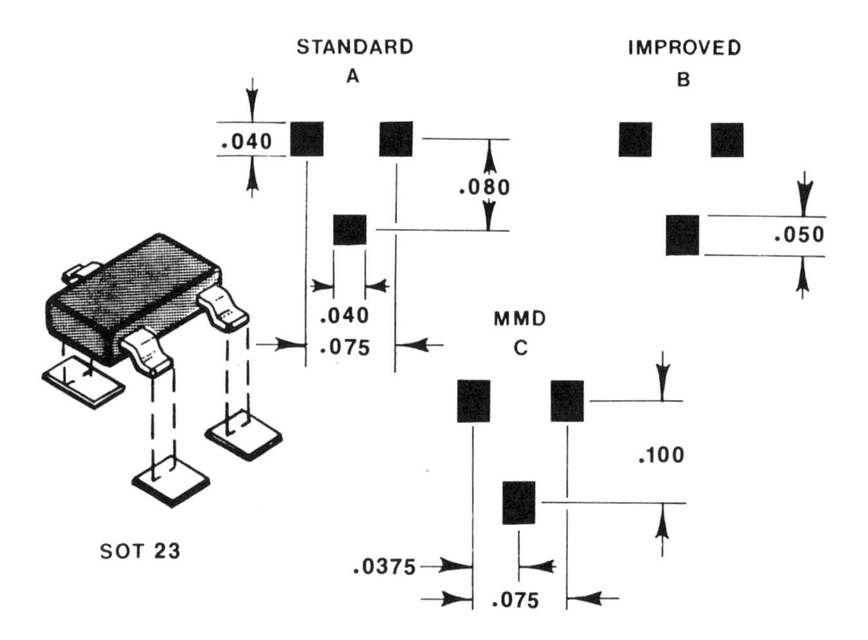

Fig. 5-43: Pattern for a MMD/MMT device that can be used to accommodate the SOT-23.

Plastic Chip Carrier (PCC) Contact Geometry

The size of the plastic chip carrier and the leadless ceramic carrier are very similar when lead spacing is .050″.

Caution: Pin 1 on the contact area of the ceramic device is longer than the others for easy identification. If the mating contact on the substrate is also lengthened, it will act as a visual orientation guide, thus avoiding the danger of unwanted interconnection to feedthrough pads or traces exposed to the contact.

Small Outline Contact Geometry

The contact pattern for SOICs is designed to furnish a reliable electrical mechanical interface of the IC to the substrate and, at the same time, be process friendly. (The term "process friendly" simply means a predictable, controlled method of attachment, using reflow solder processes.)

Successful manufacturing is assured with successful design! Process-proven pad geometry achieves repeatable results with any technique used during the assembly process, regardless of the equipment used or reflow process chosen.

Contact geometry for the PCC and SOIC devices are detailed in Chapter 7.

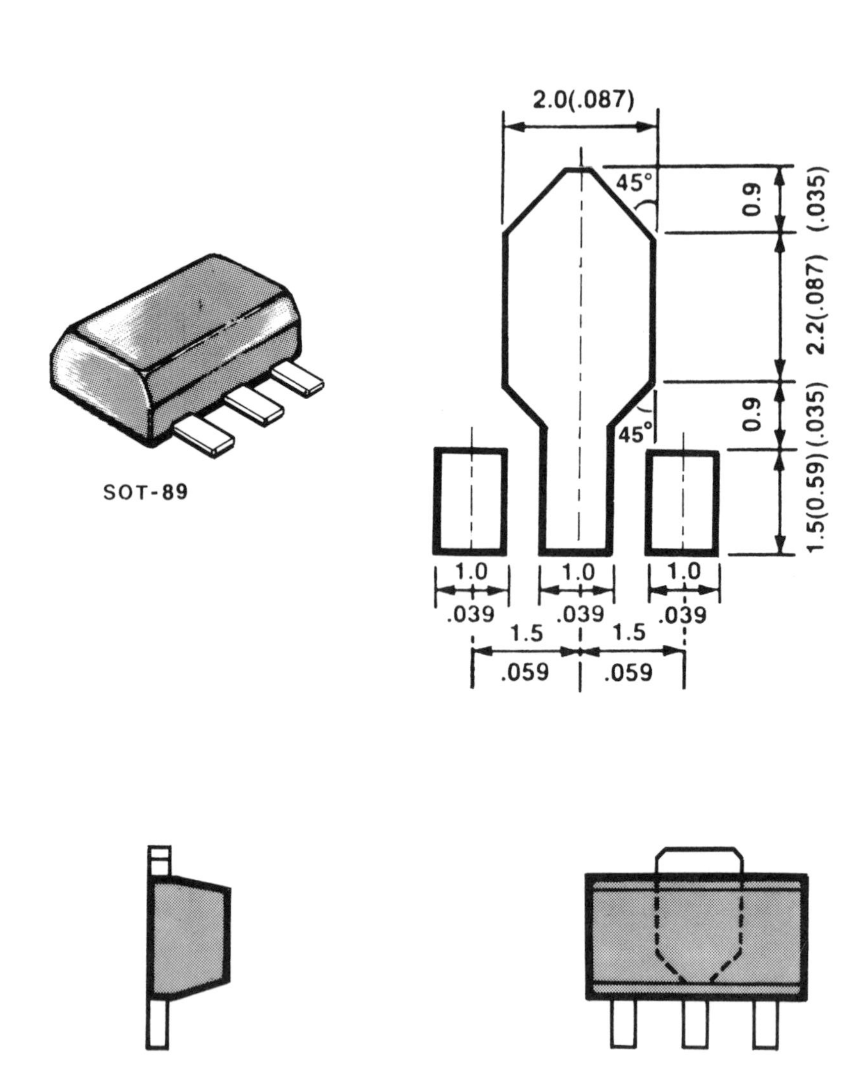

Fig. 5-44: Dimensions that do not lend themselves to grid position, but they are still recommended by the component manufacturer.

chapter 6

COMPONENT SPACING FOR SMT

Robotic assembly equipment can place SMDs on the substrate surface with the accuracy of ± .005 inch in the X and Y direction. To maximize component density, designers are tempted to pack components on the PC board with only minimum clearance between component bodies. The recommended guidelines shown in this section provide space between the components for ease of inspection and if necessary, rework.

The space between land patterns of the chip components shown in Figure 6-1 allows for a .010″ - .012″ wide trace to pass between land patterns, while providing space for rework or touch-up tools. This spacing also reduces solder bridging when the components are to be wave soldered.

Low profile, small outline and quad flat pack ICs have the gull-wing shaped leads protruding outward from the body. This lead design accommodates inspection and accessibility for touch-up or rework. When space is available, separate the parts to provide clearance for removal systems. Occasionally, when an IC is damaged, or for one reason or another has to be replaced, there should be enough space to remove it without disturbing the adjacent devices. The components should provide ideal accessibility when using removal tools.

Higher profile devices, such as plastic chip carriers with J-lead contacts, require additional component spacing. PCC components will allow for 100% visual inspection and touch-up or rework.

There is a trend, by some in the industry, to pack PCCs as close to one another as possible. This may be justified in the case of a memory module or very small assemblies. However, be aware that the assembly manufacturer is relying strictly on solder process control and cannot inspect the finished product except by cross-section or X-ray methods.

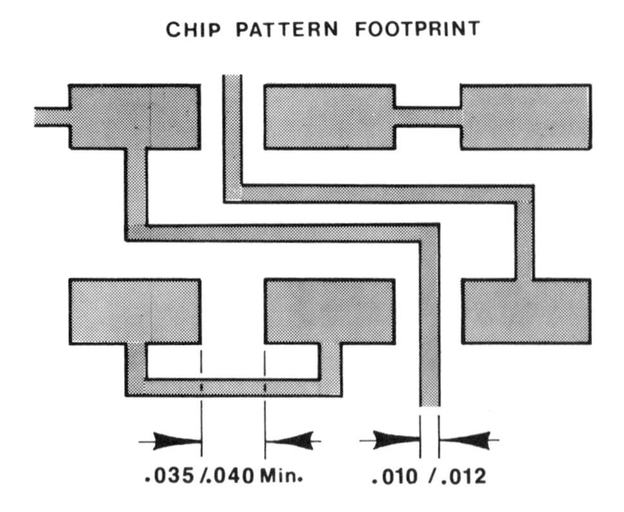

Fig. 6-1: Recommended space between land patterns.

When an IC must be replaced on a closely spaced assembly, the technician first cuts each lead from the body of the component before removing the part. Since access to contact is restricted, the entire assembly will be exposed to the solder reflow process once more.

Basic Considerations

After the basic decisions have been made, one should be ready to start the board or substrate layout. This consists of establishing the component arrangement and the circuit pad and trace patterns necessary. For this, one needs to determine the device dimensions and tolerances, as these will combine with the machine placement accuracy and the etched or printed circuit accuracy tolerances to define a set of acceptable pad layout arrangement standards.

There are many sources of suggested pad and grid, or trace, layout data available, both in the trade literature and from private publishers. Many of these are listed in Appendix I of this book. All of these will provide you with guidance and a variable degree of help. Figure 6-2 is an indication of what you will find for typical SMD components. The data generally will show the lead and body dimensions and suggested pad dimensions and locations. What is finally used, however, is highly dependent on the particular combination of components you select, the placement medium, the planned placement machines, the remaining process combinations and materials which you select for your application.

View A–For Passive Device Land Pattern Design

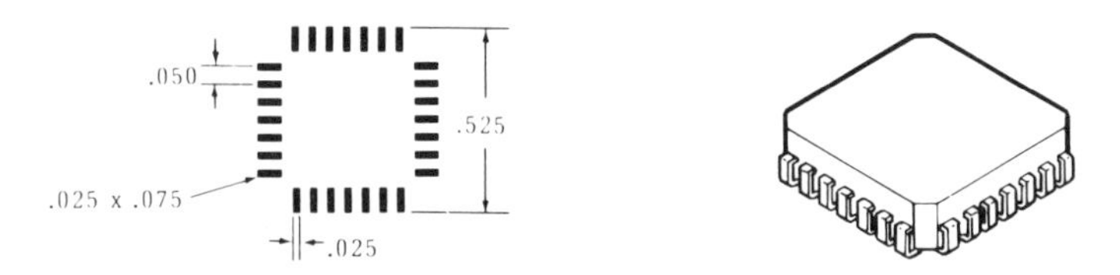

View B–For PLCC and LCC Device Land Pattern Design

View C–For SOIC Device Land Pattern Design

Fig. 6-2: Typical suggested Land Pad Pattern data for SMDs.

Before the final mounting pad arrangement standards and medium material decisions are considered acceptable, one should go through a series of controlled tests using these candidate designs, plus minor variations as well, and the proposed medium material to evaluate the soundness of the choice of design. These tests should include several soldering profiles, including flux and solder temperature evaluations. After acceptable soldering parameters have been established, thermal cycling should be done to and beyond the degree to which the design will be exposed in normal use.

With PCBs, all techniques used in the past would be applicable to any design (see Chapter 8), including double-sided and multi-layer. In both cases, "vias", or plated-thru-holes, can be used for the layer-to-layer interconnection. This technique may be used with or without the additional use of thru-hole inserted components. It does require a drilling operation, however.

Placement Accuracy Requirements

Device Function	Capacitor	Capacitor / Resistor	Resistor / Diode	Diode / Transistor	Transistor / Integrated Circuit	Integrated Circuit	Integrated Circuit	Integrated Circuit	Integrated Circuit
Available Package Forms	Polarized Chip	Chip	Cylindrical	SOT	Bare IC or Flip-Chip	SOIC	Plastic Leaded Chip Carrier	Flat-Pack	Leadless Chip Carrier
Relative Placement Accuracy Group (Note 1)	A	A	A	A	C	B	B,C	B,C	B,C
Available Handling Systems									
Tape–8mm and Larger	•	•	•	•		•	•	•	•
Sticks–Rigid and Flexible			•			•			•
Waffle Trays					•			•	
Bulk–Vibratory Feeder	•	•	• (NOTE 2)						

Notes:

1. Placement accuracy required.
   Group A–Moderate.
   Group B–Medium.
   Group C–High.
2. For Resistors only (MELF); Diodes are polarized.

Fig. 6-3: Device selection chart.

Figure 6-3 gives the relative range of placement accuracy required by device type for successful mating of the device terminal arrangement to the pad configuration on the medium.

Multi-leaded devices need a much more sophisticated machine approach for proper registration than simple two- and three-lead devices. When determining the overall placement accuracy needed, the location tolerance of the placement pads on the medium relative to the medium locating features must be taken into consideration.

Placement accuracy is a major significant characteristic of any given system and it affects machine cost, speed, and complexity. For example, those machines which simply pick up the part, orient only at 90° increments, and maybe even "center" to some simple mechanical reference, are generally adequate for the relative placement accuracy needed for the Group A devices shown in Fig. 6-3. Group A components usually need to be placed to within ±0.008″ to ±0.015″, depending on terminal and pad sizes and machine tolerances.

Components in the medium accuracy Group B are SOICs and plastic leaded chip carriers, flat packs and leadless chip carriers. Generally, those in this group with 44 leads or less, and with lead center spacing of 0.050″ (1,25mm), do not need much more care than that described above. However, devices with more leads and/or closer lead centers will typically need to have an overall placement accuracy between ±0.003″ to ±0.005″ for acceptable soldering.

For components in Group C, specifically those with many contact points and close lead or pad spacing, significantly more is needed to increase the final placement accuracy. Usually this involved the use of "Vision" for both lead location and orientation and for locating "Fiducials" on the medium, as explained in the paragraphs to follow. Generally, the larger versions of the Group C components need to be placed to within an absolute location of ±0.002″, or less, in S and Y with respect to the mating pads, and usually a rotational correction is also needed; see Fig. 6-4.

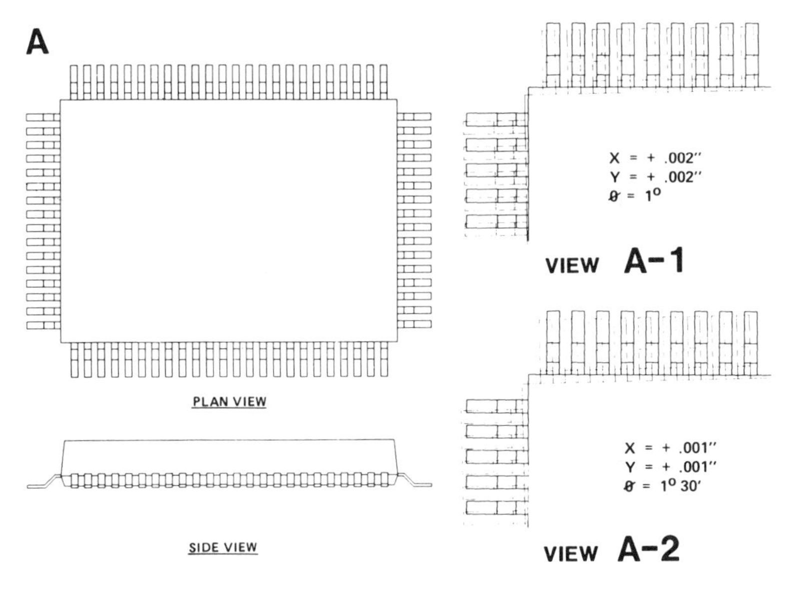

Fig. 6-4: The effect of X, Y, and an angular misalignment for the placement of a typical large, 80-lead, Flat-Pack with an overall size of approximately 1″ x 0.7″. The leads as shown are 0.013″ wide on 0.030″ centers. View A-1 shows the effect of an X and Y error of +0.002″ and an angular error of only 1°. View A-2 shows what can happen with the X and Y positioning errors reduced to 0.001″, but with the angular error increased to only 1.5°.

Placement Techniques

The technique common to virtually all SMD assembly machines is "Pick and Place." Specifically, this means "picking" the part from its input package feeder and "placing" it on the board or substrate medium. A typical SMD machine is shown, schematically in Fig. 6-5. In essence, there is a "head" and it usually carries a "vacuum spindle" which has certain common characteristics: generally, it can move up and down, or in and out. Often it can rotate the device, sometimes in 90° increments. Rotation is important for two reasons: one, having the ability to rotate the part allows the circuit designer additional freedom in arranging the artwork with 90° increments recommended. A secondary need for rotation is used with a large part with many leads with very small spacing. This, coupled with "vision" allows the necessary very small rotational corrections required to insure correct lead-to-pad alignment. Figure 6-4 shows what can happen in placing a large part with many small leads if the proper corrections are not made in both translation and rotation. In the spindle's controlled up-and-down placement motion, the down pressure is often programmable into more than one "force" level. This capability permits the device being placed to be pressed into the placement medium with a known force such that penetration into the paste deposits, flux, or glue can be controlled to give a desired and consistent result.

Another feature often found on, or in support of, many placement head assemblies is a "centering device." Generally, this will center the part on either the body outine or on the lead pattern. In most cases the pick-up positioning of devices to be placed is only approximate; that is, from a tray, a linear feeder, the open nest in a tape, etc. In these cases, the vacuum pick-up nozzle literally has only a general idea of where the part is with respect to its true center. The nozzle generally does know, however, that it has a part because most pick-and-place systems have a sensor which indicates that pick-up has been made.

The centering mechanism then "centers" either the leads of the body of the part on the nozzle,or at least moves it to a consistent poistion which becomes known as the "placement center." This operation, depending on the type of device and the centering system, can locate well defined parts with a repeatability on the placement head to as close as $\pm 0.001''$ (0.025mm). This, however, does not mean that these same parts will be placed on the mounting surface to that precision, as a number of other factors affect the final placement accuracy. Since the mechanical centering function uses the edge surfaces of the part, it is totally dependent on the part edge-to-lead tolerances. Next, one must consider some of the other factors of the process: the stretch/shrink of the circuit itself; the pad-to-pad tolerances within the part to be placed; the adhesive or solder paste influence; and finally, the remaining placement machine tolerances, X and Y resolution, calibration, angular error, etc.

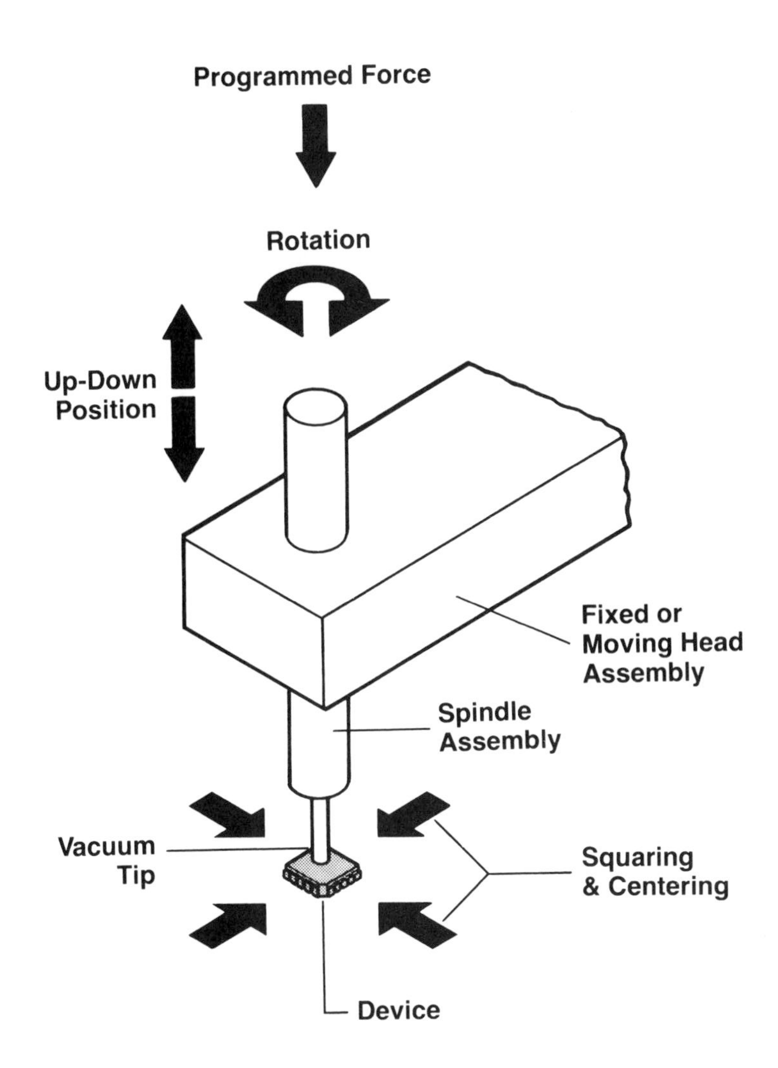

Fig. 6-5: Typical pick-and-place spindle function.

There is also a technique available called "vision" and it can help with those applications which have very tight location and placement requirements. It uses optical recognition techniques coupled with the use of optically recognizable "fiducials" to find and correct rotational and positional errors between the component and the circuit pattern on which it will be placed. Fiducials are specially shaped reference marks incorporated in the circuit pattern which provide enough information to allow recognition. Using this technique, significant improvements in placement location can be achieved, saving the manufacturer both money and time.

This placement location improvement results from reducing or eliminating many of the typical tolerance items usually found in non-vision corrected component placement. These errors generally result from: locating hole or surface references, absolute machine error in areas other than the specific placement zone, "pad" location tolerances, plus other tolerance items such as those resulting from the drawing, screening and reference processes. Generally from one to three fiducials are used to establish the true location of the pad/grid pattern on a circuit medium. With one fiducial, one can establish a second order level of correction: both an X-Y offset, an angular correction, and accomplish a linear "stretch/shrink" compensation for the medium. Adding a third fiducial, properly placed, can improve the two-fiducial correction slightly. The actual correcting is done by an algorithm in the machine software. Through the use of one or more types of fiducials several levels of corrections can be obtained: panel-to-panel, circuit-to-circuit, or on a component-by-component basis. If the overall circuit medium has been initially corrected with two or three fiducials, with the component-by-component correction, only a single fiducial is generally required and this is usually placed at the placement center of the device.

The accuracy of optical "vision" location of a fiducial can be as close as ±0.007″ (0,02mm). However, this will vary with the chosen system paraeters such as field-of-view, the optical system "gain" and magnification characteristics, and the specific fiducial size and shape selected. When using multifiducials, angular errors can be corrected to as close as 0.2°, depending on the specific machine capability. From a time standpoint, fiducials can be searched at a rate of approximately one each second, or better. This assumes that they fall within the standard field of view, which is usually ±0.100″. The field of view also sets the general size limitation on fiducials; that is, generally between 0.100″ and 0.125″.

Figure 6-6 shows several suggested shapes for fiducials. Another important concern is the color and/or contrast of the fiducial compared to the circuit medium. Also, one must take care not to have similar-appearing artwork arrangements within the "search area" of a fiducial. These and other recommended fiducial details, such as size, shape and other important characteristics may be found in manufacturer's specifications.

The best machine or combination of machines for a given application can be determined by analyzing all of the basic ingredients of the particular design approach and the expected product mix. The selection of the placement machine(s) and the design of the placement medium depends on a number of factors, some of which are discussed at the end of this chapter. These factors provide a sound basis for the detail planning, final machine selection and they will also aid in the medium design.

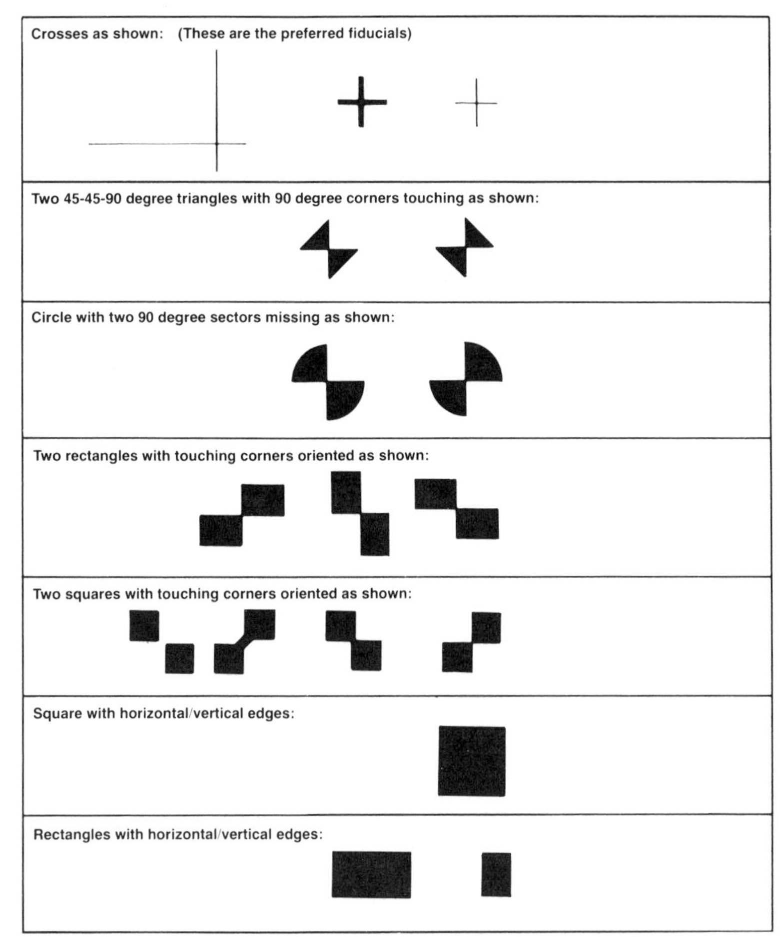

Fig. 6-6: Acceptable fiducial types.

Trace-to-Trace Guidelines

Clearance between conductor traces and trace-to-contact patterns is a matter of practical vs technological capability. The high tech board manufacturers can do just about anything to provide a quality product.

Trace width of .003 inch with equal air gap is an everyday occurrence for some fabricators, but remember, high tech boards are closely associated with high cost. The mainstream quality shop, doing multilayer PC boards recommends a more conservative approach, whenever possible. Trace-to-trace and trace-to-contact patterns shown in Figure 6-7 will allow for consistent etch quality on outer layers, while insuring thorough coverage by the solder mask over-trace. An exposed edge of the conductor passing near or between contact areas will attract solder particles and result in bridging.

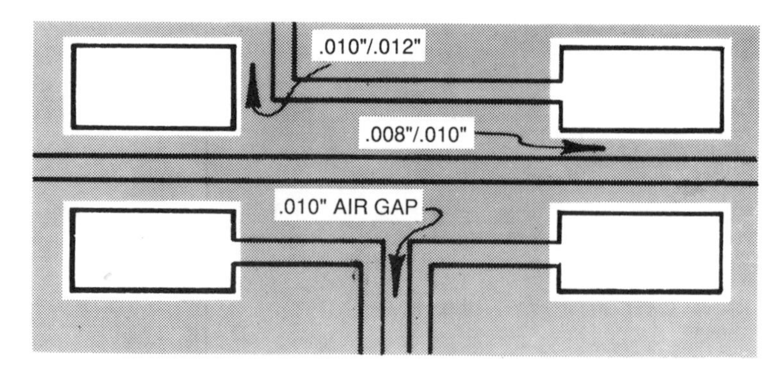

Fig. 6-7: Patterns that allow for consistent etch quality on outer layers, while insuring thorough coverage by the solder mask over-trace.

Removal of the component is the only method of eliminating the short when it occurs on an IC contact that folds under the component, such as the J-lead PCC.

Connection of the trace or conductor to the contact area is another subtle method to reduce the need for rework.

If a very wide trace or copper area is connected to the contact pad by an equally wide trace, two reactions are possible during reflow solder:

1. The component may be drawn off the liquid solder since the solder contacts adjoining the large metal masses cool the solder quickly. In the case of chip components, tombstoning is unavoidable.
2. The other reaction on tin/lead plated PC boards is a migration of the liquid solder paste during reflow processing into the liquid plating on the conductor traces. This migration from one or more leads or contacts of the SMDs will require additional touchup. Figure 6-8 illustrates a few Do's and Don'ts to help avoid these solder defects.

Contact (Footprint) to Via Pad

The space between a via pad and the component contact area should provide for a solder mask barrier. This barrier will contain the liquid solder paste in the contact area during the reflow process.

The configuration shown in Figure 6-9 provides a .008″ to .010″ wide solder strip separating the contact from the via or feedthrough pad and hole.

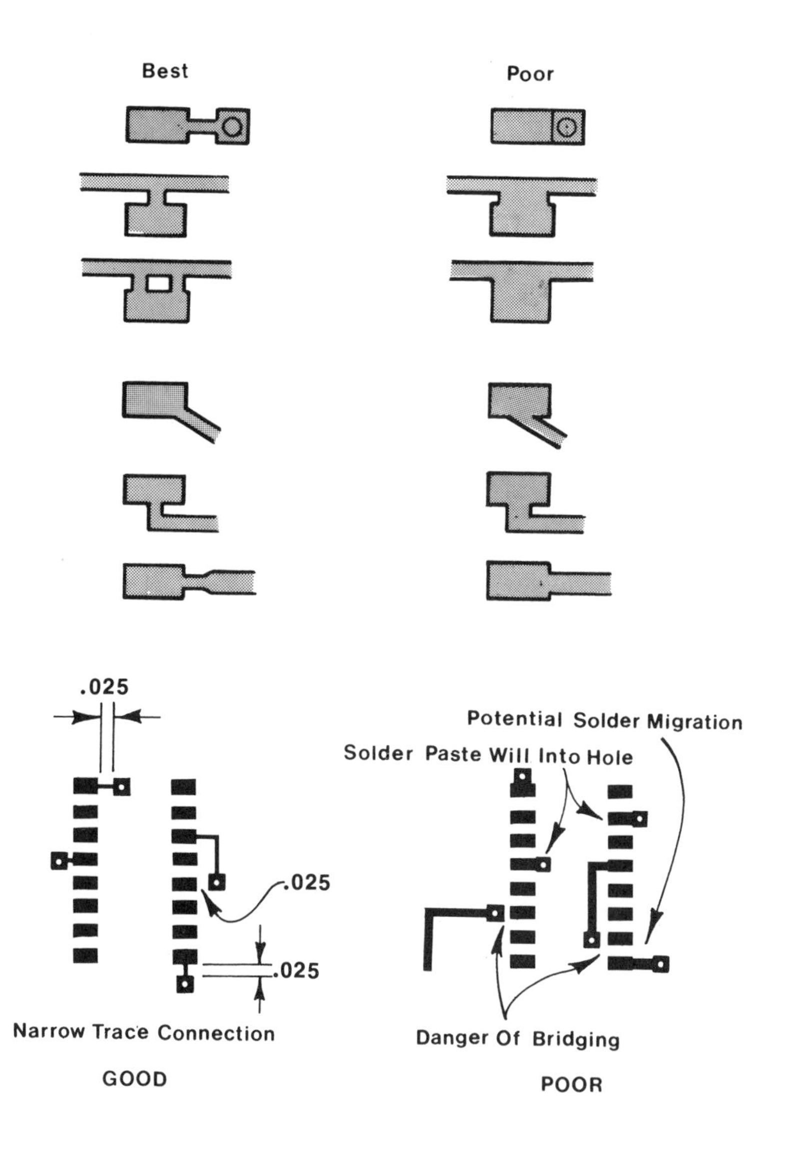

Fig. 6-8: Trace-to-contact guidelines.

If it is necessary to join the contact footprint to a via pad without adequate space, it will be necessary to cover the via pad with solder mask as shown in Figure 6-10.

Note: If the via hole is not covered the liquid solder will flow down through the hole and away from the intended component contact.

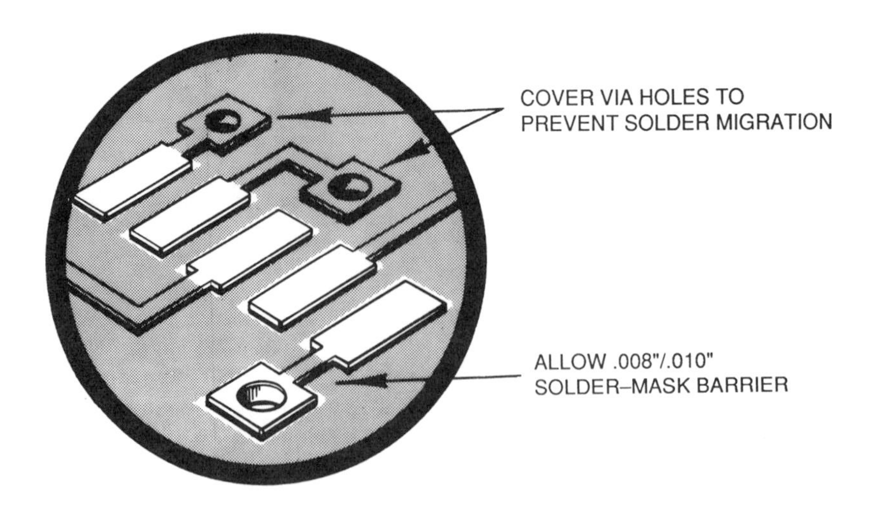

Fig. 6-9: A solder mask strip separates the contact from the via pad and hole.

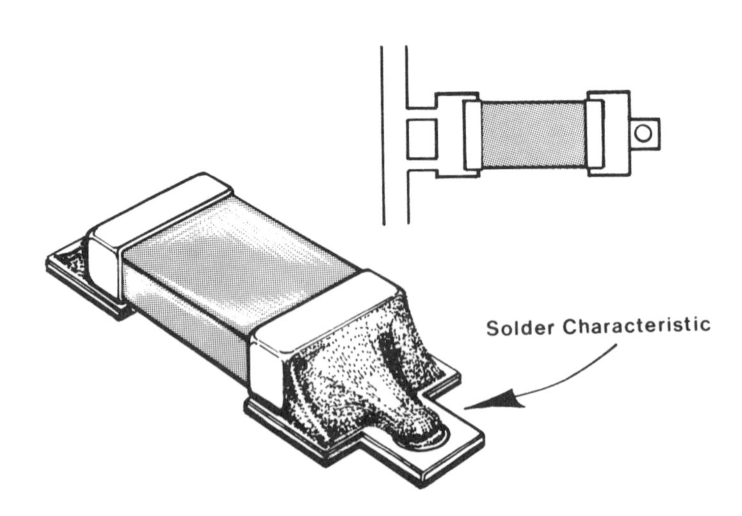

Fig. 6-10: When adequate space in not provided it is necessary to cover the via pad with a solder mask.

Solder Mask for Solder Control

Solder mask can be a valuable ally in reflow assembly of surface mounted products. The coating is either applied with:

1. A dry film lamination process.
2. A wet application—using a screening process.
3. Other more "exotic" techniques.

In any case, except for a very small PC board, the designer should specify a photo imaged mask coating.

Clearance of solder mask can be at zero expansion or as great as .005 inch or .010 inch overall expansion. This clearance can be controlled by furnishing an expanded pad master to the board fabricator with instructions not to expand further.

The three examples shown in Figure 6-11 compares the ideal solder mask clearance to the unacceptable.

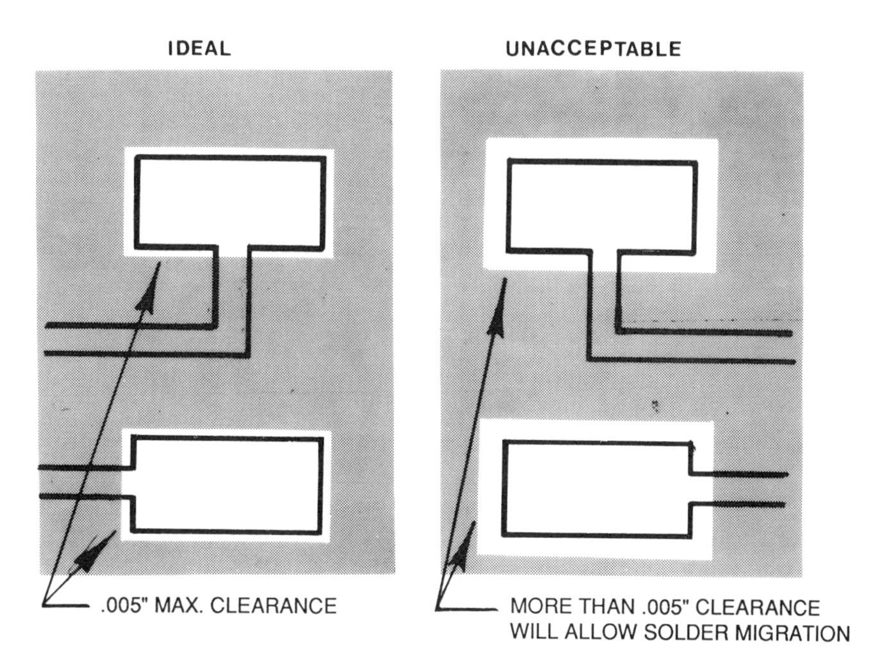

Fig. 6-11: Comparison of ideal solder masks to the unacceptable.

If solder mask overlaps the contact area or solder mask residue is left on the same area, reflow soldering will not be satisfactory. Likewise, when too great a clearance is allowed, the liquid solder will spread over too far on the area promoting an unreliable solder connection.

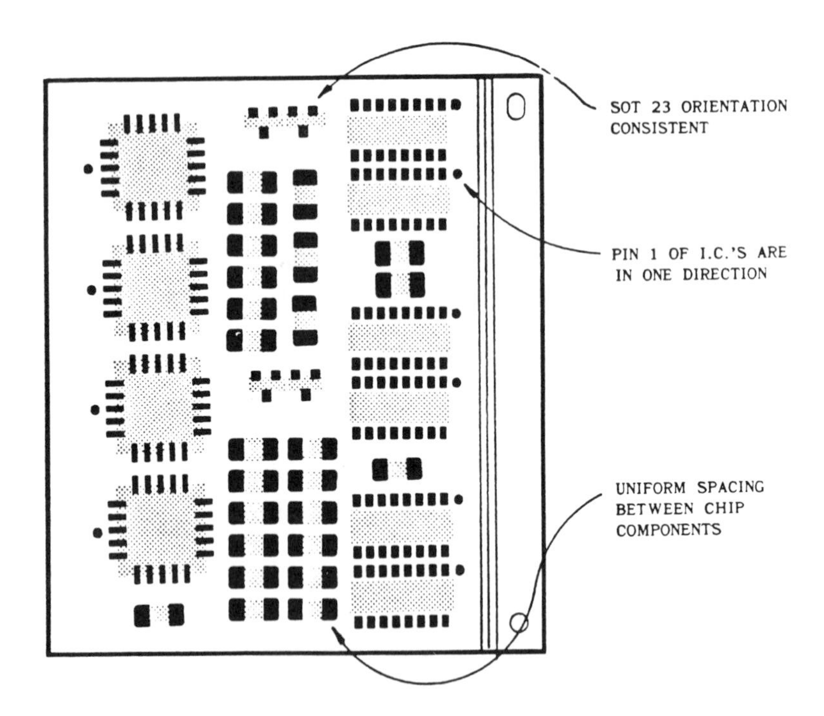

Fig. 6-12: Typical well-planned PCB.

Automatic Assembly and Testing

Component orientation and alignment will affect the assembly process in several ways. Polarized parts—such as diodes, capacitors, transistors and ICs—will require careful planning.

The equipment used to place these parts on the board's surface must be programmed to pick-up, rotate, center and transfer each device to the desired location. With consistent direction and a symmetrical component arrangement, programming is simplified and inspection or rework can be more efficient. The illustration in Figure 6-12 is typical of what the designer can achieve if the layout is well planned.

Each product requires a certain level of testing before it is passed on to the next assembly level. The product is typically designed to meet some environmental applications. A bench test is generally applied to low volume or extremely sensitive assemblies. As volume increases, it becomes necessary to employ more refined test methods to reduce labor intensive hand probing. Automatic test equipment is being developed and continually improved upon for SMT applications. Spring loaded test probes are now available for .050 inch grid spacing. Because these probes are miniature, they are more delicate than the larger probes used on PC boards with leaded through-hole technology.

As the SMT assemblies become more densely packaged, testing with probes on fixtures will become more common. The quantity of test points required on each circuit will be determined by how refined the system is and the type of test necessary. On small modules in high volume assembly operations, functional testing is the most popular. With more complex assemblies, it may require refined test methods. In all circumstances, contact area should be provided for the probe contact. It is NOT advisable to rely on component lead contacts as a test point. It is safer to use a feedthrough pad or add a contact area specifically for the point tested. See Figure 6-13.

Location of the test point should be clear of the component body and contact. Any side loading of the fragile test probe may damage the spring action required for proper contact pressure. When adding all the tolerance limits of the PC board and test fixture elements, we have concluded that a pad area of .025 inch diameter or would be adequate.

A Larger pad, .035 inch square, for test points would be better if space permits. The illustration in Figure 6-14, represents a typical circuit with a test pad for each "net" or node. Note that all test points are clear of the component body, thereby reducing the danger of interference.

Planning for automatic testing of assembled PC boards must start in the early stages of the board design. Each assembly will require a unique degree of engineering depending on the test equipment capability. Depending on the volume and product life of assemblies that must be handled, each company will have to judge the investment necessary for efficiency and economy. The automated test equipment being developed for SMT applications must accommodate both higher component density and simultaneous testing of both sides of the assembly. Many of the test equipment manufacturers are developing the "dual bed of nails" fixtures necessary to make this contact. See Figure 6-15.

There are many test levels possible depending on the environment in which the product operates. Three test methods popular for commercial and consumer applications are:

Device: Making test probe connection to each contact of every component on the board.

Functional Cluster: Partitioning the PC board in a modular fashion, with test points outside the component area.

Board Level: Using very refined test programs, the entire board is analyzed from net or the common connect between two or more components.

Test probing each contact point of every device can be expensive and will require additional real estate. Typically the "net" or common mode of several components is more practical. Functional test of a partitioned cluster will speed up the test cycle and isolate problem areas to a specific area or component.

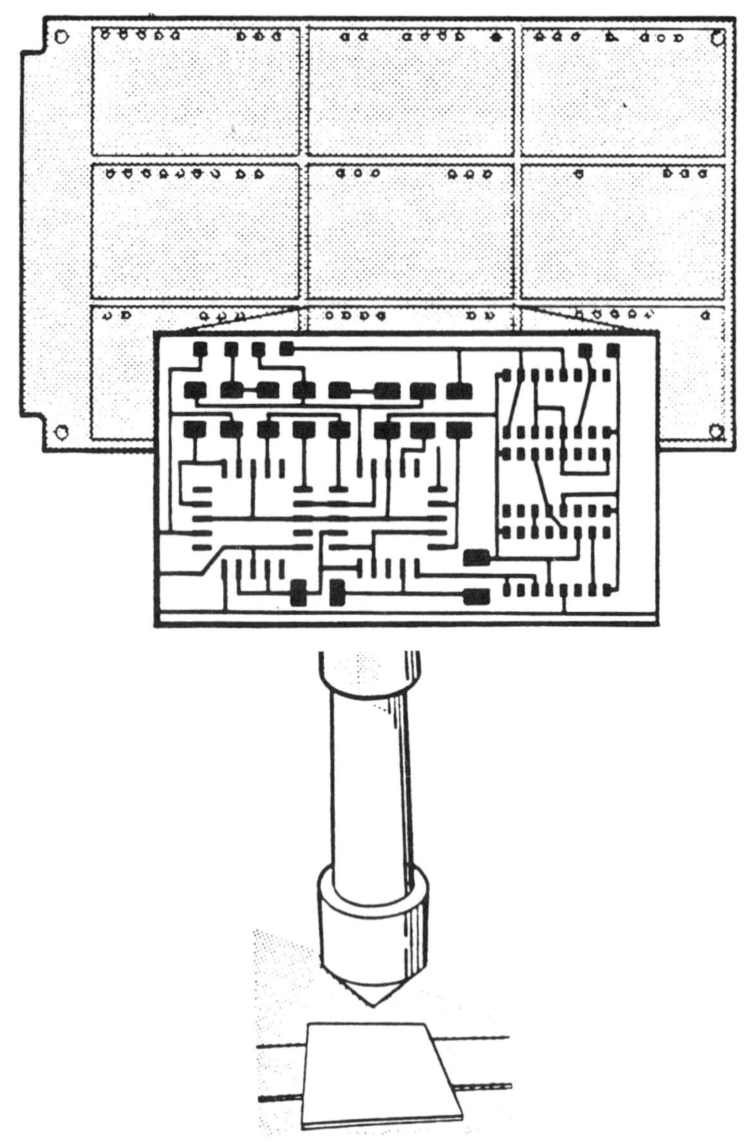

Fig. 6-13: It is safer to use a feedthrough pad for a test point rather than the component lead contacts.

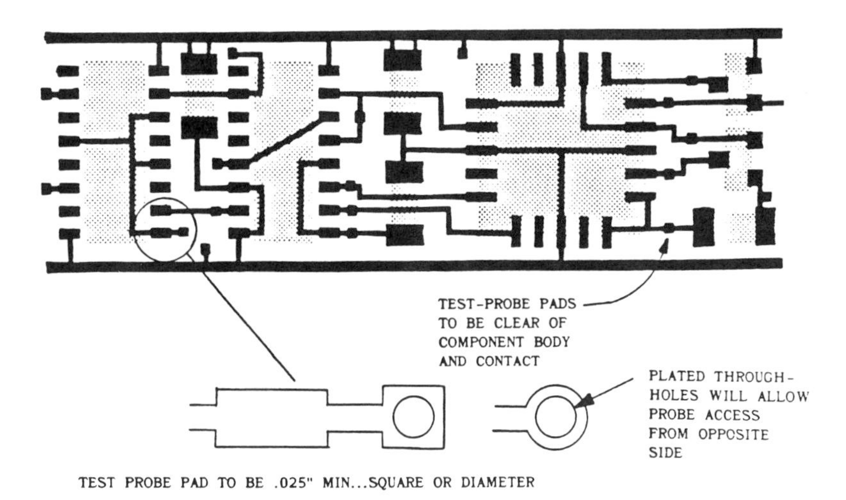

Fig. 6-14: Typical circuit with a test pad for each "net" or node.

Board level testing will require more exotic programming to isolate the trouble area or the component not meeting the specification. This method will require more refined equipment and engineering efforts, but the hardware complexity can be greatly reduced.

Tooling and handling fixtures for test equipment are a significant investment. Standardizing the board size for products will reduce costs for each process that the assemblies pass through; that does not mean the final outline of the board must be common, only that the rectangular blank should be of uniform size whenever possible. The designer and process engineer can work as a team to keep costs in control and plan efficient universal fixtures. See Figure 6-16. In some cases, testing is done by a subcontractor or service. In the case of Government contracts, the agency will normally furnish their own inspection team along with their own equipment.

Mixed Technology, Through-hole and Surface Mounted

Not all components are available or economical for surface mounting, thus, the need to continue using leaded through-hole devices on the same substrate may be necessary.

Mixed technology, as it is referred to, is usually unavoidable in one way or another. Many PC board assemblies require a connector interface to another segment of the system. The connector products, many of which are presently available in surface mounted configuration, do not always have a multiple

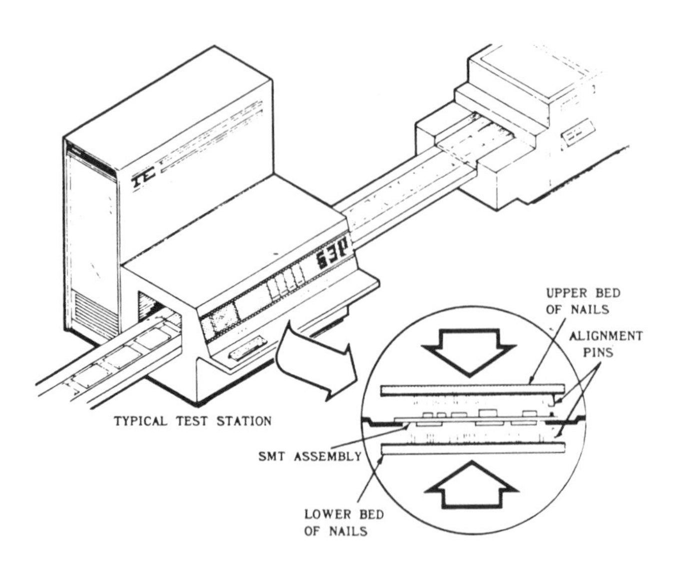

Fig. 6-15: Typical testing station for testing both sides of the SMD assembly.

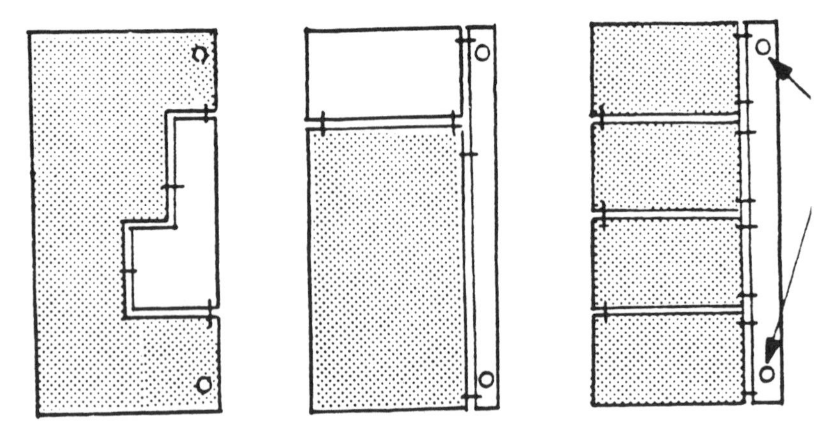

Fig. 6-16: Common tooling hole location and panel size will maximize product throughput.

source. This is also true for larger value capacitors, resistors or potentiometer found in the leaded packages. Many programmable devices are also more available in a DIP style carrier.

The designer, looking at the layout in one dimension, may fail to consider the component height when arranging surface mounted and leaded parts on the same board.

The illustration in Figure 6-17 details the profile of the PCC and the typical leaded through-hole device.

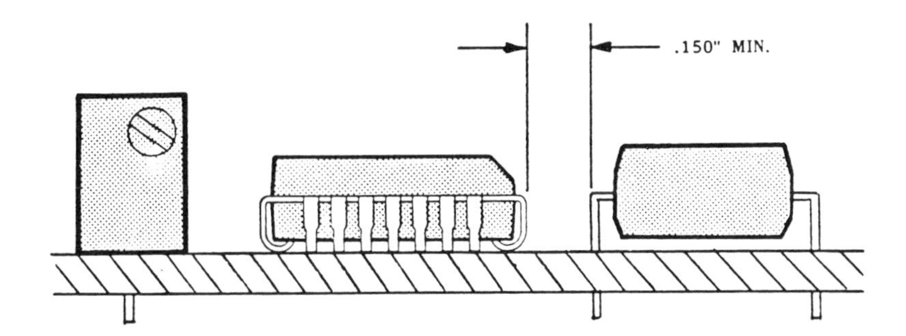

Fig. 6-17: Detail of PCC and leaded through-hole device.

The same spacing recommendations referred to earlier in this chapter must be provided for inspection and rework access.

The visual angle on the J-lead part is just as important here, as previously mentioned. Touch-up or rework tools require a minimum space to access the solder area without disturbing the adjacent component.

Component-to-Board Edge Requirement

Surface mounted devices can be mounted very closely to the substrate edge, however, the designer should avoid close edge placement, especially if the assembly is to pass through a wave solder process.

When wave soldering is to be part of the assembly sequence, break off strips could be added to the long edges of the board.

Figure 6-18 is a typical application of a break away that also includes tooling holes used in other machine handling. Generally, an edge-to-body clearance of .125 inch or .118 inch is preferred. This allows for solder mask isolation and clamping into rework stations or test fixtures.

Tolerance capability of precision machinery and commercial PC board fabrication is quite different. It is common for PC board shops to profile the etched circuit with high speed routing, using a tracing template registered with locating pins.

The accuracy of the outside dimensions may vary from the tooling hole locations up to .010 inch. On high volume boards, it may be wise to consider die punching the PC boards with hard tooling. The cost of this tooling is far greater than a routing template, but the accuracy and price per board will easily compensate for the expenditure. Machine tooling and fixtures are a more accurate process and afford closer tolerance capability.

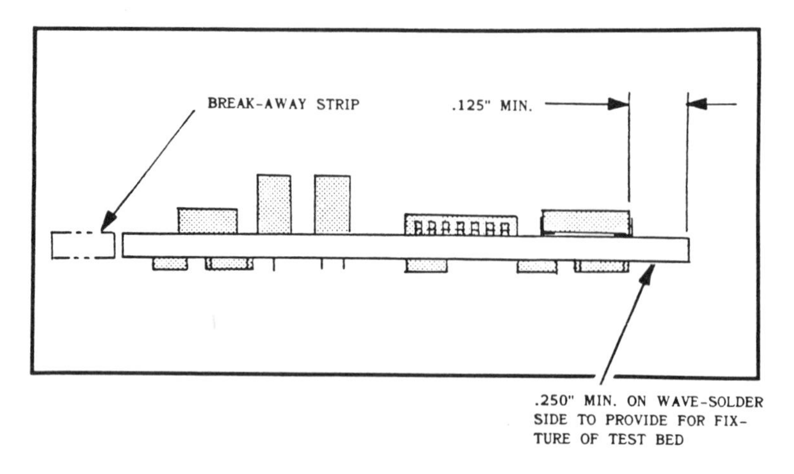

Fig. 6-18: Typical application of break-away tabs.

Knowing what is possible and what to expect from machine services will help determine the location accuracy of both component placement and test probe contacts. The test probe or "pogo" contact can be mounted on .050 inch grid centers. The location accuracy is governed by the technical quality of the machine that fabricates the holding bed of the probes and their location respective to the tooling holes. Add that location tolerance to the tolerance of the PC board fabrication method and these define the limits for the test point pad size itself.

Component placement equipment will have tolerance variables as well. Some equipment makers claim ±.002 inch placement accuracy. However, typical location accuracy on higher speed, chip placers that are now used in the industry is ±.006 inch to ±.010 inch. By allowing for these tolerance limits and the possible extremes of the PC board itself, the designer can make these appropriate considerations.

Designing the SMT board for producibility is achieved through careful planning and close communication with all segments of manufacturing.

The test engineer would be involved with the identification and numbering of test nodes on the schematic before beginning the PC board layout. The X - Y position of each node could be furnished to the test department with the IC number, thus greatly reducing fixture preparation and software development time.

The test point contact area should be well clear of any component body and preferably would be opposite from the side of the greater component population. To easily distinguish the test pad area from other via pads, consider using different sizes or shapes.

For example, a round via pad would be used for general front-to-back interconnections and a square pad would be reserved for test probe points only, as shown in Figure 6-19.

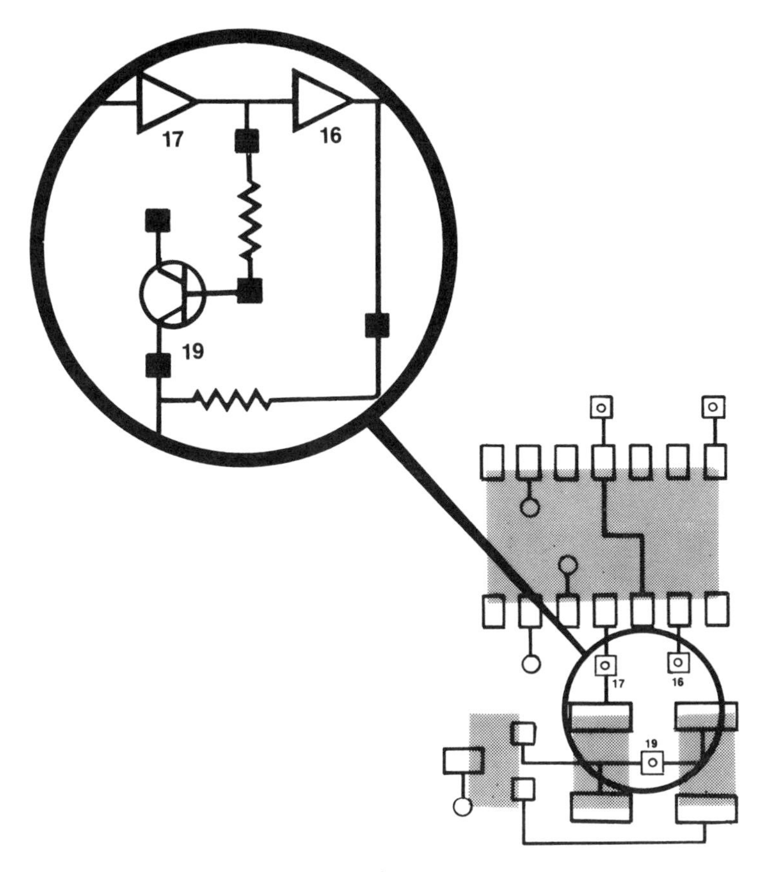

Fig. 6-19: A round via pad is normally used for general front-to-back interconnections, while a square pad would be reserved for test points only.

For those using CAD to design the board, a layer would be reserved to provide for the test contact.

Since a number is assigned to each node, the test point will become a part of the net and verified as a specific physical component.

chapter 7

ARTWORK GENERATION

When placing plated through-hole pads on your artwork, do not allow the pad to make direct contact with the component contact area. The clearance between the footprint pattern and feedthrough pads will allow a desirable dam of solder mask coating that will stop the migration of liquid solder during the reflow process. The conductor trace connecting the contact area with the feedthrough pad should be approximately .010 inch to .012 inch wide. See Figure 7-1.

As mentioned in Section 5, feedthrough pads against or within the footprint pattern as shown in Figure 7-2 will cause migration of the liquid solder from the contact area during reflow. Feedthrough pads under chip components cannot be seen and may cause failure during additional processes. For the wave solder process, feedthrough pads and heavy traces adjoining the contact area will not have significant negative results.

The added cost of the small hole size and plugging processes may not be a desirable alternative to the separation of contact and feedthrough pads. If it is necessary to design feedthrough pads against contact areas or connect feedthrough pads with contact areas with heavy traces, consider the following options:

1. Solder mask over bare copper.
2. Solder mask over feedthrough pads.
3. Plug feedthrough hole.

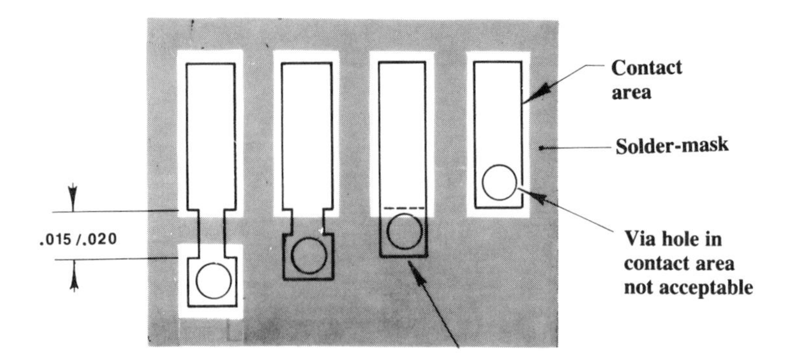

Fig. 7-1: Cover or "tent" via pads and holes when adequate solder-mask barrier is not possible.

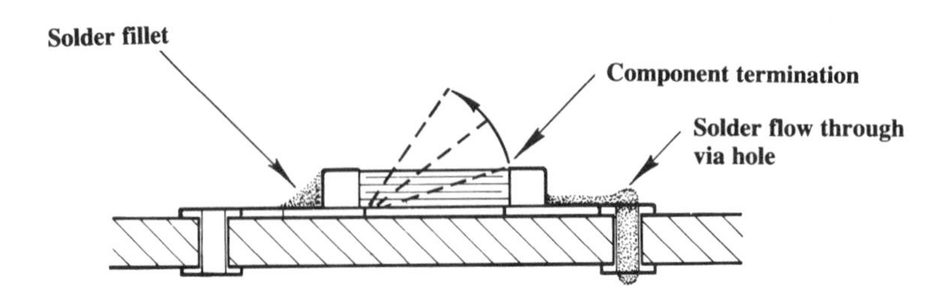

Fig. 7-2: Feedthrough pads, against or within the footprint pattern, will cause migration of the liquid solder during reflow.

To further provide for routing conductor traces while insuring an acceptable air gap, you may choose to use a .035 inch to .037 inch square pad for feedthrough holes. This square configuration will furnish more than enough metal in the diagonal corners to compensate for the reduced annular cross section at the sides of the square. The .035 - .037 inch square pad is spaced at .100 inch grid. See Figure 7-3.

The migration of solder during the reflow process will cause the liquid metal to draw away from the contact area. The details shown in Figure 7-4 further illustrate the danger of solder bridging under the body of the chip component.

Contact patterns too close to feedthrough pads will always cause production problems. The solder paste when heated to a liquid will flow into the feedthrough hole and away from the contact area, so avoid hidden feedthrough pads near unrelated contact patterns. The only way to correct a bridge or short is to remove the IC. See Fig. 7-5 for recommendations.

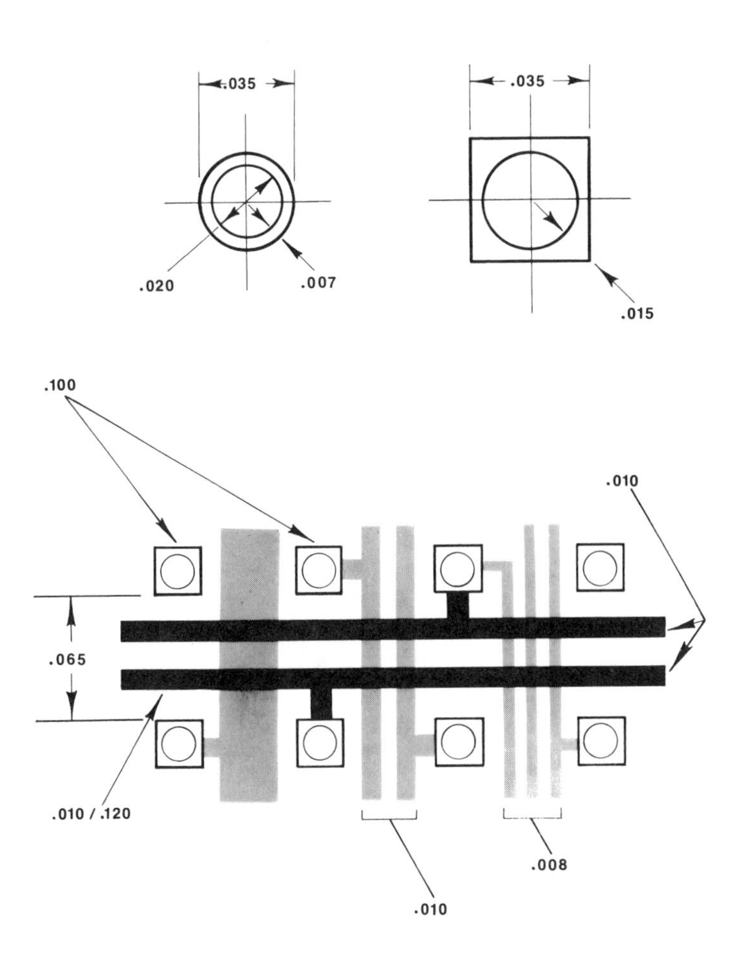

Fig. 7-3: Using .035″ to .037″ square pads for feedthrough holes will furnish enough metal in the diagonal corners to compensate for the reduced annular cross-section at the sides of the square.

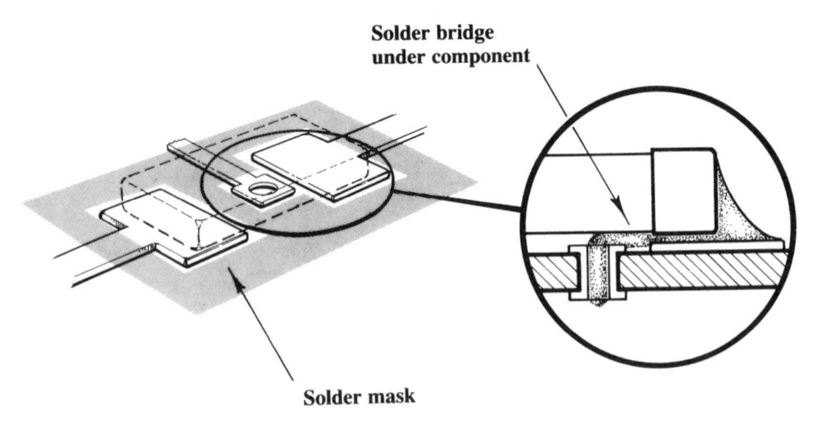

Fig. 7-4: The migration of solder during the reflow process will cause the liquid metal to draw away from the contact area.

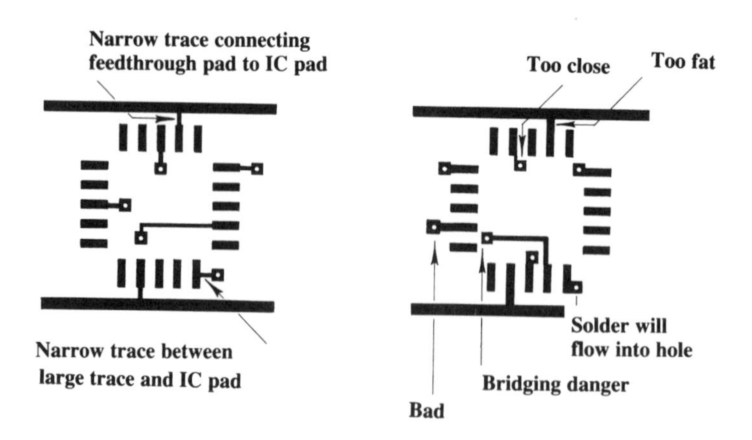

Fig. 7-5: The drawing on the left utilizes a narrow trace between the large trace and IC pad— represented good design practice. The drawing on the right, however, uses a wide trace which is poor practice.

Adding feedthrough holes directly onto the contact areas is allowable if the hole itself can be plugged to stop solder paste migration. Solder paste migration and bridging are to be reduced when possible. Rework and touch-up are an unwanted detour in the high-volume manufacturing operation.

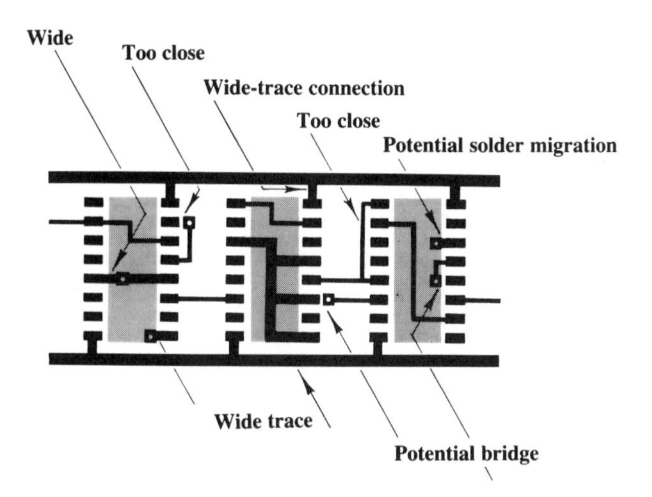

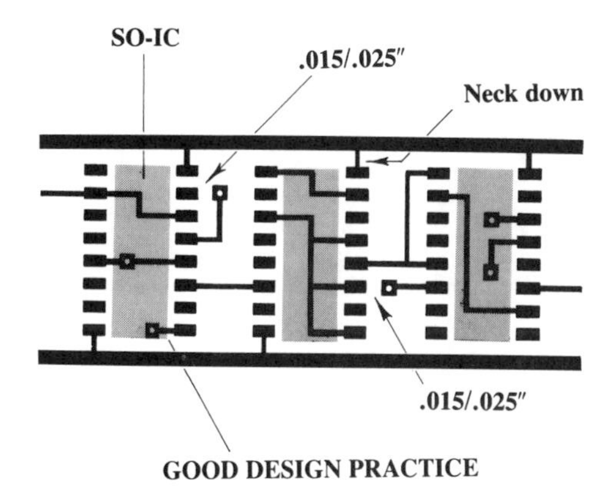

Fig. 7-6: A: potential problems with feedthrough pads on SOICs; B: compares good design practice with poorer design practice in A.

On the more traditional .100 inch hole to hole or when necessary, it is possible to route two conductor traces between via pads. This density is only possible on internal layers of multilayer boards with leaded through-hole boards. The smaller .035 inch via pads will easily allow two or three conductor traces. Using multilayer for surface mounted applications will further increase density possibilities.

The reduced size of the plated feedthrough hole diameter will easily allow three conductor traces on internal layers without resorting to "fine line" (.006 inch wide traces). As component density increases further, it may be necessary to limit the outside layer surfaces to component mounting contact patterns and feedthrough pads only. See Figure 7-7.

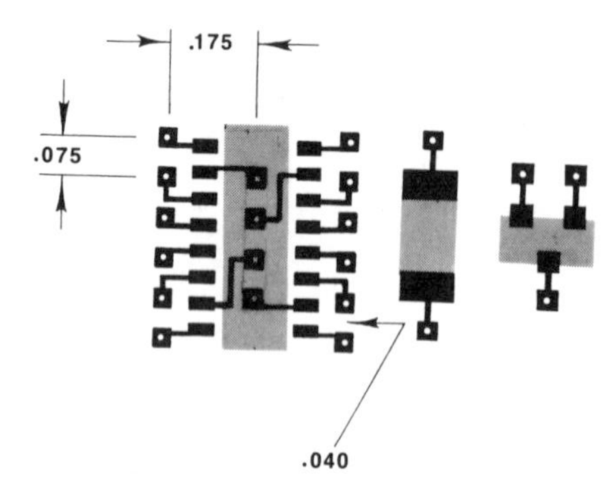

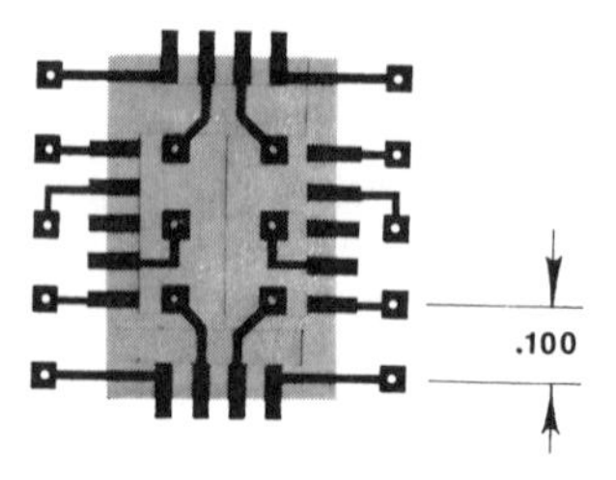

Fig. 7-7: Typical patterns of "pads only" when using internal routing on multilater PCBs requiring maximum density.

Footprint (Contact) Pattern For Passive Devices

The chip component patterns shown in Section 5 are recommended for reflow-solder processes, but also work well on dual wave-solder machinery. Because wave soldering is one of the most popular processes in the world, many companies choose to use leaded through-hole components for all ICs and larger passive devices. By mounting most of the resistors, capacitors and SOT components on the wave solder side, the board size can be further reduced. However, mounting the chip components on the wave solder side does restrict the conductor routing paths. See Figure 7-8.

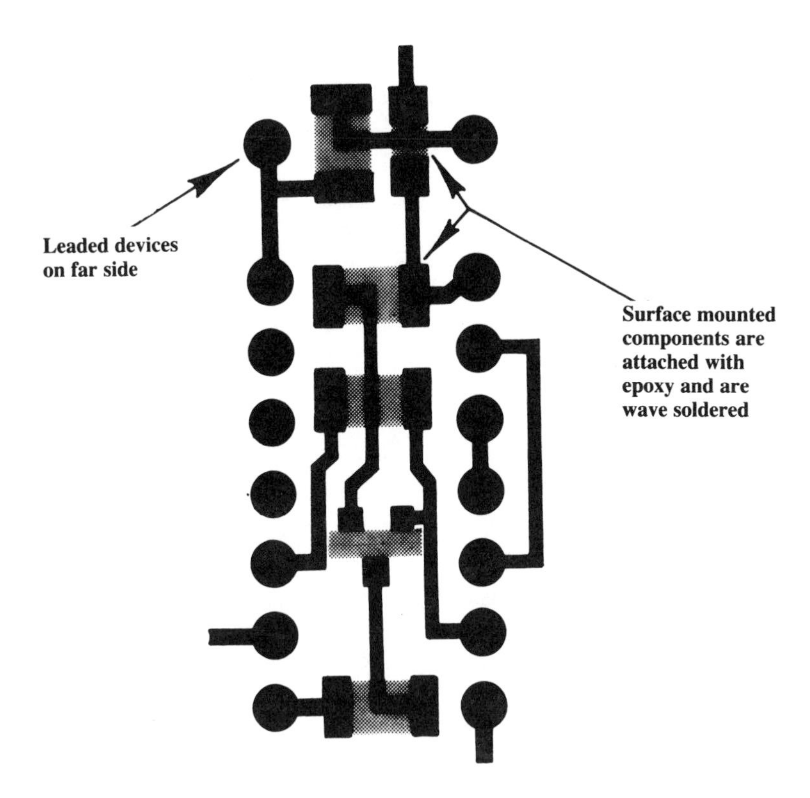

Fig. 7-8: Mounting the chip components on the wave solder side restricts the conductor routing paths.

Component footprint details for the wave solder process are based on guidelines noted in Chapter 4. The footprint patterns in Fig. 7-9 allow for placement accuracy of the assembly equipment, as well as possible component shift during adhesive cure. When using the narrower contact pattern, be sure to allow adequate clearance between the component body and the conductor trace. Keep in mind that the component body will overlap both sides of the footprint pattern when using this design. Many possibilities may exist in footprint design to further reduce secondary rework procedures. The component footprint patterns were designed to reduce or eliminate rework and touch-up of boards after assembly procedures. Figure 7-9 compares different contact geometry used for wave solder application.

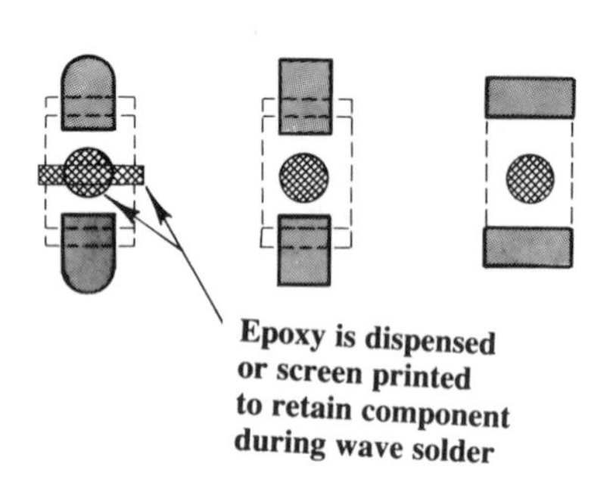

Fig. 7-9: A comparison of different contact geometry used for wave solder applications.

Footprint Planning For Active (IC) Components

The JEDEC registered IC packages available today have .050 inch space between lead centers. Component manufacturers have recommended standard limitations for the contact pattern (footprint). Many manufacturers are not familiar with the technology for creating artmasters for the PC board itself and are unaware of the importance of grid location in design. The contact patterns shown in Figures 7-10 and 7-11 will allow grid center placement when creating the footprint library in a CAD system or when autorouting. Figure 7-10 describes the footprint pattern used for the JEDEC plastic chip carrier. Leadless ceramic chip carriers may require a change in size and center position. The contact .025 inch width will allow for the routing of .008 inch wide conductor traces between leads when needed. If length is added to the contact pattern, it may be necessary to reposition the grid in order to maintain the desired solder reflow characteristics.

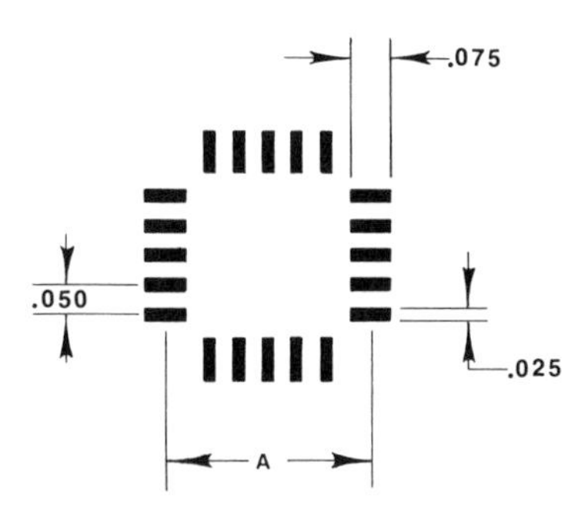

Fig. 7-10: Footprint pattern used for the JEDEC PPC (plastic chip carrier).

The JEDEC-registered small outline, or SOIC contact patterns is easily adaptable for the designer. The two parallel rows of contacts have the same pin assignment as the dual inline, through-hole IC it replaces. For most logic devices, the .200 inch spacing between contact-row centers will be consistent. However, as mentioned before, some 16-pin devices require a wider lead frame to provide for the die size or heat dissipation. A PC board may have a mix of the SO-16L (wide) devices. Always check the component manufacturer's specifications. SO-IC's greater than 16-leads will always be in the wide lead frame but they will be limited to a total of 28-leads. Beyond 28- leads, most active components will be supplied in the PCC (plastic-chip carrier) or other quad configurations. See Figure 7-11.

As mentioned previously, the worldwide electronics industry is cooperating in the development of standardization. In most cases, an active component can be obtained from multiple sources and will be dimensionally and pin-for-pin compatible with a device manufactured in another part of the world.

The first suppliers of the small outline (SOIC) chose the .050 inch lead spacing to be compatible with the accepted grid system. As illustrated in Fig. 7-11, even the quad arrangement on plastic chip carriers can be located on a grid pattern.

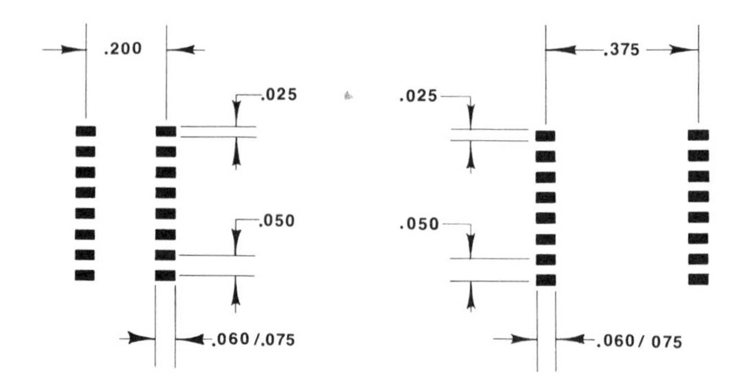

Fig. 7-11: Beyond 28 leads, most active components will be supplied in the PCC or other quad configurations.

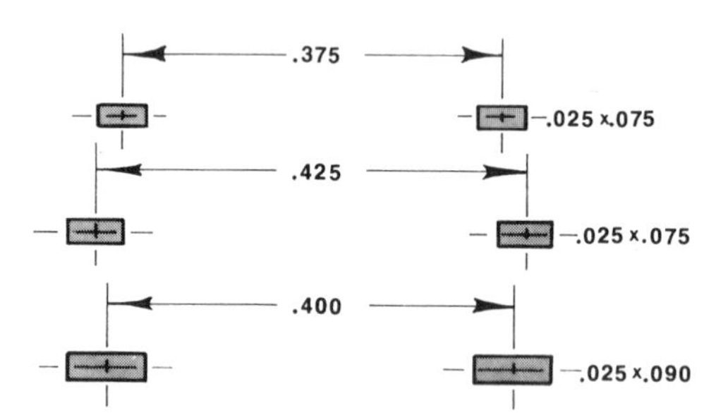

Fig. 7-12: Footprint size and row spacing will vary between different manufacturers.

While the Japanese use the metric system of measurement, they do maintain the .050 inch lead spacing on many components to be compatible with the world market. Among the products available with .050 inch lead spacing, there are significant differences. The distance between the rows of contacts on the EIAF-SOP IC is not the same as on the JEDEC-SO IC. If it is necessary to use IC devices from multiple sources, it may not be the component manufacturer.

Figure 7-12 shows manufactured differences in footprint size and row spacing. Even though each device is pin-for-pin compatible, the designer must choose one or the other, or develop a universal pattern to accommodate them.

The SO/SOP universal pattern is a configuration that can be considered. Beacause of the excess length of the contact area, the IC may shift to one side during the solder reflow process. If this becomes a problem, it may be necessary to add adhesive to the component when it is placed on the board surface. Other options have been developed to eliminate this difference. Many Asian component suppliers comply with the JEDEC standards and offer direct compatibility.

Hand-Taped Artmaster Preparation

Generating the artwork for two-sided SMT PC boards is similar to those supplied for conventional leaded boards. The major difference is the separation of the SMT footprint patterns. When choosing preprinted footprint patterns for surface mounted designs, make sure the pad geometry is process-proven or meets the recommendation of the component manufacturer. The scale for preparing SMT artwork is 4-to-1 or greater to insure accuracy and maximize density.

One sheet of mylar will be dedicated to the footprint patterns of the SMT components. The footprint pattern sheet, when photo reduced to 1:1 size, will be the master for preparing the solder-paste screen. This film will not be required if the wave solder process is used.

ARTWORK LAYERS: (2-SIDED PCB, WITH SMT ON ONE SIDE)

1. Pad master (all common vias, boarder, etc.)
2. Footprint master (all SMT component patterns)
3. Side one conductor
4. Side two conductor
5. Side one legend (screen master)

When ordering working film, instruct the photography service to furnish the following composite film set:

1. Composite sheet 1, 2, 3 (side one)
2. Composite sheet 1, 4 (side two)
3. Composite sheet 1, 2 expanded .010 inch (side one solder mask)
4. Sheet 1 only expanded .020 inch (side two solder mask)
5. Sheet 2 only (solder-paste screen master)
6. Sheet 5 only (legend screen master)

All the film is sent with a fabrication drawing to the board house, except sheet 2, the solder paste master. The solder paste master is only used to prepare the screen or stencil used in applying solder paste for reflow process.

All areas of the board except the contact areas and feedthrough pads will be coated with solder mask. The solder mask is necessary to contain solder paste on the contact pads.

Tape and Reel

Today's systems are requiring the use of more complex high leadcount ICs with increasing density. Cost pressures dictate the need for automated assembly. As IC packages become smaller, their lead spacings become closer and, as a result, more fragile. This causes more difficulty for automated equipment to effectively handle and test these packages. Today's surface mount packages, consisting mainly of the small outline (SO) and plastic leaded chip carrier (PLCC), solve some of the density requirements (up to about 84 leads), but testing and handling difficulty is still encountered.

Tape & Reel packaging was developed to meet the high density package handling and testing issues lacking in other packaging methods. Tape & Reel is a method in which the SMDs are placed in embossed pockets located along the length of a plastic carrier tape. A second plastic tape, called a cover tape, is then sealed over the carrier tape. The loaded and sealed tape is then wound onto a plastic reel, labeled, put in foil-lined boxes, and shipped to the user.

Tape & reel packages have been designed with leadcounts ranging from 20 to more than 400, all within standard package body sizes.

Computer-Aided Design

When preparing for photoplotting of closely spaced conductor traces on a SMT board, it will be necessary to change trace width. Routing .010/.012 inch wide signal lines between the.025 inch wide contact patterns used on most SMT ICs will require "necking down". Reducing the conductor width to .008 inch as it passes between contact patterns will still provide a reasonable air gap. Care must be taken to overlap the start and stop position of the line aperture when photoplotting. Line apertures are generally a circular opening, so the end of any line run will have a full radius end. When necking down, a smooth transition from one width to the other can be achieved if each starts and stops at the same grid point. This stopping may not be obvious when looking at a pin plot, but the careless overlap of line joints can cause potential problems when the photoplot is generated.

Because CAD systems snap to a grid point, it is customary to use a 45-degree angle when traces must divert from a continuous line but remain traveling in the same direction. Offset-stepping several conductor traces on close

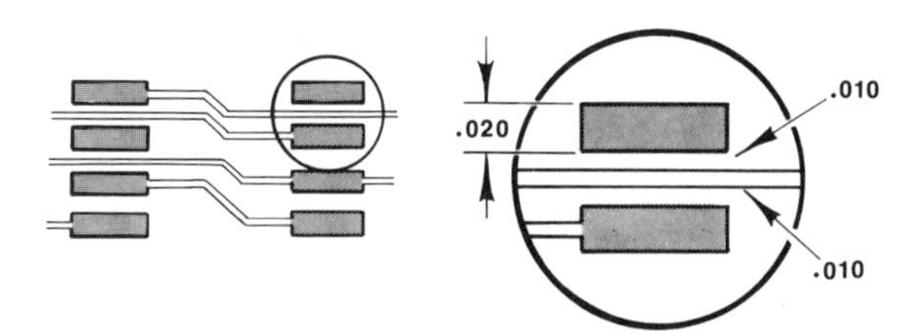

Fig. 7-13: Some designers reduce the contact pattern for SMD ICs to .020″ width, allowing the use of .010″ conductor trace widths throughout the board.

spacing will require attention to maintain a proper air gap. The start and stop points of these aperture runs must be very carefully executed to reduce the chance of overlap and shorting. If laser plotting is used, many of the customary limitations of aperture photoplotting are eliminated.

Some designers may choose to reduce the contact patterns for SMT ICs to .020 inch width. See Figure 7-13. This will allow the use of .010 inch conductor trace width throughout the board and necking down can be reduced to a minimum. While digitizing and photoplotting time is reduced a great deal, the cost of the PC board may increase due to the finer line etch control necessary. The .020 inch wide contact may cause additional difficulty in placing larger PCC ICs on older robotic equipment. With this narrow target, placement and rotational accuracy must be 25% greater due to the reduced contact area.

Autorouting SMT

High density, autorouting for surface mount PC boards cannot be accomplished by most CAD systems available today. The systems are primarily hole intelligent through all layers of the multilayer board.

To maximize the efficiency of your existing system, create a footprint library that will include the component contact pattern and a uniform, preassigned via pad connected by a narrow trace(.010inch/.012 inch wide). The via pad will appear on all layers, but the contact pattern and trace will only appear on the outside layer. Figure 7-14 illustrates the advantage of the pads-only configuration. The inside layers will provide unrestricted routing paths between vias and under contact areas. The router can now ignore the footprint of the component and address only the pre-assigned via pads. Of course, via pads added by the router program, must be kept out of the footprint zone. To further increase density, especially when components are mounted closely on both sides of the board surface, the designer may have to revert to blind or buried vias. Figure 7-15. This procedure is to be avoided, if possible, because of the difficulty and cost added to board fabrication.

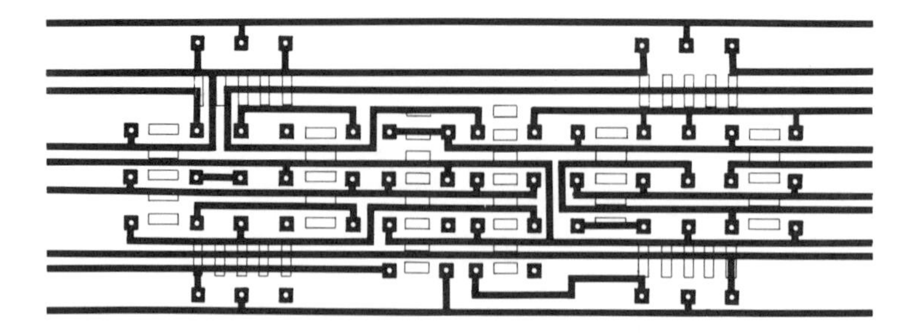

Fig. 7-14: Pads-only configurations have certain advantages that makes their use desirable in all types of SMD design work.

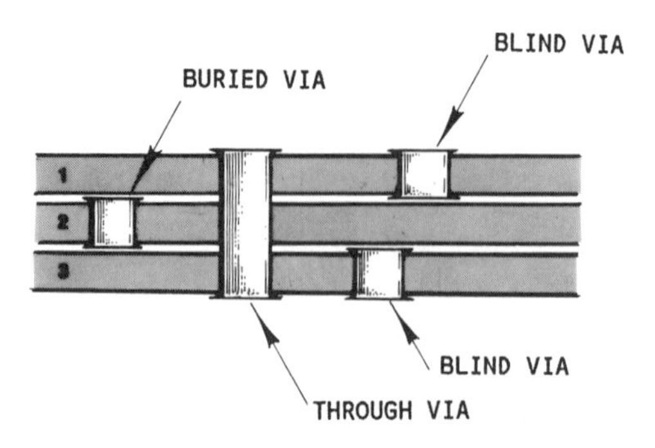

Fig. 7-15: In some cases, the designer may have to revert to blind or buried vias.

Preparing For Future Assembly Methods

To comply with the accuracy requirement forced on the industry by new, fine pitch (.020 inch and .025 inch lead spacing), assembly equipment is becoming more sophisticated. With more precise positional accuracy and optical camera systems, what was considered impossible yesterday will be tomorrow's normal way of doing business. To assist in the future use of optical techniques of component placement, the addition of fiducial targets (patterns) at two or three locations on the component side of the board is recommended.

Figure 7-16 illustrates a few pattern configurations. The optical system compares the fiducial location to the component as the machine is placing. Because tooling holes will vary in size and true position from board to board, the machine will be looking at the fiducial patterns etched into the board; adjust the X-Y and rotational position of the component to the footprint pattern on the board surface; then place the component.

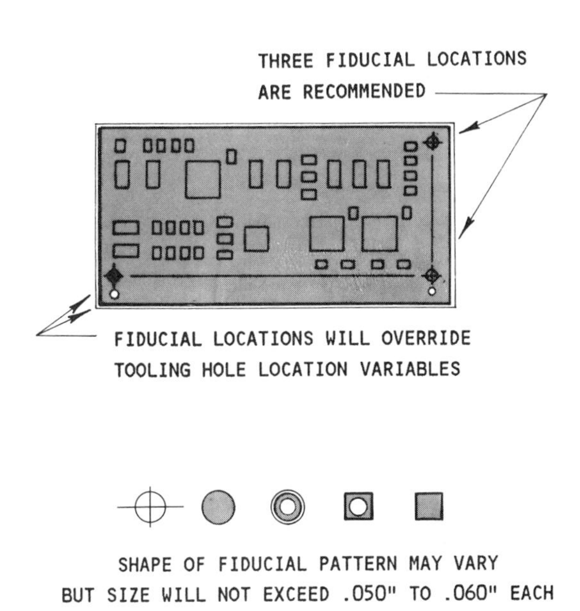

Fig. 7-16: Three fiducial locations are recommended; fiducial locations will override tooling-hole location variables.

NOTE:

1. Fiducials must be present to the vision system a clear-cut image with its bright part well-contrasted photographically against the dark part.
2. For the bright, a surface of copper, gold, or tin is acceptable. If you have to tin-plate fiducials, be sure the top surface is flat and the outlines are smooth and unbroken.
3. The dark part of the fiducial should not have any bright spots in it.

Furnishing the manufacturing personnel with X -Y positional data will accelerate the implementation of CAD/CAM necessary for automation.

As the footprint library is being prepared for each device, add the center point of the component on one of the pad stack layers. This center-point position will be referenced from the zero datum of the PC board. That datum will easily reference the offset of any individual assembly system.

When the orientation of the component can also be supplied, the programming of the equipment will take only a fraction of the time for the "on machine" or X -Y calculation methods. See details in Figure 7-17.

Providing usable data to the machine programmer is the most important factor. Find out what CAD/CAM output is really needed before spending time creating unnecessary data.

As we evolve into new assembly technology, the designer's role in the manufacturing processes becomes more interactive.

New components, introduced each year, will change the way we design products which will, in turn, reflect the way we do business. As the component evolution continues, the equipment and processes will be modified and refined in response. A flexible response to this new technology, combined with motivation and cooperation, will change problems into solutions.

Remember, what appears to be a very unique method of assembly today will be tomorrow's business as usual.

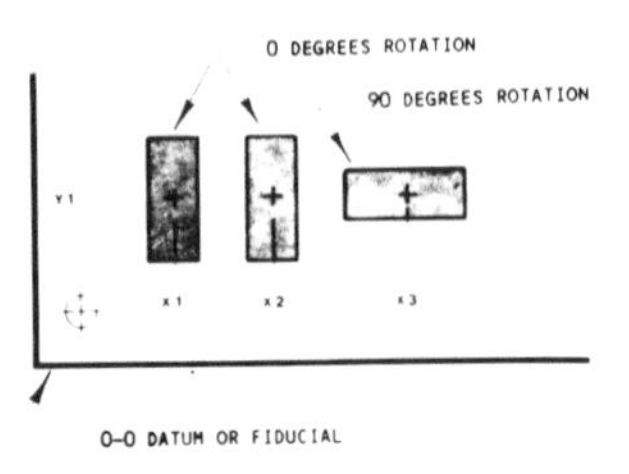

Fig. 7-17: As the footprint library is being prepared for each device, add the center point of the component on one of the ad stack layers.

CAD Applications

This section illustrates the use of a computer aided design system to generate a rectangular connector.

To begin the exercise, install the ProDesign II program. When the *Drawing Screen* appears, set the *Drawing Units* by selecting an appropriate linear dimension to establish the proportional relationship. Do this by depressing the "0" key to set a point at one extremity of a proposed line. Then move the cursor to the opposite end of the intended line and set the second point to define the chosen linear measurement. Now depress the "U" key to display the *Units* command. At the prompt enter the correct distance completing the dimensional definition (in this illustrated example—Fig. 7-18—the Units should be 208.0 and are recorded to three decimal places when applicable).

STANDARD SHELL

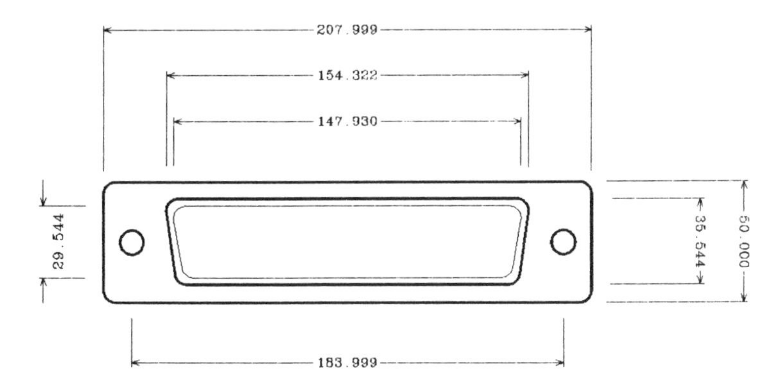

Fig. 7-18: Typical CAD generated SMD drawing.

Depress the "Q" key to access the *Drawing Parameter Menu*. This will cause the screen to display 11 default values. However, at this time we need to be concerned with only three. Both the *Cursor Step Size* and the *Line Width* should be set to 1.000 and the *Fillet Radius* to 5.000. Press the F1 key to use these changed values and terminate the *Drawing Parameter* command. Next, depress the Alt-Q key to display the *System Parameter Menu* screen showing 12 default values. At the first prompt change the Character Font to 3. Then for the third value chose the mathematical option and change selection 6, 11 and 12 to yes. Now terminate the*System Parameter* command with the F1 key.

Before you begin drawing, depress the “J” key to display the *Status Line* options and select format 2. This will activate a display of both distance and angle measurements plus X & Y co-ordinates.

To begin generating the outer perimeter of the connector case, set a line value of 1.000 cursor step and then try to visualize the finished component to assist you in initial lay-out and positioning. After this preliminary consideration, set the first point to create a line for the top edge of the connector case by pressing the “0” key. Now move the cursor laterally to the opposite extremity of the intended line while observing the *Status Line Measurement* until it reaches 208.0. At this time set the second point and generate a line with the *Vector* command by pressing the “V” key. Now press the “.” key to set a *Gravity Point*. This command causes the cursor to move to the nearest existing point and set a point on that exact location. This insures that you can access the precise end point to extend or intersect that line with accuracy.

Using the *Status Line Measurement,* move the cursor vertically to 50.0 and set a second point and generate the intersecting line using the *Vector* command. Again, use the *Gravity Point* feature to re-establish the last point location. Then move the cursor laterally to 208.0 and set the second point before generating the third line with the *Vector* command. Complete the figure by first evoking the *Gravity Point* command and then move the cursor near the first point set in the first line and again depress the “.” key. This ensures closure when you use the Vector command to complete the figure.

Radius the four corners by again evoking the *Gravity Point* command to ensure that this point is set on the apex of the corner. Access the *Root Menu* and toggle down to *Edit*, press return and highlight the *Fillet* command. Press the Return key and since the point has been defined, the corner will be rounded by an arc with a radius previously set in the *Drawing Parameter* command. Move the cursor near the next corner and set a point on the apex by again using the “.” command. Next, depress the Return key to access the *Fillet* command and once again to execute it. Move successively to each corner and repeat the operation to complete the task.

NOTE: The above section purposely contains repetitious functions to familiarize the new user with a keyboard-controlled CAD System. For example, an advanced user could generate the above with far fewer commands, utilizing abbreviated keystrokes and advanced functions to produce the same component in a much more expeditious manner. The following series indicates the required keystrokes with brief explanations enclosed in parenthesis.

After setting the required parameters and forming a visual concept to determine a starting position execute the following commands: **“0”** (set point), **“>”** (cursor to 208.0), **“0”** (set point), **“/”** (cursor to .50), **“0”** (set point), **“<”** (cursor to .208), **“0”** (set point), **“/”** (cursor to .50), **“.”** (GP to insure closure), and **"V"** (generates a rectangular shape using a line one cursor step wide).

The next series of keystrokes will radius the corners: **"."** (GP to mark upper left vertex), **"F"** (radius corner), **"CTRL >"** (cursor to right margin), **"."** (GP to mark upper right vertex), **"F"** (radius corner), **"CTRL /"** (cursor to bottom margin), **"."** (GP to mark lower right vertex), **"F"** (radius corner), **"CTRL <"** (cursor to left margin), **"."** (GP to mark lower left vertex), **"F"** (radius corner).

Note: This brief series of keystroke commands has effectively, yet quickly, generated the previously defined figure. Both the time and effort required to generate this image may be further reduced with the use of a supplementary input device such as a digitizer or mouse.

To establish the upper edge of the inter flange, move the cursor near the middle of the line defining the top of the connector. Depress the Return key to access the command line and type MID and press enter. This will cause the cursor to snap to the center of that line and set a point there. Now move the cursor down and locate a point in the middle of what will be the top of the inter flange line. Next, while observing the measurement on the Status Line, move the cursor until you achieve one half the measurement of the top flange line.

Note: The first two points are for reference only and should now be deleted by pressing the ESC key twice. Now set a point at this location and move the cursor laterally toward the other extremity. When the correct measurement is reached set the second point to define the line. This line, however, is to have a different value. Therefore, the Drawing Parameter screen must be accessed by typing "Q". At the prompt, change the Line Width to .700 and enter F1 to return to the drawing screen. With the changed line value, enter the Vector command to create the line.

The angular side of the flange will be generated at a 10 degree angle. To accomplish this, establish the end-point of the top by using the Gravity Point command. Depress the Return key to access the *Command Line* and enter the keystroke command **";"** to evoke the **Set Point** using angle and distance. The prompt will first request a distance which should be entered as 35.5. Next enter the mathematical angle as 280 degrees which will cause a second point to be set and a line can now be generated using the Vector command. Now move the cursor near the opposite extremity of the top inter flange line and repeat the operation using the co-ordinates 35.5 and 260 degrees to create the opposing side. The bottom side may now be added utilizing the Gravity Point and Vector commands.

Finally the corners may be rounded using the same technique previously employed to create the contour of the outer case.

Next, the inside flange line should be created by first re-defining the Line Width on the *Drawing Parameter Menu*. Access this screen by depressing the

"Q" key and setting the width parameter tc 200 and then return to the drawing screen with the F1 keystroke command. The Parallel Line function may now be used to generate a series of lines to define the inter flange. This action will require that this area be enlarged to permit visual identification of the center points used to radius the four corners. To accomplish this, move the cursor near the center of the shape now defining the outer flange and set two points in the same location. Next, evoke the Zoom function by pressing the "Z" key and then enter a value of 3 at the prompt. This will cause the designated portion of the screen to be re-generated on a scale three times greater. Now, with the enlarged view displayed, move the cursor to a point near the left end of the line defining the top of the flange and press the "." key to establish the end point of that line. Then move the cursor down five steps and set the second point using the "0" key. Next, use the "=" key to generate a parallel line through that point. Now, to locate the upper end of the line defining the left side of the flange, move the cursor near that area and evoke the Gravity Point command once more. Then, move the cursor laterally, five steps to the right and set the second point. Again, evoke the Parallel Line command to create the line forming the left side. Now, move the cursor near the center point used to generate the outside radius and evoke the Gravity Point command to establish the exact location. Then, move the cursor to the left end of the first inside line formed and press the "." key. This establishes the correct radius and the Parallel Line function may be used to generate the line. The remainder of the inside flange may now be created using the same procedure.

Before continuing to the next stage, again set two points near the center of the component and evoke a Zoom factor of .25 to return the drawing to the original scale.

Next, create the mounting holes in the flange by first locating the mid-point of a perimeter side line and with the aid of the *Status Line Measurement*, move the cursor horizontally to a co-ordinate of .140. Now delete the first reference point with the ESC key before setting the center point for the hole. Next move the cursor three steps laterally and set a second point to establish the radius dimension. Access the *Drawing Parameter Menu* and set the Line Width to .700; return to the drawing screen with the F1 key. Next evoke the "O" keystroke command which will generate a circle using the first point as a center and the second as a radius dimension. Move to the opposite side of the flange and create a second circle in the same manner.

Only dimensioning and labeling remain to complete this view.

chapter 8

PCB DESIGN CONSIDERATIONS

The design criteria described in this chapter establish the considerations for maximizing machine insertion of leaded electronic components into Prined Circuit Boards (PCBs).

The purpose of this chapter is to aid the designer of PCBs by detailing the considerations and some of the trade-offs which will provide reliable insertion in a production environment.

Techniques that apply to conventional circuit boards are also applicable to SMD technology, to a certain extent. There are also many applications where both SMD and leaded components are used on the same board or substrate. Therefore, a review of circuit board design considerations are in order.

Printed Circuit Board Design

In the application of machine insertion to Printed Circuit Board (PCB) assembly, the following general rules should be applied. Each of these is described in subsequent paragraphs of this chapter.

- Standardize tooling for Printed Circuit Board mounting.
- Utilize known references.
- Orient components properly.
- Select standardized components whenever possible.
- Select proper PCB hole diameters.
- Provide proper clearances between components.

- Provide proper clearances between tooling holes and components.
- Provide proper clearances for Insertion Head and Cut and Clinch.

Board Considerations

Material: Most insertion principles do not limit PCB construction or material. Components may be inserted in single or multiple layer constructed circuit boards.

Size: The maximum size of the PCB is generally related to the insertable area of the type insertion machine used and/or by any automatic board handling system. Standard insertable areas vary according to the machine selected.

Most manually fed, processor controlled machines have a standard insertable area of 18 x 18 inch (457,2mm x 457,2mm). Automatic board handling machines have a maximum board size of 16 x 18 inch (406,4mm x 457,2mm) and with certain special overall dimensional tolerancing. For special situations, contact the manufacturer's sales representative.

Where small PCBs are required, it is sometimes advantageous to use a "break-away" or "biscuit" board. This configuration consists of a series of small boards which are punched or drilled within a single larger board. Once inserted, they can be separated on perforated or grooved edges, or they may be routed or sheared. A technique called "press-back" may also be used when small boards are punched out of a larger sheet. After punching the outline, it is pressed back into a larger board and held by friction until the assembly and soldering process is complete. "Break-away boards", or boards sheared after assembly, minimize board handling throughout all phases of board production, particularly when only a few components are inserted into each PCB.

Of prime importance in PCB design and construction is the establishment of accurate reference points, or location references, to which all holes are drilled or punched. Locating references generally take the form of two holes. Other possible approaches are "V" notches, multiple slots, or straight surfaces. These references are used to position the PCB on the workboard holder for automatic insertion. It is recommended that the distance between the locating references be the maximum separation permitted by the length of the board. Also, it is recommended that the location reference hole diameter tolerance be held to ±.001″ (0,025mm). See Figure 8-1. The lower left reference hole is generally given the location terminology of X_0, Y_0.

The locations of all insertion holes should be defined relative to the board locating references. The accuracy of the hole locations with respect to the locating references will affect insertion reliability and hole diameter requirements. With drilled boards, the hole tolerance from the board references should be equal to or less than 0.003″ (0,076mm). With punched boards, the reference holes or surfaces should be punched using the same die-set used for the insertion holes and not a part of the blanking die.

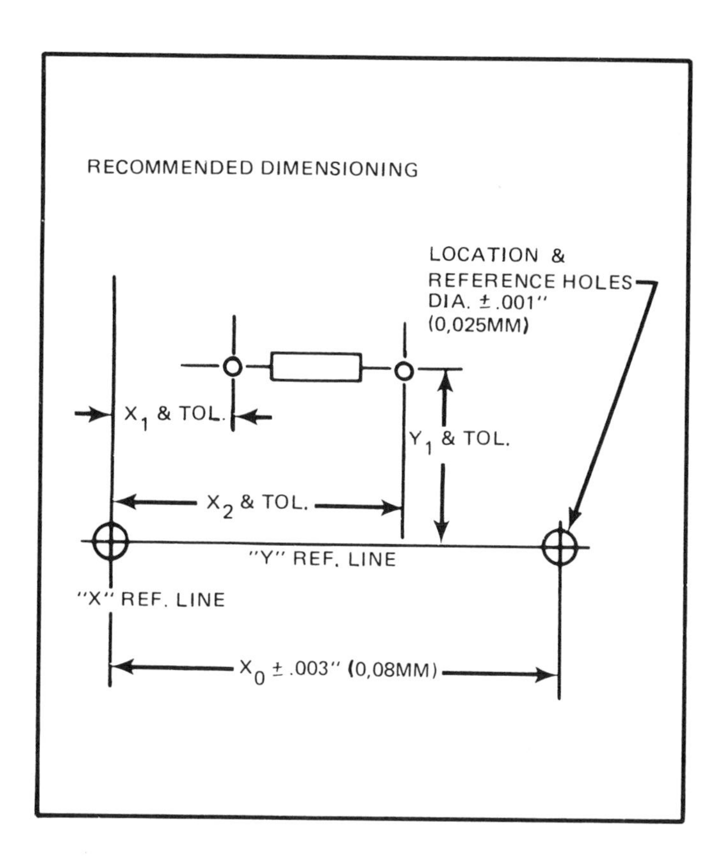

Fig. 8-1: Recommended methods of locating PCB holes and dimensions.

Care must be exercised in the initial board layout to ensure that the insertion hole locations and the board locating references are only "one tolerance" apart—particularly for boards with drilled holes. If this consideration is not met, tolerance build-up during insertion will reduce insertion reliability. The "one tolerance" requirement can easily be accomplished by establishing X-Axis and Y-axis dimensional reference lines through the centerlines of the location references and using these pre-drilled holes to locate the board for insertion hole drilling. If both the reference holes and insertion holes are drilled at the same setup, the net accuracy is two times the normal drilling machine tolerance. Once these lines are established, they may be used as the references for drilling or punching of the PCB, as well as for insertion. See Figure 8-1.

Component Location Objectives

Axis Considerations: The most efficient use of automatic insertion equipment and the greatest component population density can be achieved by inserting components in only one or two orthogonal axes.

Use of only one axis, as shown in Figure 8-2, minimizes the amount of board handling and machine operations.

Two axes insertion, shown in Figure 8-3, is an acceptable and efficient way of inserting components. There are two common methods of performing two axes insertion. One method, used on earlier machines without a Rotary Table,

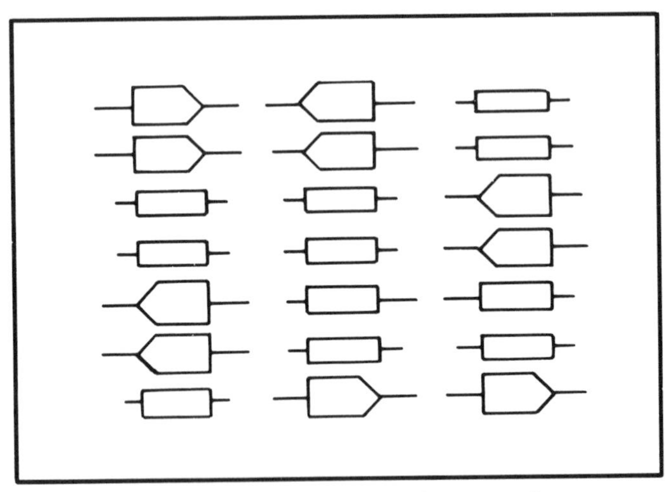

Fig. 8-2: Acceptable arrangement for one-axis insertion on PCBs.

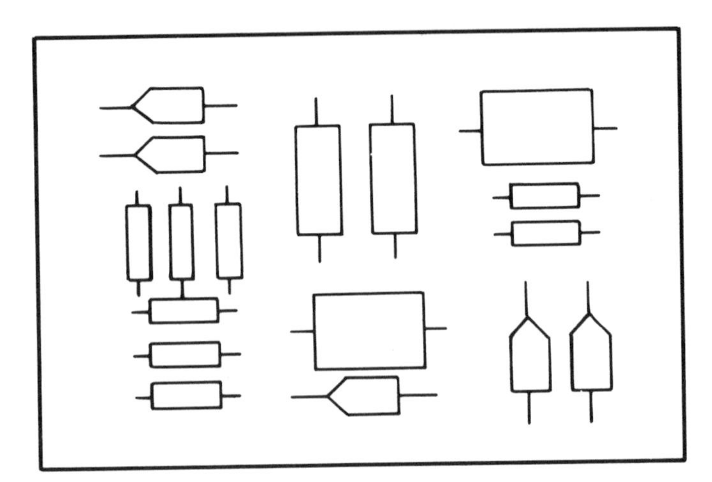

Fig. 8-3: Acceptable arrangement for two-axes insertion on PCBs.

involves designing the workboard holder, which supports the PCB on the machine table, such that one window (PCB location) supports the PCB in an X-Axis while another window supports it in the Y-Axis. The operator must load each PCB on the workboard holder twice. Using the second method, a properly designed workboard holder is rotated in 90 degree increments. Using the Rotary Table, now standard, rotation can be automatically accomplished as part of the insertion process. However, one must recognize that the use of the Rotary Table contributes to a tolerance build-up which can lower insertion reliability. With Automatic Board Handling, a Rotary Table is required for axes insertion.

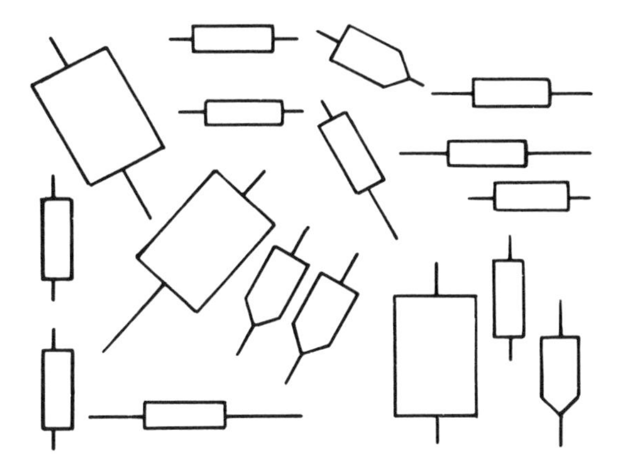

Fig. 8-4: Many axes insertions are unacceptable for automatic machines.

Applications which use more than two axes, as shown in Figure 8-4, are generally inefficient and difficult to implement.

Component Selection

Universal Insertion Machines allow a wide range of components to be inserted which minimum tooling changes. Careful component selection, however, can provide improved automatic insertion reliability. It is wise, whenever possible, to limit the range of component body sizes and lead diameters used. Doing this will decrease the need for editing of established programs and generally reduce the need for time consuming tooling changes.

Lead Diameter Considerations

Each general type of component insertion has its own rule for establishing hole diameter. Hole diameter is basically a function of the machine, board accuracy, and lead diameter.

While some hole diameters may seem somewhat large in size, they take into account typical conditions on the production floor. They deal with nominal dimensions, plus the tolerances of all elements of the insertion system regardless of type. It is one thing to say components can be inserted into smaller holes and it is another thing to accomplish it in a production environment.

There have been occasions where manufacturers have successfully inserted components into holes smaller than those recommended. Usually there are some other compensating factors involved, such as a very small board size, better hole location tolerance, a recognized lower yield, etc.

Location Considerations: Automatic insertion of PCBs requires that there be sufficient clearances around components and locating references to allow for the Insertion Head and Cut and Clinch tooling. The clearances required will vary with the type of insertion machine.

Board Holder Design

It is necessary to firmly retain and support the PCB on the insertion machine during insertion, and to provide an easy method of board loading and unloading. This is done by means of a workboard holder, a metal plate with specially designed opening and accurate locating means. See Figure 8-5.

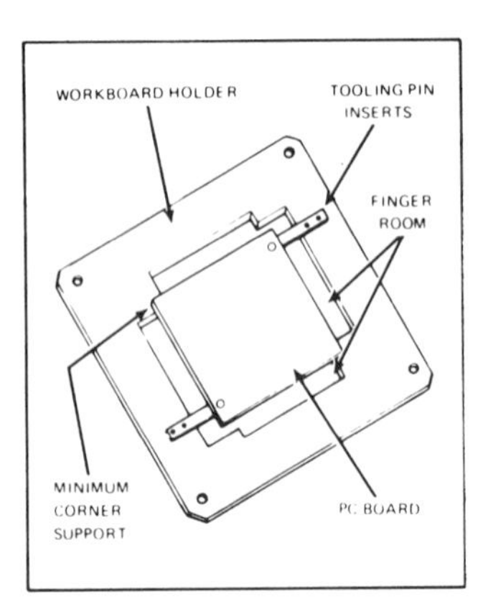

Fig. 8-5: Typical board holder opening.

When designing a workboard holder, the following general rules should be applied.

- Support the PCB as much as possible.
- Provide finger room for loading and unloading for manually loaded machines.
- Utilize known references.
- Provide proper clearances around location references.
- Provide board hold down during insertion.

Blank workboard holders are available from manufacturers for all machines. Many also provide workboard holder designs and build these to owner's specifications.

Workboard Holder Considerations

Board Openings: Board openings, or windows, must be positioned so that all components to be inserted are located within the insertable area of the insertion machine. Maximum insertable area begins at machine zero, located a defined distance from the lower left table reference bushing, and proceeds in a positive X and Y direction from that point.

When designing workboard holder openings, it is desirable to support the PCB as much as possible with at least .125 inch (3,18 mm) of workboard holder surface extending beneath the board at all support areas. Supporting the board at all four corners is preferable; larger boards may require additional support near the middle on long sides.

A certain amount of edge clearance is required for "Cut and Clinch" clearance when inserting components close to the edge of the board. Where manual board loading and unloading is used, additional "finger room" should be provided at the right and left sides of the windows to allow easy loading and unloading of the PCB.

The acceptable distance between the edge of the board opening and an insertion varies with the type of component, machine, etc. In general, printed circuit designs requiring components inserted close to locating references and board edges may result in loss of automatic insertion of these components due to board support requirements.

Locating Methods: Locating holes "V" notches, slots or other accurate surfaces may be used for locating the PCB on the workboard holder for insertion. Regardless of the method used, it is required that two locations be utilized, and that the distance between them be the maximum permitted by the length of the board. The same location references from which all the holes are drilled or punched should be employed for this purpose.

Pins through two locating holes are the most common and preferred locating method; they may be mounted directly into the workboard holder or into inserts which are mounted onto the workboard holder. Inserts allow insertions to be made closer to the locating references than do workboard holder-mounted locating techniques limited support area. See Figure 8-6.

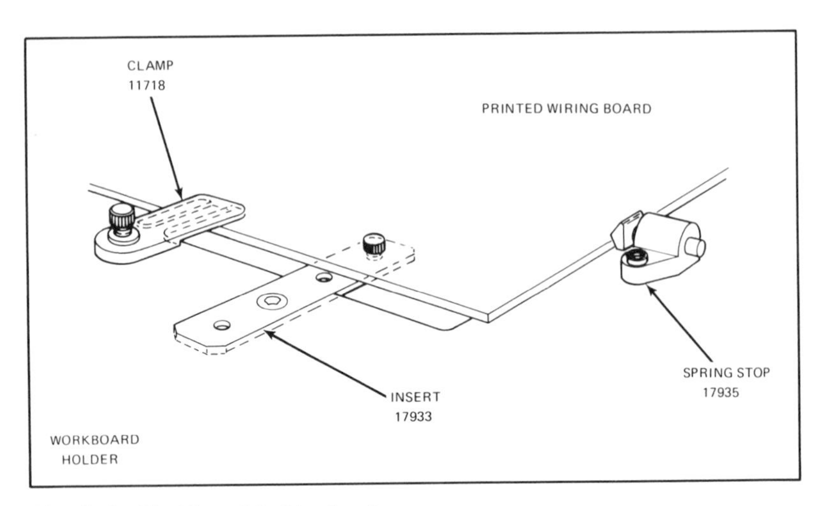

Fig. 8-6: Workboard holder hardware.

Hold-Down Devices: Hold down devices are used to maintain board position during table movement and component insertion. The most common of these are clamp assemblies which must be manually positioned during board load and unload or spring stops which use spring loading plungers to apply pressure to the edges of the board to hold it in position. Both are shown in Figure 8-6.

Hold-down devices are mounted at any place where the board is supported and, most advantageously, at opposite sides of the PCB near the locating pins.

Rotary Workboard Holder Design

Rotary workboard holders usually have one, two, or four board openings. These locations are usually arranged in a symmetrical pattern about the center of the workboard holder to allow component insertion in both 0 degrees or 90 degree axis positions. (Figure 8-7).

For manually loaded machines, "finger room" for board loading and unloading should be provided in both the X and Y-Axes if the workboard holder is to be rotated.

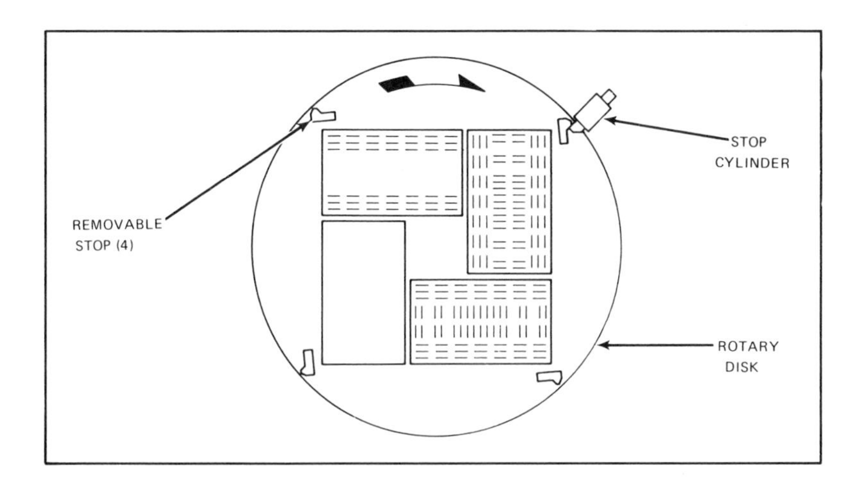

Fig. 8-7: Four window, two-axes insertion.

Automatic Board Handling Workboard Holders

Two types of board locations are available with most types of automatic board handling equipment: front edge justified or centerline justified. Front edge justified means the front edge of the board is fixed and the board "grows" to the rear of the machine. Centerline justified means that the centerline of the board on the table is fixed and everything "grows" symmetrically. Front edge justified automatic board handling is not compatible with some earlier machines now in use.

The major advantage of the front edge justification approach is that a standard, adjustable, workboard holder may be used.

If centerline justification is used, the clearances for support and locating are similar, but because of the fixed front and rear edge support and the automatic locating and hold-down means, the uninsertable areas are usually greater. A typical automatic board handling workboard is shown schematically in Figure 8-8.

Most automatic board handling machines utilize the standard rotary table. From a board design standpoint, edge clearances on certain components, primarily DIPs, vary with the table rotation. Specifically, with 180 degree rotation, DIPs vary with the table rotation. Again, foot-print considerations and precedence are of extreme importance. For wide boards, say 8 inches or more, and/or for boards which have been weakened, or for those which will have a heavy component load by the end of the last automatic inserter, some form of board support at or near the centerline may be required to control "sag".

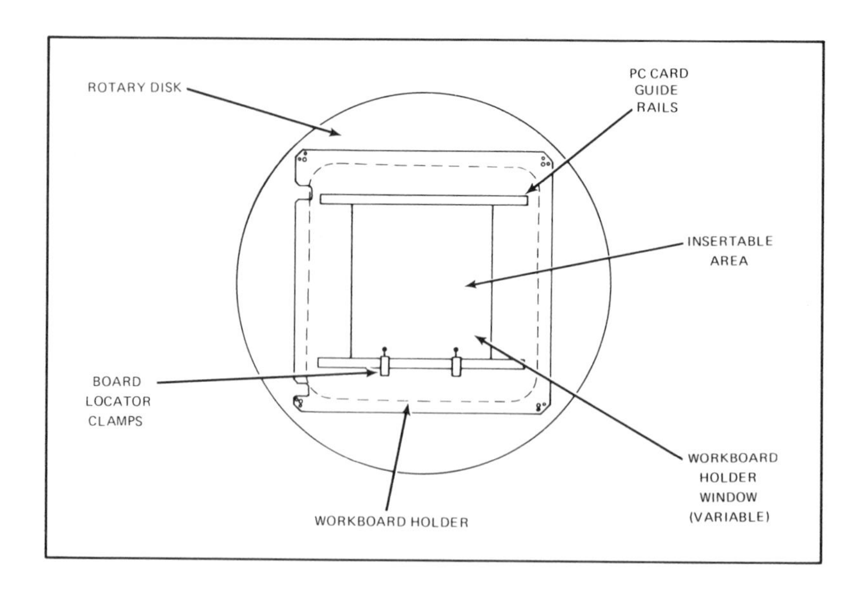

Fig. 8-8: Automatic machine workboard holder with rotary positioning system.

board handling machine is 16.00 inches wide x 18.00 inches long (406,4mm x 457,2mm) and the minimum size is 4.25 inches wide x 5.90 inches long (108,0mm x 149,9mm).

Additional requirements: When designing a PCB for any automatic board handling inserter, certain special requirements must be considered. Precision positioning of a PCB on the workboard holder is accomplished by locator holes. In addition, the width of the board, "W", and the length, "L", must be maintained. This is to ensure proper transfer of the board from the magazine (or stack) through the machine and into the output magazine as edge support is used throughout. This also limits the insertable area.

Programming Considerations

Processor controlled machines require that a pattern program be entered into the controller memory. The pattern program contains all the information required to automatically populate a PCB.

The information required in the pattern data will vary according to the type of insertion machine. The format used to input the required information to the controller memory is relatively fixed depending on the type of software being used. Earlier processor controlled machines used a basic pattern format using software defined as Satelite which goes up to version 2E. The newer machines use updated software compatible to the individual machines.

Programming Considerations

Processor controlled machines require that a pattern program be entered into the controller memory. The pattern program contains all the information required to automatically populate a Printed Circuit Board.

This chapter has been prepared to assist in the understanding of pattern programming which will also aid in the understanding of automatic insertion (population) requirements discussed in other chapters of these guidelines.

Insertion Machine Pattern Program Format

The information required in the pattern data will vary according to the type of insertion machine. The format used to input the required information to the controller memory is relatively fixed depending on the type of software being used. Earlier processor controlled machines used a basic pattern format using software defined as Satellite which goes up to version 2E. The newer machines use UICS, Universal Instruments Control Software. This has a much more expanded format and can contain additional information beyond the basic insertion data.

The actual format and arrangement will be found in the manual for the specific machine you are programming.

Generally, a pattern program consists of three basic sections. These are the heading, the offsets, and the pattern steps. Each of these are handled somewhat differently depending on the software being used.

Pattern Heading: The pattern program heading serves to identify the pattern program.

In most cases, each Printed Circuit Board to be inserted or populated is considered one complete pattern. As a result, it is necessary to define the insertion locations only once, regardless of the number of boards that are mounted on the workboard holder, also in the case of "break-away" boards. All insertion locations are defined relative to one point on the board, hereafter referred to as the Pattern Program Zero reference.

The pattern offsets then define the distance between the machine zero and the Printed Circuit Board datum reference. One offset is entered for each Printed Circuit Board location (window) on the workboard holder and/or each Printed Circuit Board on a "break-away" board.

The offsets for a typical Printer Circuit Board are shown on Figure 8-9.

The pattern steps give the specific insertion instructions for each component to be inserted. There is basically one pattern step for each component to be inserted. Each pattern step is typed on a single line. A pattern step consists of the following minimum information.

X and Y coordinates: These coordinates define the location of the component insertion reference point relative to the PCB Pattern Program Zero reference or offset. This will vary with the type of component used.

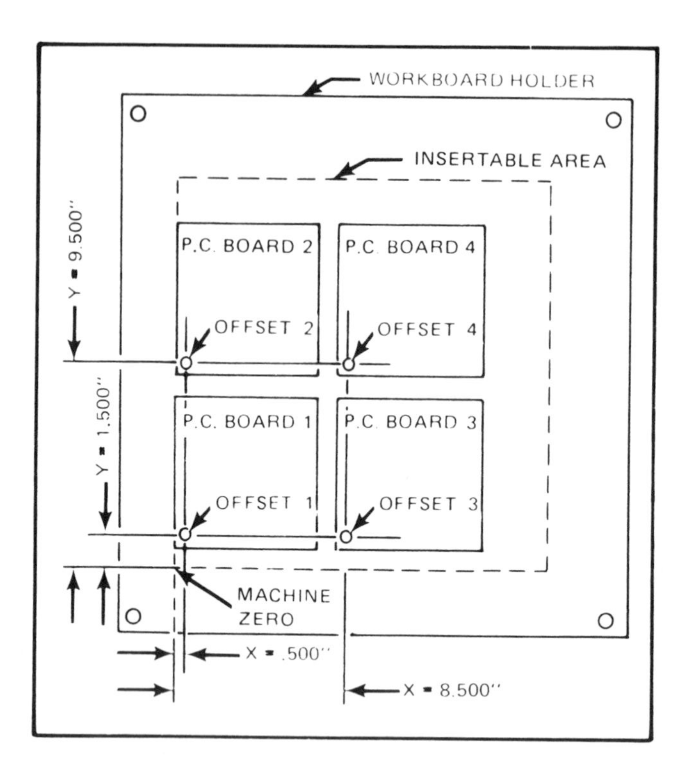

Fig. 8-9: Offsets for a typical PCB.

The X and Y coordinates consist of up to five digits which describe a dimension either in milli-inches or thousandths of a centimeter. Leading zeroes need not be programmed; training zeroes are required.

Z axis data and/or other data functions (third data field): These functions vary with the machine type. Examples are as follows:

Axial Lead Insertion Machines—Insertion Center Distance
DIP Insertion Machines—Magazine number
Radial Lead Insertion Machine—Dispensing Head Number

The functions are fully described in the individual Programming Appendix for each machine.

Single Letter or Modifier Fields: The first of two single character instructions is the auxiliary or modifier function. This field was originally the depth-stop field and is now often referred to as the MODIFIER field. For example, on Axial Lead Insertion Machines, this character represents the depth-stop, or the depth to which the component body is assembled to the Printed Circuit Board.

On Radial Lead Insertion Machines, the auxiliary function defines the Insertion and Clinch orientation.

Insertion or operation instruction: The last character on a pattern step line is also an alphabetic character whose purpose is to indicate the function to be performed by the machine. Typically these mean *insert and check continuity, index rotary table, park, etc*. See the data for the specific machine for a full listing.

There are four more 5-digit fields available for use. On some machines, one or more of these are pre-assigned. On others they are totally available at the user's discretion for, say, part numbers, part values, error count, second Z-axis, depth stop, etc.

Board Error Correction

The Board Error Correction (B.E.C.) option is offered with most insertion machines (Universal is one brand). B.E.C. permits the positioning system to compensate for lead hole location errors that contribute to insertion failures. Its main purpose is to detect displacement, stretch, and skew of Printed Circuit Board lead hole patterns in comparison to programmed hole positions. This option includes provisions for use as a Bad Board Sensor, whereby pre-inspected, faulty board segments may be bypassed during the processing multi-segmented circuit boards.

Because the B.E.C. sensor is not on the insertion center line, there is generally some area on each machine which cannot be scanned. These limitations are usually detailed in the manufacturer's specifications for each individual machine.

The generation of the pattern program must take into consideration many things in addition to the above basics to produce optimum results.
Some of these other considerations are:

a. Path — a minimum travel path to cover the board in the least total time. There are suggestions to optimize this in each chapter.
b. Clearances — including both Head, Anvil and workboard holder clearances.
c. Precedence—which component should be programmed first? Generally, the least expensive are often done first to minimize possible scrap costs, however, sometimes size requires that certain components be done last like radials and square-wire pins. Other considerations are ease of repair and tooling footprint.
d. Magazine Selection vs. Speed—On DIP inserters, where shuttle travel is a factor in establishing insertion rate, component values used most frequently should be located closest to the Insertion Head.

All in all, programming and planning go hand in hand in a straightforward, logical manner to completely populate as many parts by automatic means as is possible.

Axial Lead Component Insertion

Manufacturers offer wide varieties of equipment for automatic and semi-automatic insertion of axial lead components: from a manually loaded, single head, Variable Center Distance (VCD) Inserter, through several sizes of Dual Head VCD's plus the ability to incorporate full automatic board handling on single head pass-thru versions. All standard machines are equipped with rotary tables and they can automatically insert a wide range of presequenced component types.

These guidelines are intended to provide near optimum insertion reliability of axial lead processing equipment in the production environment. Where tradeoffs can be made to further increase reliability, these are described. Optional tooling and equipment is available for certain special applications. Consult your nearest sales engineer for applications not defined herein.

Component Input Taping Considerations

Component taping requirements for axial lead processing equipment generally comply with EIA Standard RS-296-D which is accepted as the industry standard. Figure 8-10 represents the adaptation of this specification as it applies to some manufacturer's products, specifically as the input to a sequencer, or directly to a VCD, where non-sequenced parts are used.

When used as the input to a VCD, the sequencer output must meet the VCD Head input specifications as well. These requirements are shown on Figure 8-12. If non-sequenced, taped components are to be used in a VCD, they must also meet this same specification.

Axial Lead Sequencing

Sequencing, which in this case is the preparation of axial leaded components prior to insertion on a VCD, usually permits complete or nearly complete insertion of this type of a component on a given Printed Circuit Board.

In a Satellite High Speed Expandable Sequencer, the components are cut from the input tape, dispensed onto a moving conveyor and fed into a special taping mechanism where they are accurately centered, re-pitched and retaped to a distance between tapes somewhat smaller than the original input re. As a consequence, sequenced components do not use the same maximum insertion spans obtained with standard taped components.

Figure 8-11 shows the taping/sequencing relationship in tabular form. It shows the four input classes for a sequencer, as defined by the distance between tapes, that may be fed into the machine. Most sequencers are set up for Class I or Class II input. Class III is generally used for very large component input and Class SI is used for very small ones.

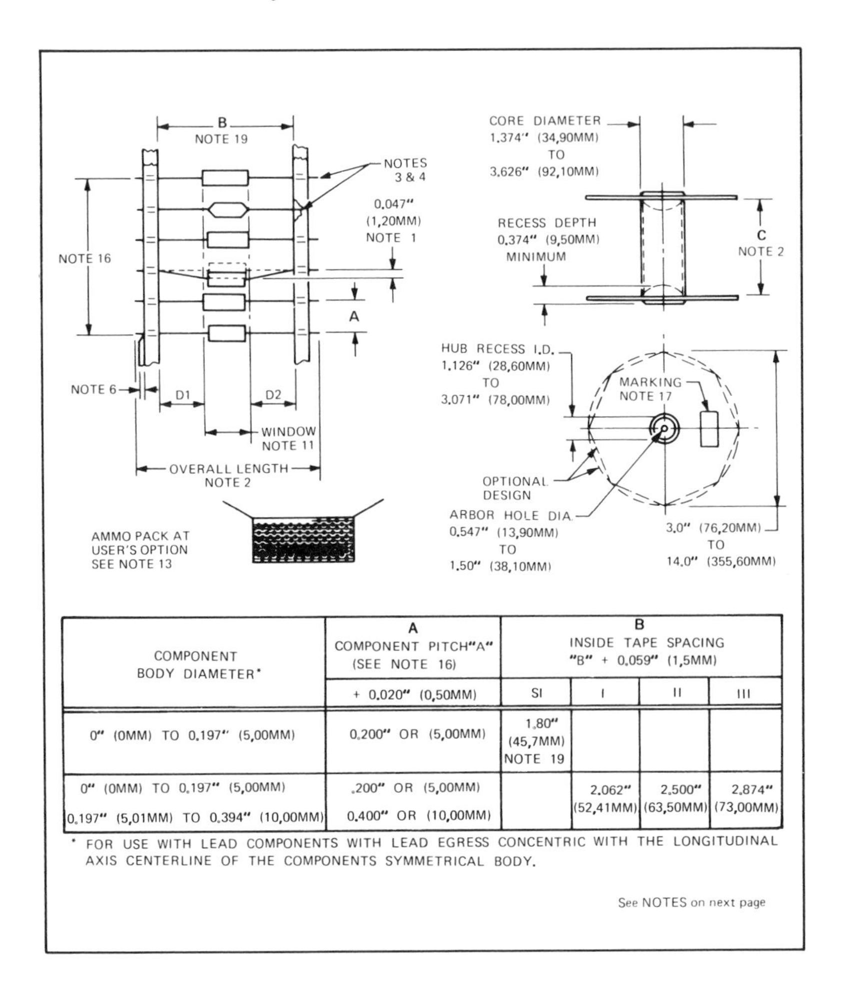

COMPONENT BODY DIAMETER*	A COMPONENT PITCH "A" (SEE NOTE 16)	B INSIDE TAPE SPACING "B" + 0.059" (1,5MM)			
	+ 0.020" (0,50MM)	SI	I	II	III
0" (0MM) TO 0.197" (5,00MM)	0.200" OR (5,00MM)	1.80" (45,7MM) NOTE 19			
0" (0MM) TO 0.197" (5,00MM) 0.197" (5,01MM) TO 0.394" (10,00MM)	.200" OR (5,00MM) 0.400" OR (10,00MM)		2.062" (52,41MM)	2.500" (63,50MM)	2.874" (73,00MM)

* FOR USE WITH LEAD COMPONENTS WITH LEAD EGRESS CONCENTRIC WITH THE LONGITUDINAL AXIS CENTERLINE OF THE COMPONENTS SYMMETRICAL BODY.

See NOTES on next page

Fig. 8-10: Axial lead component specifications.

Once the input class has been tentatively selected, the next task is to check the Insertion Head specifications to see if this class will allow the desired insertion spans. It is not generally recommended that class changes be planned as part of an operation routing because too many items must be reset and readjusted.

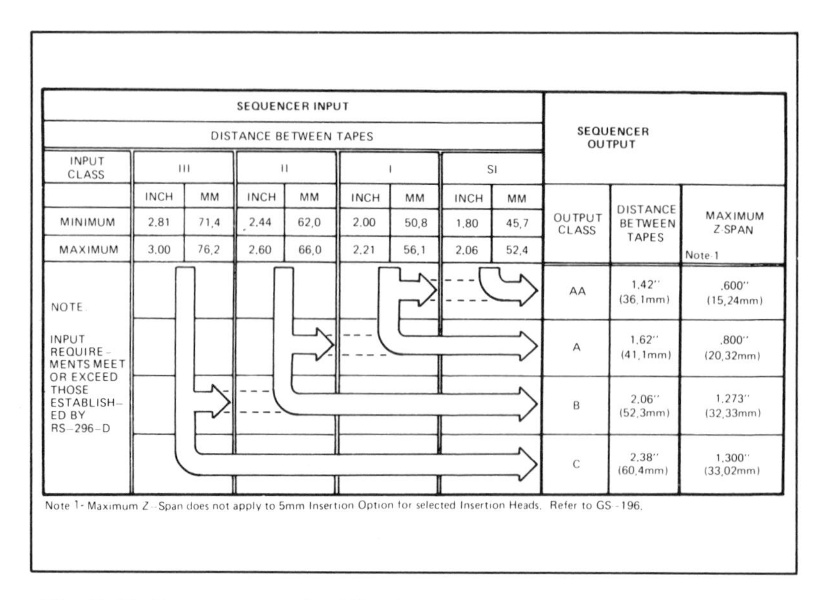

SEQUENCER INPUT — DISTANCE BETWEEN TAPES

INPUT CLASS	III		II		I		SI	
	INCH	MM	INCH	MM	INCH	MM	INCH	MM
MINIMUM	2.81	71,4	2.44	62,0	2.00	50,8	1.80	45,7
MAXIMUM	3.00	76,2	2.60	66,0	2.21	56,1	2.06	52,4

NOTE:

INPUT REQUIREMENTS MEET OR EXCEED THOSE ESTABLISHED BY RS-296-D

SEQUENCER OUTPUT

OUTPUT CLASS	DISTANCE BETWEEN TAPES	MAXIMUM Z-SPAN Note 1
AA	1.42" (36,1mm)	.600" (15,24mm)
A	1.62" (41,1mm)	.800" (20,32mm)
B	2.06" (52,3mm)	1.273" (32,33mm)
C	2.38" (60,4mm)	1.300" (33,02mm)

Note 1- Maximum Z-Span does not apply to 5mm Insertion Option for selected Insertion Heads. Refer to GS-196.

Fig. 8-11: Input-output classifications—sequences

Insertion Center Considerations

Some manufacturer's Axial Lead Insertion Heads are similar in design and function. The maximum spans available for each class are shown in Figure 8-11. Figure 8-12 shows the input and output characteristics of the VCD Insertion Head.

Minimum VCD spans are further limited by component body length. See Figure 8-13 for a typical situation. To be more precise, one may use the following:

Min. Insertion Span = Body length + 0.062" + (inside former width x 2) + one wire diameter

or, for metric:

Min. Insertion Span = Body length + 1,57mm + (inside former width x 2) + one wire diameter

In the above formula, the 0.062" (1,57mm) value accounts for component centering variations, which is two times 0.031" (0,79mm)(Note 1, Figure 8-12), and the "one" wire diameter used is really the two inside halves. The width of the standard inside former is 0.032" (0,81mm).

Note that the programmed Z-Axis span is the selected inserted span plus one wire diameter (the two outside halved).

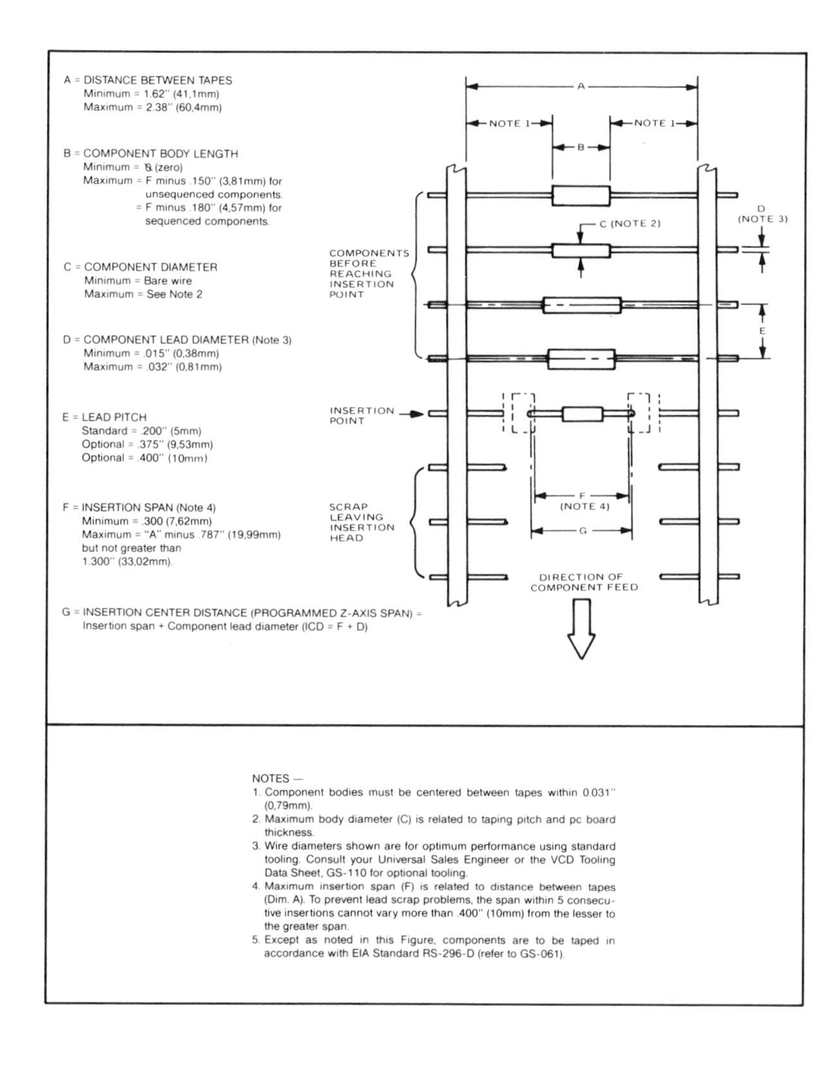

Fig. 8-12: Input specificatoins—VCD insertion head.

Lead Form and Tooling

The VCD Head is tooled for the lead form shown in Figure 8-13. The "X" dimension (end of component body to inside of form in lead) given is a function of the Driver Tips and Inside Formers selected. Typically this is approximately 0.080″ (2mm) minimum. Larger values of "X" will give more reliable insertion. With special attention to component selection and body centering during sequencing, the .075″ (1,91mm) dimension can be used as the minimum.

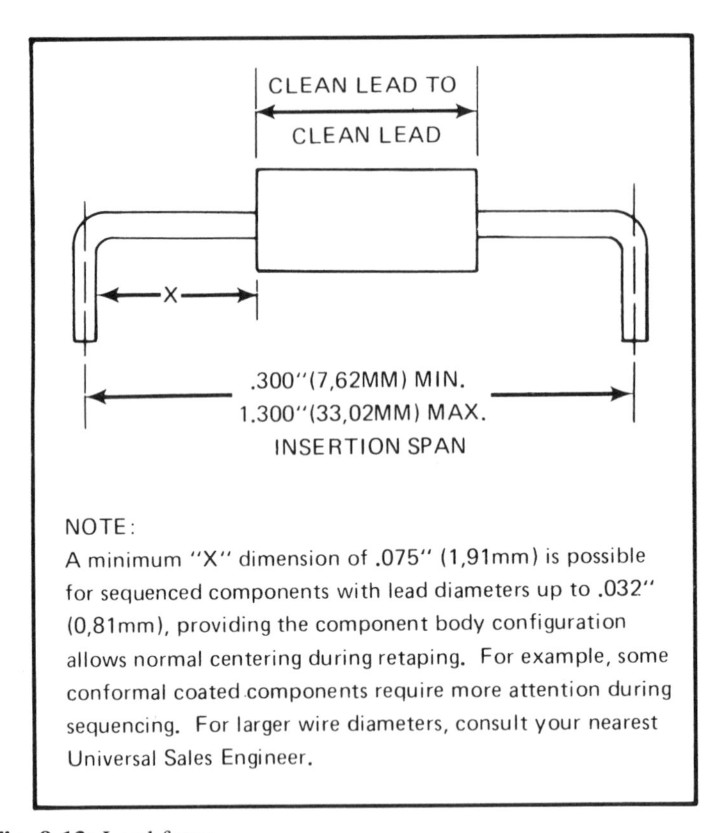

Fig. 8-13: Lead form.

Printed Circuit Board Thickness Versus Body Diameter

Another factor to be considered for the VCD head is the board thickness versus maximum body diameter, since overall cut-off length is fixed. Taking board thickness into consideration, the maximum body diameters that are insertable with sufficient lead remaining for cutting and clinching are shown in the chart in Fig. 8-14.

MAXIMUM BODY DIAMETER		
Board Thickness	Standard Tooling	Optional Tooling
0.093" (2,36mm)	0.234" (5,94mm)	0.354" (9,0mm)
0.062" (1,57mm)	0.296" (7,52mm)	0.394" (10mm)*
0.032" (0,81mm)	0.356" (9,04mm)	0.394 (10mm)*

Fig. 8-14: Maximum body diameter is limited by the tape pitch.

Hole Diameter Requirements

In general, the hole size necessary for reliable component insertion is a function of the following:
a. Component lead diameter
b. Machine Insertion Head tolerances
c. Table positioning accuracy
d. Board holder accuracy
e. Printed Circuit Board hole pattern and tooling reference accuracy
f. Component lead length desired below the board

For some Axial Lead Insertion Machines with Rotary Tables, the maximum machine tolerances for the head, X-Y table, Rotary Table and workboard holder is a true position accuracy of .005" (0,12mm). This value requires a starting hole diameter of .010" (0,25mm) larger than the lead diameter being inserted. The manufacturing tolerance of the Printed Circuit Board insertion hole pattern must be added to this hole diameter.

The formula for determining hole diameter is:

Minimum Hole Diameter = Lead Diameter + 0.010" (0,25mm) + Hole Location Tolerance

Example:

A 1/4 watt resistor with 0.025" (0,64mm) lead diameter, boards with hole location accuracy of within 0.003" (0,08mm) of true position, and the standard machine tolerance of 0.010" (0,25mm).

Solution:

Minimum Hole Diameter = 0.025" + 0.010" + 0.006" = 0.041"

In metrics, the solution will be:

Minimum Hole Diameter = 0,64mm + 0,25mm + 0,15mm = 1,04mm

The above formula normally produces insertion reliability of 99.9 percent. Reduced clearance from wire diameter to hole diameter to as low as 0.005″ (0,13mm), may be used with small, high accuracy boards and still maintain reasonably successful insertion. However, the insertion reliability factor will be reduced.

Conversely, an even higher reliability factor can be achieved by increasing hole diameter by 0.001″ or 0.002″, by better control of sequencer centering, by increasing the span related to body length, and also by using more accurate boards. The maximum hole diameter is usually limited by soldering or good clinching requirements.

When the Rotary Table feature is not being used, the machine tolerance of 0.010″ (0,25mm), above, may safely by reduced to 0.008″ (0,20mm).

Component Body Configuration

Normally body configuration is not critical unless the extreme limitations of the equipment are being used. The components are handled by the leads so that component bodies may be round, oval, dog boned in shape, etc.

The Depth Stop is the depth to which the component body is assembled to the Printed Circuit Board. Due to the lead form requirements of axial lead components, shown in Figure 8-13, the depth stop should be selected so that the component is held securely to the board without deforming the lead.

The VCD Insertion Head has an electrically driven Depth Stop cam with twenty-six program selected letter positions in 0.008″ (0,20mm) increments in the modifier field. With newer single head, UICS programmed machines, the Depth Stop cam can be programmed in increments of 0.001″ (0,025mm).

A reasonably close approach to the depth stop dimension to be programmed has been found to be:

DS = Body Dia. + Lead Dia./2x 0.936 - 0.028″

All values are shown in inches. For metric values, subtract 0,71mm. This accounts for most factors, including the groove in the Drive Tip.

This value can be modified in either direction to get the desired results. A slight looseness is recommended to avoid residual internal stresses after soldering.

Location Considerations

Some Axial Lead Insertion Heads are equipped with Outside Formers which guide the leads to the point of insertion on the printed circuit board. Clearance around a given hole must be taken into consideration to allow the equipment to function properly. Figure 8-16 shows the top view of a standard Outside Former of an Insertion Head at the point of insertion, illustrating the clearances required between the lead being inserted and any adjacent component body or lead.

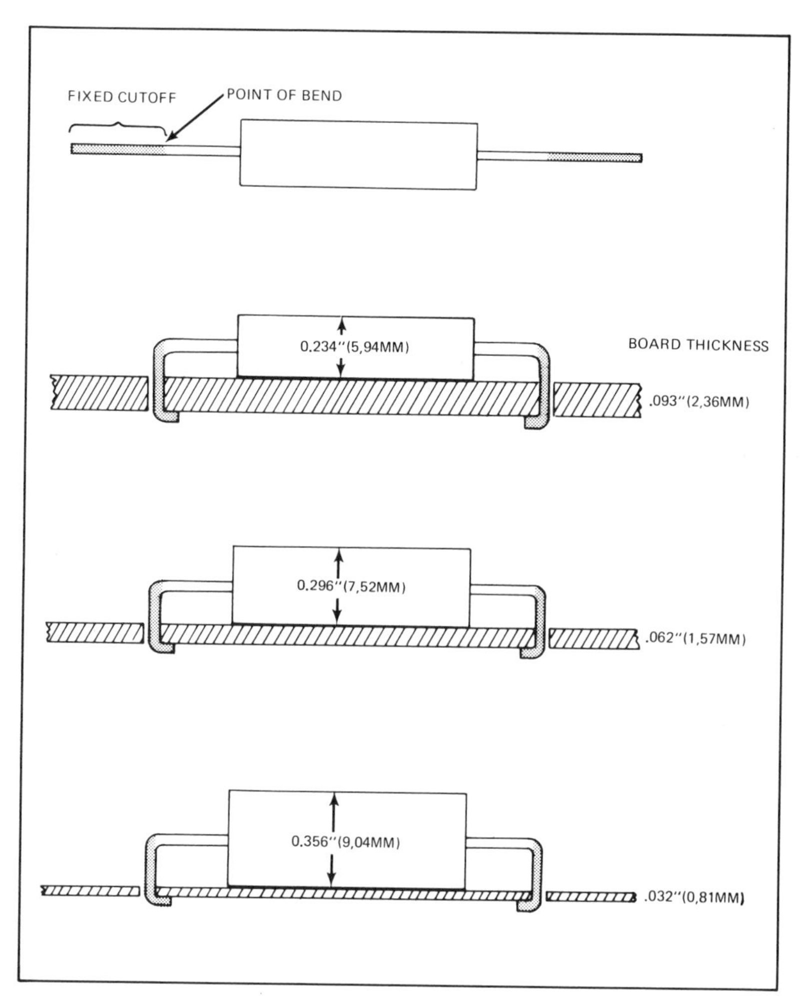

Fig. 8-15: Board thickness versus body diameter, using standard tooling.

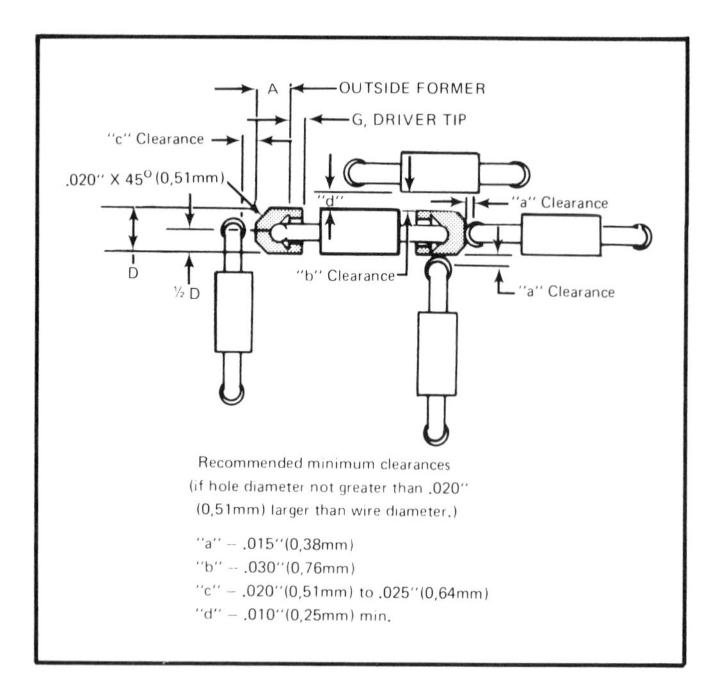

Fig. 8-16: Plan view of insertion head showing tooling clearances.

The VCD Insertion Head employs an Outside Former with a single rounded "Vee" groove. The centerline of lead to former clearance line varies as a function of lead diameter. Figure 8-15 shows the largest wire diameter being inserted with standard tooling.

For applications involving dense component assembly of small components (diodes, or resistors of ¼ watt and smaller), Outside Formers are available with smaller "footprint" dimensions. This, however, results in a minor loss or durability of the tool.

The Cut and Clinch is normally positioned approximately 0.600″ below the board to clear the workboard holder and the rotary disk. As the Head begins the insertion cycle, the Cut and Clinch Anvil is raised to a theoretical 0.005″ to 0.010″ (0,12mm to 0,25mm) below the board to support it and provide lead clearances. At the bottom of the Head stroke, the Cutters are driven upwards to cut and clinch the component leads. See Figure 8-17 for minimum side-to-side and end-to-end clearances for the Axial Lead Cut-and-Clinch to previously inserted components of both axials and DIP's.

The various methods used to locate boards for insertion exclude mounting of components, in most cases, in areas around reference points. This varies greatly between the different board configurations.

As a general rule, the minimum uninsertable are in the vicinity of any reference hole is approximately 0.5″ (12,7mm) radius from the edge of the workboard holder which supports the locating pin. Edge locating and support methods usually require 0.5″ (12,7mm) along the edge guides. With the use of inserts, these requirements can be reduced. See Figure 8-18.

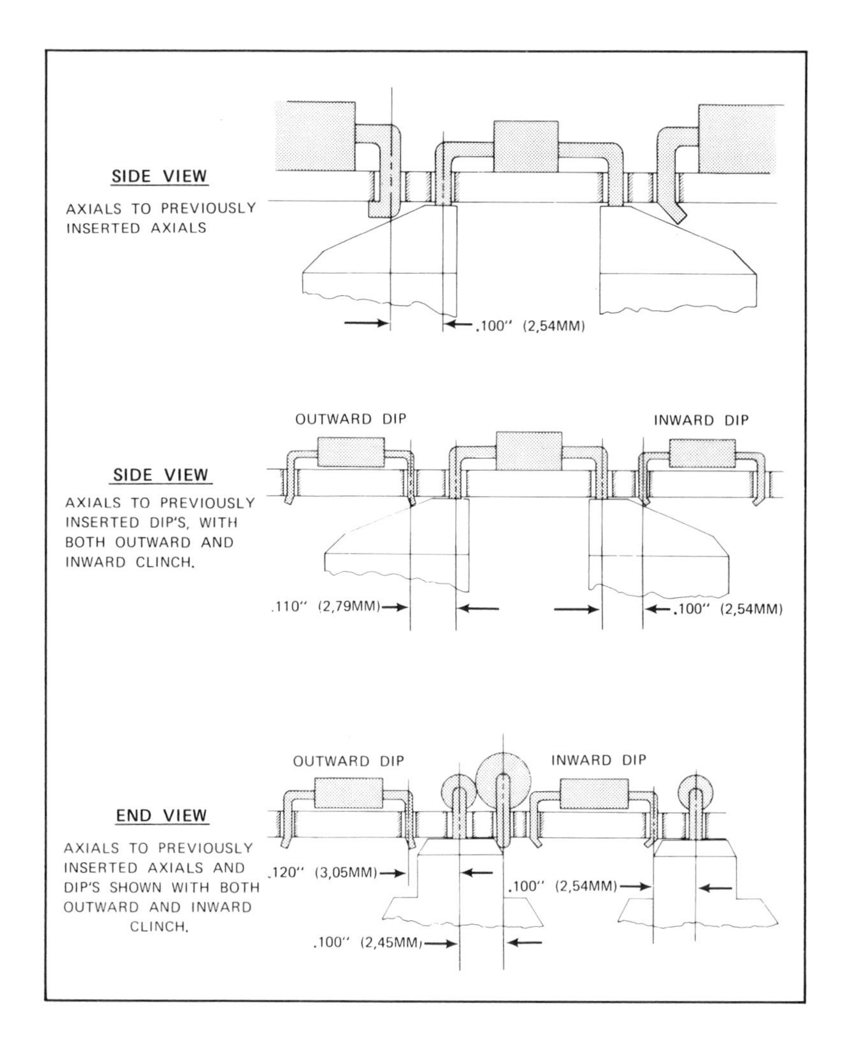

Fig. 8-17: Minimum clinch clearances.

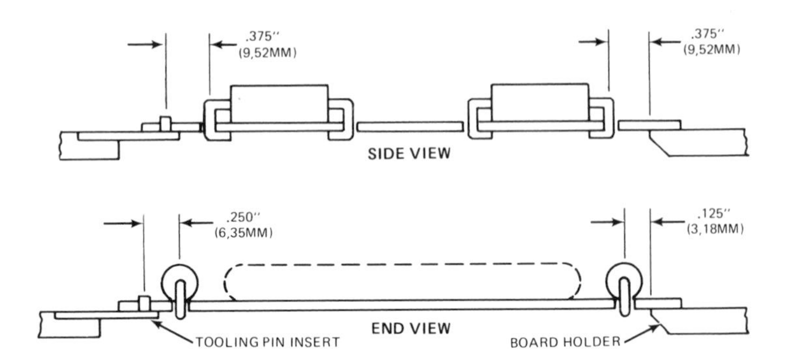

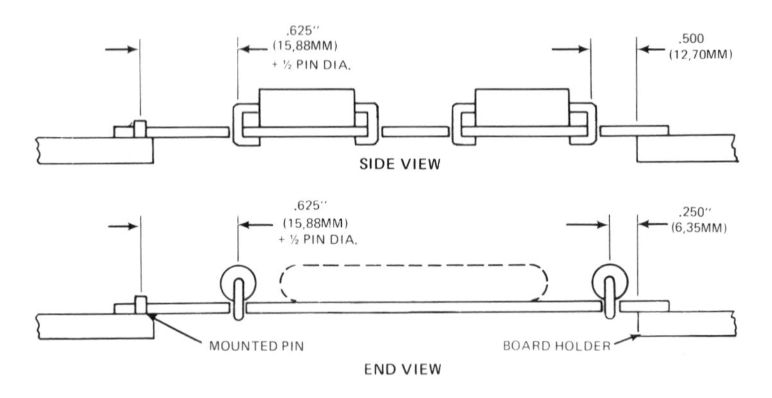

Fig. 8-18: Uninsertable areas—clinch to work-board holder, for VCD insertion machines.

Certain Rotary Tables have additional uninsertable areas in the four corners; refer to the specific machine data.

Clinch Patterns

Lead forming for axial and other similar components is along the centerline of the component, parallel to the component leads, and inward toward the body. Although the component is cut from the tap by the Insertion Head before insertion, a second and more precise cut occurs beneath the board. The minimum length of the lead under the board is a function of lead diameter. Lead length is measured parallel to the board after clinch. See Figure 8-19.

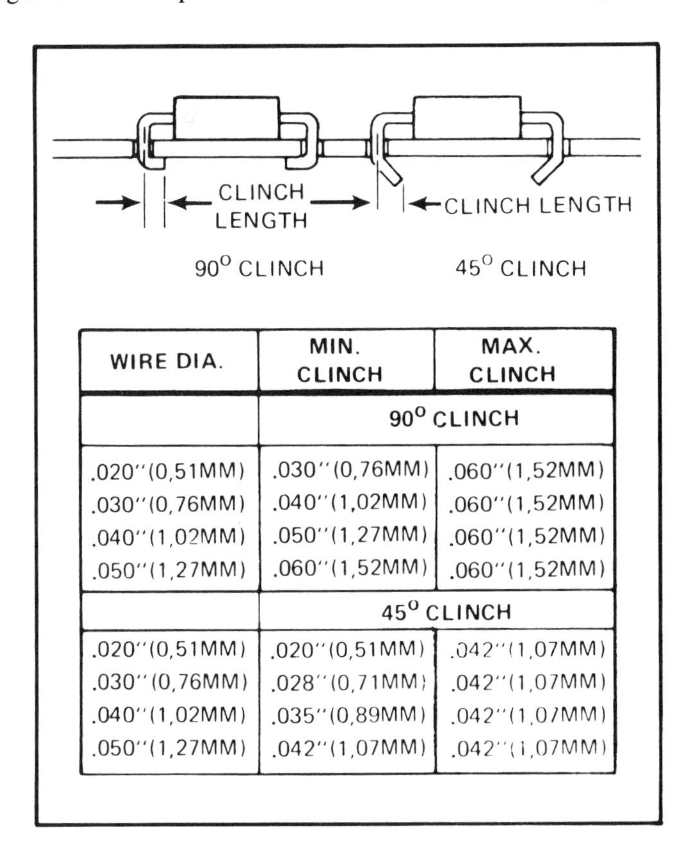

WIRE DIA.	MIN. CLINCH	MAX. CLINCH
	90° CLINCH	
.020"(0,51MM)	.030"(0,76MM)	.060"(1,52MM)
.030"(0,76MM)	.040"(1,02MM)	.060"(1,52MM)
.040"(1,02MM)	.050"(1,27MM)	.060"(1,52MM)
.050"(1,27MM)	.060"(1,52MM)	.060"(1,52MM)
	45° CLINCH	
.020"(0,51MM)	.020"(0,51MM)	.042"(1,07MM)
.030"(0,76MM)	.028"(0,71MM)	.042"(1,07MM)
.040"(1,02MM)	.035"(0,89MM)	.042"(1,07MM)
.050"(1,27MM)	.042"(1,07MM)	.042"(1,07MM)

Fig. 8-19: Clinch pattern options.

The clinch length is measured from the centerline of the lead as it extends through the hole in the Printed Circuit Board. Minimum clinch lengths for smaller wire diameters are somewhat dependent on the hole diameter in the board. To maintain these minimum clinch lengths, hole size must not be more than 0.015" (0,38mm) larger than the lead diameter. Clinch length is adjusted by changing anvil span relative to head span.

Once the Clinch Unit is set for a lead length, the repeatability is ±0.005″ (0,13mm) or less depending on the lead material consistency, with all other insertion parameters remaining constant.

The clinch may be adjusted to give any desired clinch angle from 30 degrees to 90 degrees. Figure 8-19 shows a 45- and 90-degree clinch. When selecting the clinch angle for a given application, hole diameter and lead length should also be considered. Board thickness variations, lead diameter, and material will also have an effect on the clinch angle.

One of the major concerns in selecting an appropriate clinch angle is the solderability of the lead to the Printed Circuit Board. Generally, the 45 degree clinch is preferred because it is easier to repair than the 90 degree (tight) clinch, however there are certain limitations to consider. With single-sided boards which have no holes plated through, the 45 degree clinch is usually good only if the hole diameter does not exceed the lead wire diameter by, say 0.015″ (0,4mm). Generally, this is about the limit for achieving a good solder joint without a void.

On the other hand, there are some users who have used axial components on single-sided boards having no plated-through holes, with hole diameters of 0.060″ (1,5mm) and a wire diameter of 0.020″ (0,5mm) with a tight 90 degrees clinch and have felt they have had totally acceptable soldering.

With plated-through holes, the above soldering limitations practically vanish regardless of clinch angle and wire-to-hole clearance.

Pattern Program Considerations

Processor controlled Axial Lead Insertion Machines require that a pattern program be entered into the controller memory. The pattern program contains all the information required to automatically populate a Printed Circuit Board including component location, insertion span and body diameter.

When generating a pattern program, it is necessary to know the insertion reference point of the component to be inserted. The component insertion reference point of axial lead components is the intersection of the X (distance between hole centers) and the Y (line running through the center of the component leads) center-lines of the component as inserted into the Printed Circuit Board. In this respect, it is important that initial board layout include positioning holes with respect to a broad datum reference. See Figure 8-20.

The Z-axis of processor controlled Axial Lead Insertion Machines defines the insertion center distance of the component insertion tooling. The insertion center distance (programmed Z-span) is the distance between component lead hole centers plus one component wire diameter.

The pattern program can make optimum use of the Axial Lead Insertion equipment by minimizing X-axis movement. Whenever possible, programming should proceed in a plus or minus Y direction. See Figure 8-21.

It is obvious that the distance from PARK to the first component in the "string" and from the last component to PARK again should be minimized. This would be true for both manually loaded machines and PASS-THRU machines.

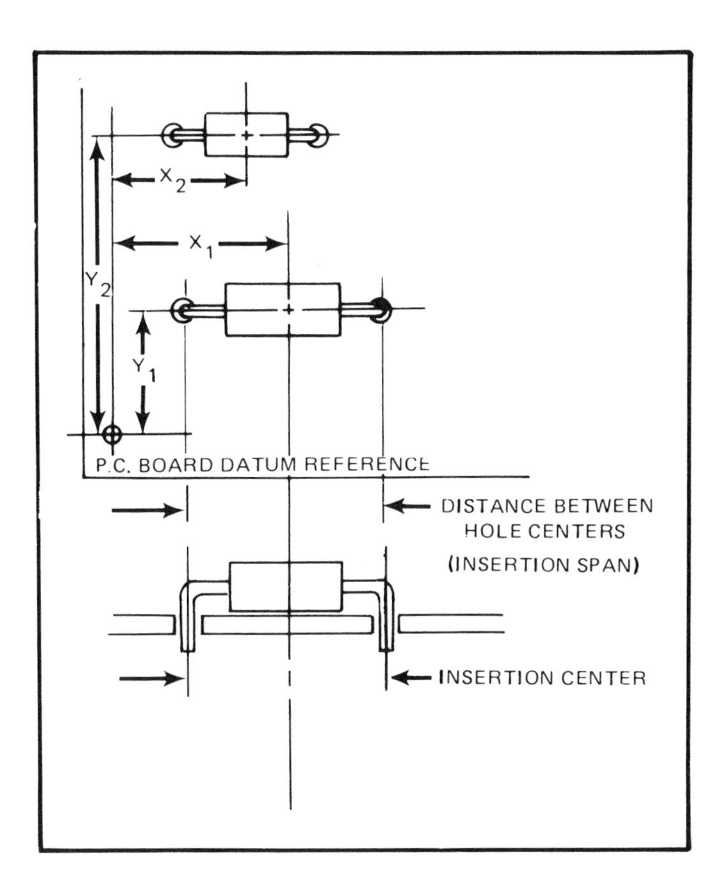

Fig. 8-20: Axial lead insertion reference point.

Where there are a large number of variations in insertion spans (Z moves) in a given board, because the Z-axis drive is slower than the X and Y-axis drives, this will sometimes lower the yield significantly. Some programmers have found that they can actually increase the overall rate by making two passes over the board: one for the large span (or spans), and the second for the small (or smaller) spans. For example, if the span differences are over 0.200″ (5mm) and you have a long production run, this is worth a try.

Radial Lead Component Insertion

Manufactuers have offered several variations of automatic equipment for radial lead component insertion. The descriptions to follow are based on Universal's Model 6346A Radial Lead Sequence Insertion machine. The 6346A may be used as a stand-alone machine, or Pass-Thru II Automatic Board Handling equipment may be added.

These machines are designed to automatically insert most randomly sequenced radial lead devices with two leads and also three lead transistor devices. Universal's Radial Lead Sequence Inserters are capable of processing disc and electrolytic capacitors, coils, resistors, thermistors and similar two leaded devices packaged in a radial configuration. Other components which can be processed and inserted include axial lead components that have been prepared for processing as radial lead devices by "hairpin," lead forming, or with the use of a new input forming station can be taken directly from standard axial lead tape. In addition, three lead transistors that have been prepped for inline lead taping may also be used.

Taping Considerations

Two and three lead components prepped and taped in a radial lead configuration which conform to the specifications described in Figure 8-22 can be processed by Universal's Radial Lead Sequence Inserters.

Component input for the Model 6346A is through a component sequencing module. Properly prepared components may be dispensed to the Dispenser Heads from reels and "ammo pack" cartons, which are the preferred methods, or from cassettes. Radial lead taped components should meet the lead-to-tape adhesion tests and the taped-component-removal pull test.

Tape Splicing Specification

Tape may be spliced using either an acceptable splicing tape and tool, or by using staples. Universal splicing tape is recommended for its superior adhesion and adaptability to automatic machine processing. If staples are used, they cannot be place along the centerline of the feed holes. Placement of staples in the feed hole area will interfere with carrier tape cutting during automatic processing, and may damage the Dispensing Head Cutters. For maximum feed reliability, tape feed holes should remain free of splice interference, all feed holes must remain clear of punched material and the overall tape thickness should not exceed 0.056″ (1,4mm).

Hole Diameter Requirements

The minimum hole diameter for radial lead component insertion should equal the lead diameter plus 0.013″ (0,33mm). The maximum hole diameter

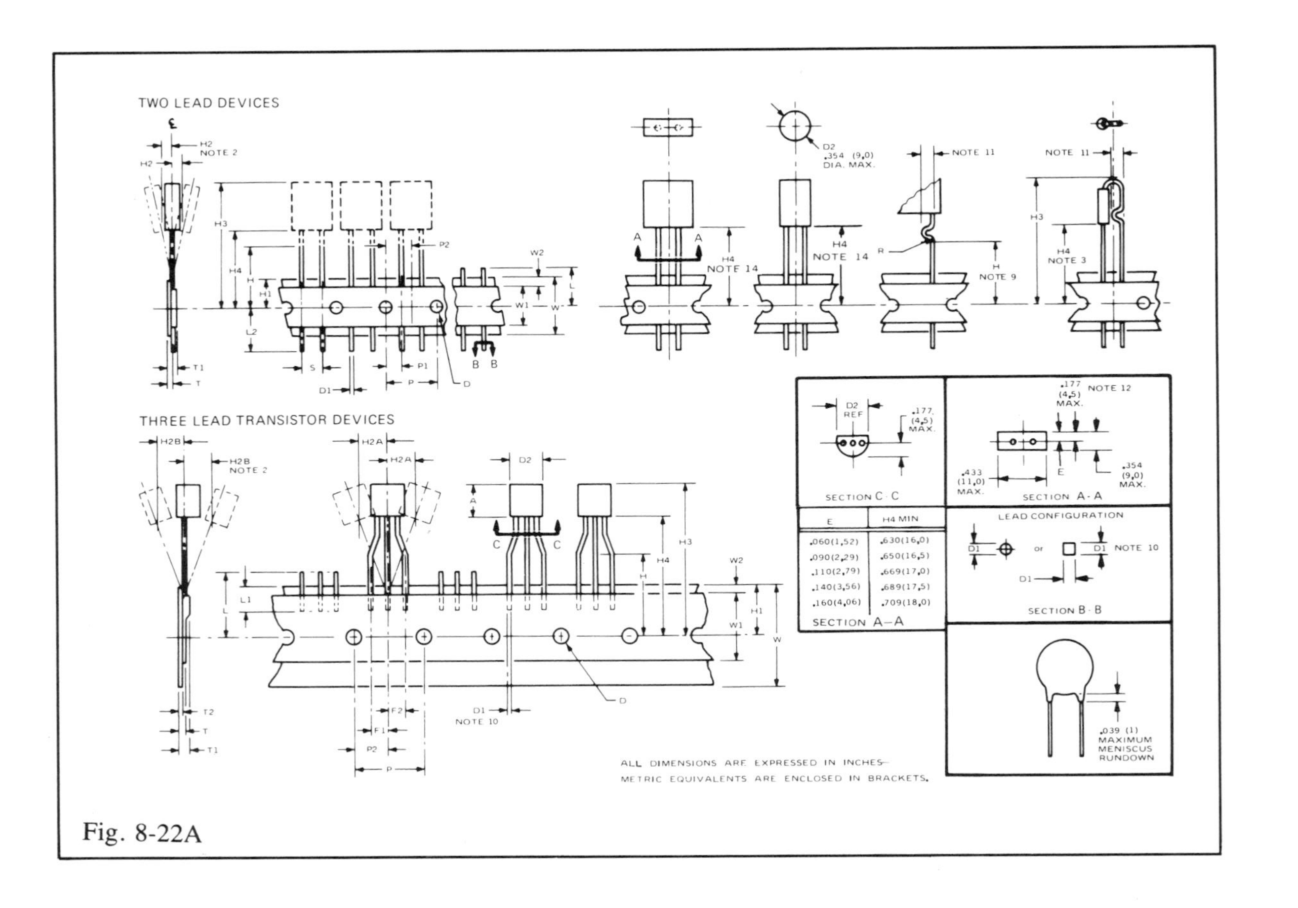
TWO LEAD DEVICES
THREE LEAD TRANSISTOR DEVICES
D2
.354 (9,0)
DIA. MAX.
NOTE 2
NOTE 3
NOTE 9
NOTE 10
NOTE 11
NOTE 12
NOTE 14
SECTION C-C
.177
(4,5)
MAX.
SECTION A-A
.433
(11,0)
MAX.
.354
(9,0)
MAX.
LEAD CONFIGURATION
SECTION B-B
.039 (1)
MAXIMUM
MENISCUS
RUNDOWN
E
H4 MIN
.060(1,52)
.630(16,0)
.090(2,29)
.650(16,5)
.110(2,79)
.669(17,0)
.140(3,56)
.689(17,5)
.160(4,06)
.709(18,0)
SECTION A—A
ALL DIMENSIONS ARE EXPRESSED IN INCHES—
METRIC EQUIVALENTS ARE ENCLOSED IN BRACKETS.

Fig. 8-22A

SYMBOL	ITEM	TWO LEAD SPECIFICATIONS					THREE LEAD SPECIFICATIONS				
		MINIMUM		MAXIMUM		NOTES	MINIMUM		MAXIMUM		NOTES
		INCH	MM	INCH	MM		INCH	MM	INCH	MM	
A	COMPONENT BODY HEIGHT						.125	3,18	.472	12	
D	FEED HOLE DIAMETER	.146	3,7	.169	4,3		.146	3,7	.169	4,3	
D1	LEAD DIAMETER	.014	0,36	.028	0,7	10	.014	0,36	.021	0,53	10
D2	COMPONENT BODY DIAMETER			.354	9,0				.354	9,0	
F1,F2	COMPONENT LEAD PITCH						.094	2,4	.114	2,9	2
H	HEIGHT OF SEATING PLANE	.610	15,5	.650	16,5		610	15,5	—	—	9
H1	FEED HOLE LOCATION	.335	8,5	.384	9,75		.335	8,5	.384	9,75	
H2	FRONT-TO-REAR DEFLECTION	.000	0	.031	0,8	1,2	0	0	.039	1	1,2
H2A	DEFLECTION LEFT OR RIGHT						0	0	.039	1	1,2
H3	COMPONENT HEIGHT	—	—	1.26	32		—	—	1.26	32	
H4	FEED HOLE TO BOTTOM OF COMPONENT	.610	15,5	.886	22,5	14	.610	15,5	—	—	14
L	LEAD LENGTH AFTER COMPONENT REMOVAL	H1	H1	.433	11	4,13	H1	H1	.433	11	4,13
L1	LEAD WIRE ENCLOSURE						.098	2,5	—	—	
L2	LEAD PROTRUSION	.000	0	.44	11,2						
P	FEED HOLE PITCH	.488	12,4	.512	13	5	.488	12,4	.512	13	5
P1	LEAD LOCATION	.124	3,15	.179	4,55	6					
P2	CENTER OF SEATING PLANE LOCATION	.234	5,95	.266	6,75	6	.234	5,95	.266	6,75	6
S	COMPONENT LEAD SPAN	.165	4,2	.228	5,8	6,2					
T	OVERALL TAPE THICKNESS	—	—	—	—	7					7
T1	TOTAL TAPED PACKAGE THICKNESS	—	—	.056	1,42	7	—	—	.056	1,42	7
T2	CARRIER TAPE THICKNESS						.015	0,38	.027	0,68	7
W	TAPE WIDTH	.689	17,5	.748	19	8	.689	17,5	.748	19	8
W1	ADHESIVE TAPE WIDTH	.216	5,5	.748	19	8	.216	5,5	.748	19	8
W2	ADHESIVE TAPE POSITION	.000	0	.236	6	8	0	0	.020	0,5	8

Fig. 8-22B

should not exceed the sum of the lead diameter plus 0.015″ (0,38mm). The maximum would be recommended for highest reliability.

Hole Span Considerations

When PCB density makes it necessary to insert component leads into PCB holes where the two diameters are close (such as 0.014″ [0,36mm] leads into 0.027″ [0,69mm] holes or 0.028″ [0,71mm] leads into 0.041″ [1.04mm] holes, for instance), reliability will be enhanced by considering the jaw tooling design when laying out the PCB hole drilling centers.

During the insertion process two outer spring clamps are used to grasp the component leads. These clamps secure the outside component leads against two fixed surfaces of the Jaw Guide. As a consequence of this clamping method, the actual component lead span will increase or decrease from the minimum insertion span by a factor of one lead diameter plus 0.177″ (4,50mmm) as the lead diameter increased or decreased from optimum.

Board Sizing Considerations

The maximum standard board size which can be processed by the Model 6346A without Pass-Thru is 18″ x 18″ (457.2mm x 457.2mm). When used in the Pass-Thru II version, the maximum board size is 16.00″ wide x 18.00″ long (406,4mm x 457.2mm) and the minimum size is 4.25″ wide x 5.90″ long (108,0mm x 149,9mm).

Thickness: The thickness range of board material for Radial Insertion is basically identical to that for axial leaded components. The recommended minimum Printed Circuit Board thickness is 0.032″ (0,81mm), and the maximum allowable is 0.093″ (2,36mm).

Warp and Sag: Printed Circuit Board warp is not easy to define, and its effect varies widely with the type of board handling that is used. For a bare board, warp generally creates a problem during feeding from a stack where Automatic Board Handling is used. With Automatic Board Handling, the problem continues to increase as the weight of inserted components causes a sag because of the typical front and back edge support used in Pass-Thru machines. For a Radial Insertion Machine, the combination of board warp and sag may not exceed ±0.004″ (0,1mm) with a maximum center deflection in either direction of ±0.062″ (1,6mm). It may be necessary to provide some form of board support at or near the centerline to meet these requirements.

For a manually fed machine the same combination of sag and warp still applies, since neither the Head tooling nor the Clinch is capable of correcting vertical board displacements beyond those dimensions stated. With a workboard holder giving four edge supports, this may completely eliminate any need for an intermediate board support, even on larger board sizes.

Location Considerations

Universal's Radial Lead Insertion Head consists of an Insertion Jaw which positions the component and an insert pusher unit which exerts pressure on the top of the component envelope after it is positioned at the programmed hole location. The Insert Pusher unit is designed to permit overtravel beyond the insertion jaw tip. The insertion jaws may be rotated +90 degrees/0 degrees/90 degrees, relative to the front of the machine. This motion is controlled by the pattern program. The insertion jaw retracts away from the inserted component while the part is being inserted. Special consideration should be given so that the insertion jaw does not interfere with previously inserted components.

The direction of the insertion jaw as it moves away from the component, as well as the clearance required in relation to particular standard components is also shown. These clearances when maintained, will prevent component damage during the insertion process. The width dimension of the jaw is given in Figure 8-23.

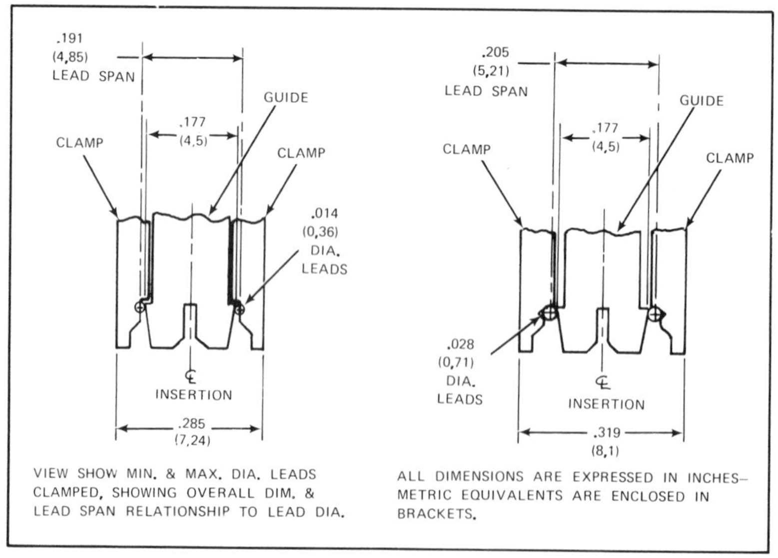

Fig. 8-23: Lead hole span to lead diameter relationship.

Below the Board: The Cut and Clinch is raised to accept the component leads as they pass through the PCB and to support the bottom of the board to prevent excessive flexing during the insertion process, thereby ensuring a uniform lead length. When the component leads are completely through the PCB, two cutters trim the leads to their finished length and clinch them to the underside of the PCB to secure the component to the board.

Two Cut and Clinch modules are available. The first is called an Adjustable Cut and Clinch. This Cut and Clinch yields a tight component with some sacrifice on longer lead lengths.

The second module is Non-Adjustable. This cutterhead provides a typically shorter lead length that does not vary with lead diameter or material consistency. The compromise involves maintaining close hole tolerances to ensure stable components after insertion. When hole sizes exceed 0.017″ (0,43mm) over the lead diameter, component stability is affected (that is, it can move slightly); hence the potential for insertion errors increases since previously inserted components tend to protrude into the Insertion Head path during the next adjacent cycle. Workboard holder clearances for both Cut and Clinch modules, with and without Pass-Thru, are shown in Figures 8-24 and 8-25.

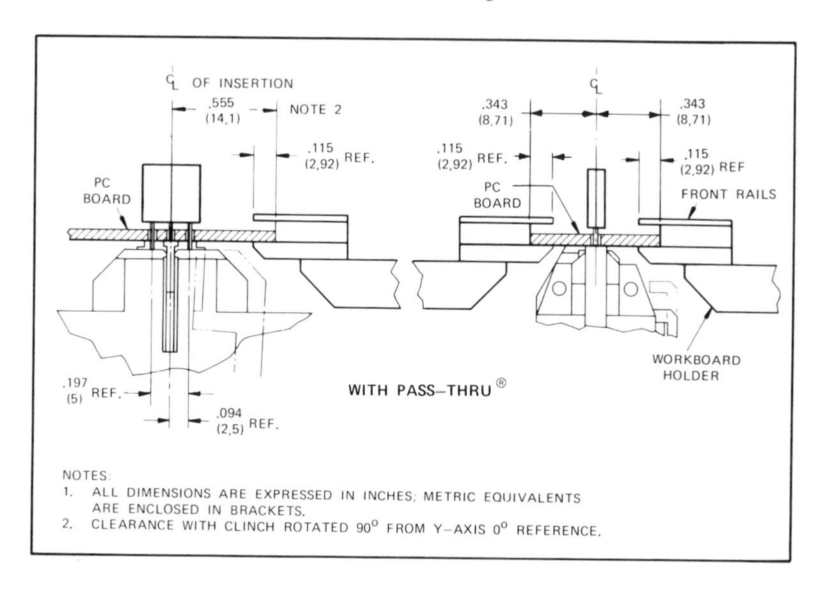

Fig. 8-24: Cut and clinch shown with pass-thru.

Uninsertable Area

The various methods used to locate boards for insertion exclude mounting of components in areas around reference points, or in the case of Pass-Thru, along the edge guides. Uninsertable area may vary significantly between different board configurations.

As a general rule, the minimum uninsertable area in the vicinity of any reference hole is approximately a 0.0625″: (15,88mm) radius for the Model 6346A.

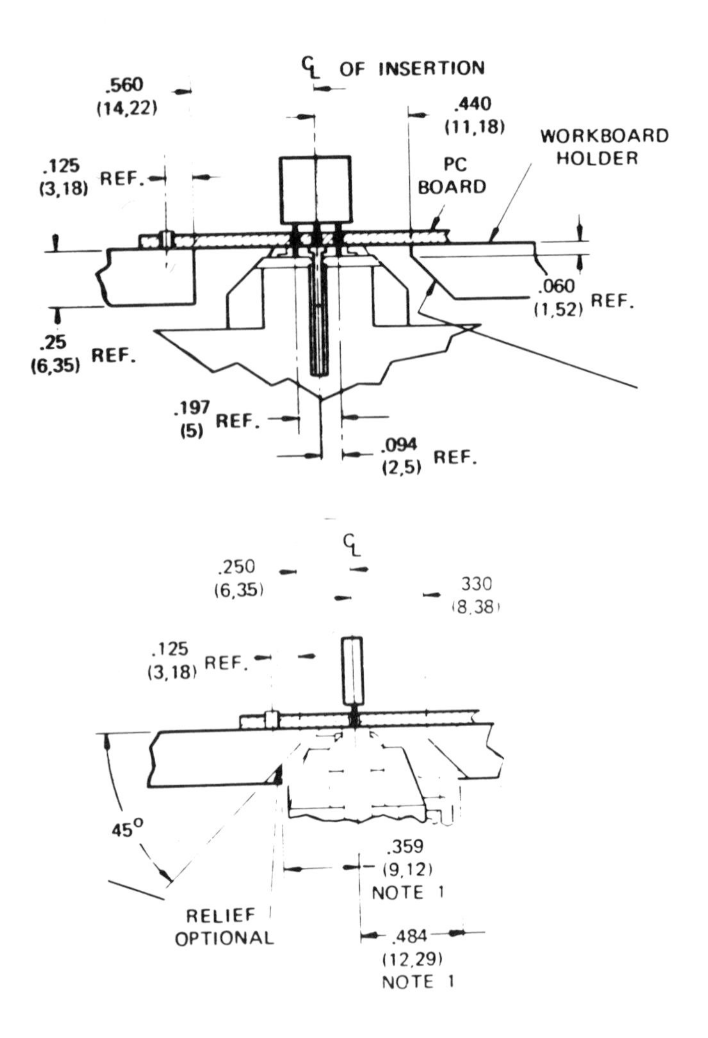

NOTES:
1. CLEARANCE REQUIRED WHEN OPTIONAL RELIEF IS NOT USED.
2. ALL DIMENSIONS ARE EXPRESSED IN INCHES; METRIC EQUIVALENTS ARE ENCLOSED IN BRACKETS.

Fig. 8-25: Cut and clinch shown without pass-thru.

The shaded areas around the locator arms indicate further possible uninsertable areas. The determination of particular dimensions is also dependent upon the locator arm assembly and the footprint of the specific machine to be used for component processing.

Cut and Clinch Patterns

The Cut and Clinch unit rotates in a 0 degrees, 90 degrees counterclockwise relationship with the insertion jaw. The leads of a two lead device are clinched in an outward pattern along the axis formed by the center lines of the leads. The lead clinch angle and trimmed lead length are a function of the clinch selected.

The Radial lead Cut and Clinch unit is mounted below the X-Y table, directly under the Insertion Head Assembly and is activated in conjunction with the Insertion Head. The Cut and Clinch unit contains Anvils which raise to support the Printed Circuit Board during component insertion. The Anvil indexing mechanism allows the Anvil to be oriented at 0 degrees or 90 degrees in the clockwise direction with reference to the front of the machine. The desired orientation is entered in the pattern program.

Pattern Program Considerations

Processor controlled Radial Lead Insertion Machines require that a pattern program be entered into the controller memory. The pattern program contains all the information required to populate a Printed Circuit Board, including component locations (X-Y table coordinates of each component insertion). Sequencer Head number, insertion orientation, and function type. When a Printed Circuit Board is densely populated, the pattern program should be written with the tooling clearances and component orientations taken into consideration. For specific programming instructions, refer to the Programming Supplement of the machine manual.

Insertion Reference Point

When generating a pattern program, it is necessary to know the insertion reference point of the component to be inserted. The insertion reference point for a two or three leaded component in a radial configuration is the point formed by the intersection of the X (insertion span) and Y (perpendicular line equidistant from the end points of the X insertion span) centerlines of the component.

Optimum Pattern Programming

In order to make optimum use of the Radial Lead Insertion Equipment and prevent tooling interference with previously inserted components, insertion machines should be programed with components in the back (rear of the machine) inserted first. Component insertion should then proceed from the back of the board to the front.

The motion of the insertion jaw should be considered when programming. The insertion jaw moves away from the inserted component as the pusher applies pressure. When the jaw is oriented in the 0 degrees position, it will move toward the front of the machine. When oriented 90 degrees in the clockwise direction, the jaw moves to the left, while a 90 degrees counterclockwise orientation results in an insertion jaw movement to the right. Minimum clearance should be considered when a pattern program is written.

Special Programming Considerations

Generally, pattern programs for the Radial Lead Insertion Machines follow the same format as other insertion machines described in this chapter.

appendix I
SURFACE MOUNT EQUIPMENT, SUPPLIES AND SERVICES

This list of surface mount equipment and material manufacturers has been compiled to facilitate obtaining additional information about Surface Mount Technology.

ADHESIVES

Amicon, Inc.
25 Hartwell Ave.
Lexington, MA 02173
Phone: 617-861-9600

Ablestik Labs
833 W. 182 St.
Gardena, CA 90248
Phone: 213-532-9341

Furane Products
5121 San Fernando Rd. West
Los Angeles, CA 90039
Phone: 818-968-6511

CERAMIC PIECE PARTS

Brush Wellman, Inc.
17876 St. Clair Ave.
Cleveland, OH 44110
Phone: 216-486-4200

Kyocera Intl.
10050 N. Wolfe Rd.
Cupertino, CA 95014
Phone: 408-257-8000

Naramics Corp.
3350 Scott, Bldg. 9
Santa Clara, CA 95051
Phone: 408-727-6550

NTK Ceramics
NGK Spark Plugs, Inc.
349 Cobalt Way, Suite 304
Sunnyvale, CA 94086
Phone: 408-736-7205

Rosenthal Technik
North America, Inc.
100 Niantic Ave.
Providence, RI 02907-3190
Phone: 401-943-2200

Spec Industries
1615 S. Rancho Santa Fe Rd.
San Marcos, CA 92069
Phone: 714-744-6692

CLIPS (PINS)

Diatech (Comatel)
87 Sand Pit Rd.
Danbury, CT 06810
Phone: 203-797-9445

DuPont Co.
Berg Electronics
Camp Hill, PA 17011
Phone: 717-975-2000

CONNECTORS

AMP, Inc.
Harrisburg, PA 17105
Phone: 717-564-0100

MALCO
12 Progress Dr.
Montgomerville, PA 18936
Phone: 215-699-5373

Texas Instruments, Incorporated
34 Forest Street, MS 14-3
Attleboro, MA 02703
Phone: 617-699-5205

Winchester Electronics
Div. of Litton
Main & Hillside Ave.
Oakville, CT 06779

CLEANING SYSTEMS

Baron-Blakeslee, Inc.
2001 N. Janice Ave.
Melrose Park, IL 60160
Phone: 312-450-3900

Corpane Industries, Inc.
Bluegrass Industrial Park
250 Production Court
Louisville, KY 40299
Phone: 502-491-4433

Crest Ultrasonics Corporation
P.O. Box 7266
Trenton, NJ 08628-0266
Phone: 609-883-4000

Delta Sonics
7309 E. Compton Blvd.
Paramount, CA 90723
Phone: 213-634-7140

Detrex Chemical Industries, Inc.
Equipment Division
P.O.Box 501
Detroit, MI 48232
Phone: 313-358-5800

Hollis Engineering
15 Charron Avenue
Nashua, NH 03063
Phone: 603-889-1121

Lanape Sales & Service, Inc.
27 Colby Ave.
Manasquan, NJ 08736
Phone: 201-223-9450

Unique Industries, Inc.
P.O. Box 1278
Sun Valley, CA 91353
Phone: 213-875-3810

FEEDERS/HANDLERS

AMI
P.O. Box 5049
North Branch, NJ 08876
Phone: 201-722-7100

Daymarc Corporation
Suite H 189
1101 South Winchester Blvd.
San Jose, CA 95128
Phone: 408-244-8297

deHaart, Inc.
P.O. Box 356
Burlington, MA 01803
Phone: 617-272-0794

Engineered Automation
152 Cregar Road
High Bridge, NJ 08829
Phone: 201-638-8555

Ismeca U.S.S., Inc.
1857 Charter Lane, Suite A
Lancaster, PA 17601
Phone: 717-397-8855-6

Lanco AG
Middlesex General Industries
6 Adele Road
Woburn, MA 01801
Phone: 617-935-8870

MCT (Micro Component Technology)
P.O. Box 43013
St. Paul, MN 55164
Phone: 612-482-5100

Sym-Tek Systems, Inc.
3912 Cale Fortunada
San Diego, CA 92123
Phone: 619-569-6800

Systemation Engineered Products,Inc.
16805 Victor Rd.
New Berlin, WI 53151
Phone: 414-785-1255

Trigon Industries Inc.
311 Ravendale Drive
Mountain View, CA 94043
Phone: 415-965-3600

Unique Industries, Inc.
11544 Sheldon Street
Sun Valley, CA 91352
Phone: 213-875-3810

INSPECTION/TEST/REPAIR

Alphatech, Inc.
P.O. Box 15564
Austin, TX 78761
Phone: 512-458-9179

Automation Unlimited, Inc.
10 Roessler Road
Woburn, MA 01801
Phone: 617-933-7288

John Chatillon & Sons, Inc.
83-30 Kew Gardens Rd.
Kew Gardens, NY 11415-1999
Phone: 212-847-5000

EFD
977 Waterman Ave.
East Providence, RI 02914
Phone: 401-434-1680

ETP (Engineered Technical Products, Inc.)
3380 U.S. Highway 22
Somerville, NJ 08876
Phone: 201-725-0330

Everett/Charles
2887 North Towne Ave.
Pomona, CA 91767
Phone: 714-612-9511

Fairchild
One Fairchild Square
Clifton Park, NY 12065
Phone: 518-877-7042

Gen Rad
300 Baker Avenue
Concord, MA 01742
Phone: 617-890-4900

Machine Intelligence
330 Protero Ave.
Sunnyvale, CA 94086
Phone: 408-737-7960

Manix (Henry Mann,Inc.)
Mann Road
Huntingdon Valley, PA 19006
Phone: 215-355-7200

Nicolet Analytical Instruments
5225-1 Verona Rd.
Madison, WI 53711
Phone: 215-822-8400

Pace, Inc.
9893 Brewers Ct.
Laurel, MD 20707
Phone: 301-490-9860

Projectina Ltd.
CH-9435 Heerbrugg, Switzerland
Dammstrasse, P.O. Box 138
Phone: 071-722-044

Sonoscan Inc.
530 East Green Street
Bensenville, IL 60106

Vanzetti Systems, Inc.
111 Island St.
Stoughton, MA 02072
Phone: 617-828-5650

PASSIVE COMPONENTS

Capacitors

AVX Corporation
Myrtle Beach, SC 29577
Phone: 803-448-9411

Centralab, Inc.
P.O. Box 2032
Milwaukee, WI 53201
Phone: 414-228-7380

Corning Glass Works
225 Great Road
Littleton, MA 01460
Phone: 617-486-3125

Murata Erie
1140 New Market Parkway,
Suite 102
Marietta, GA 30067
Phone: 404-953-1496

Johanson Dielectrics
2220 Screenland Drive]
Burbank, CA 91505
Phone: 213-848-4465

MCC (Monolithic Components Corp.)
6828 Nancy Ridge Drive
San Diego, CA 92121
Phone: 619-578-9390

Siemens Components, Inc.
186 Wood Avenue South
Iselin, NJ 08830
Phone: 201-321-3400

Sprague Electric Company
70 Pembroke Road
Concord, NH 03301
Phone: 603-224-1961

Vitramon North America
P.O. Box 544
Bridgeport, CT 06601
Phone: 203-268-6261

Tantalum Capacitors

Mepco/Electra
5900 Australian Ave.
West Palm BEach, FL 33407
Phone: 305-842-3201

Panasonic
P.O. Box 1503
Secaucus, NJ 07094
Phone: 201-348-5207

Inductors

Murate Erie
1140 New Market Parkway,
Suite 102 Marietta, GA 30067
Phone: 404-953-1496

Panasonic
P.O. Box 1503
Secaucus, NJ 07094
Phone: 201-348-5200

TDK Corp. of America
4711 W. Gulfward, Suite 300
Skokie, IL 60076
Phone: 312-679-8200

Resistors

Allen-Bradley Co.
P.O. Box 2086
Milwaukee, WI 53201

Bourns, Inc.
1200 Columbia
Riverside, CA 92507
Phone: 714-781-5050

Mepco/Electra
Columbia Road
Morristown, NJ 07960
Phone: 201-539-2000

Rohm
P.O. Box 19515
Irvine, CA 92713
Phone: 714-855-2131

Sprague Electric Company
70 Pembroke Road
Concord, NH 03301
Phone: 603-224-1961

State of the Art, Inc.
2470 Fox Hill Road
State College, PA 16801
Phone: 814-355-8547

Thermistors

CAL-R Inc.
1601 Olympic Blvd.
Santa Monica, CA 90406
Phone: 213-450-1761

Coilcraft
1102 Silver Lake Rd
Cary, IL 60013

Dale Electronics, Inc.
2064 12th Avenue
Columbus, NB 68601

Electro Films Inc.
111 Gilbane St.
Warwick, RI 02886

IMS Inc.
50 Schoolhouse Lane
Portsmouth, RI 02871
Phone: 401-683-9700

Kamaya Inc.
4000 Transporation
Ft. Wayne, IN 46808
Phone: 219-489-1533

Mini Systems Inc.
P.O. Box 69
No. Attleboro, MA 07260
Phone: 617-695-0203

PACCOM
Dept. EN073
3928 148th N.E.
Redmond, WA 98052
Phone: 206-883-9200

Semi-Films Div.
National Micronetics Inc.
P.O. Box 188
West Hurley, NY 12491
Phone: 914-338-7714

VRN International
P.O. Box 440001
St. Petersburg, FL 33743
Phone: 813-347-2181

PCB MATERIALS

Augat Pactel Prod.
2520 Turquoise Cir.
Newbury Park, CA 91320

Advance Circuits, Inc.
15102 Minnetonka Industrial Rd.
Minnetonka, MN 55343
Phone: 616-935-3311

Cirtel Corporation
Div. of Interconics
2302 Barranco Road
Irvine, CA 92714
Phone: 714-660-1510

DuPont Company
Chestnut Run
Wilington, DE 19898
Phone: 800-441-4494

EMCA
609 Center Ave.
Mamaroneck, NY 10543
Phone: 914-698-8434

Graphic Research, Inc.
9334 Maon Ave.
Chatsworth, CA 91311
Phone: 213-886-7340

Howe Industries, Inc.
13704 Saticoy St.
Van Nuys, CA 91402
Phone: 213-781-4122

Oak Materials Group, Inc.
Laminates Div.
174 N. Main St.
Franklin, NJ 03235
Phone: 603-934-5736

Parlex Corporation
145 Milk Street
Methuen, MA 018144
Phone: 617-685-4341

PCK Technology Inc.
322 So. Service Road
Melville, NY 11747
Phone: 516-454-4543

Printed Circuits Inc.
1200 West 86th Street
Minneapolis, MN 55431
Phone: 612-888-7900

Rockwell Intl.
Collins Div.
400 Collins Rd. N.E.
Cedar Rapids, IO 52498
Phone: 319-396-5787

Texas Instruments
Metallurgical Materials Div.
34 Forest St.
Attleboro, MA 02703
Phone: 616-699-1619

Zetron, Inc.
2894 Aiello Dr.
San Jose, CA 95111
Phone: 408-365-7000

PLACEMENT SYSTEMS

AMI
U.S. Route 22
P.O. Box 5049
North Branch, NJ 08876
Phone: 201-722-7100

Celmacs Corporation (Mamiya)
25067 Viking Street
Hayward, CA 94545
Phone: 415-785-3390

Dyna/Pert Division
Emhart Machinery Group
181 Elliot St.
Beverly, MA 01914
Phone: 617-927-4200

Engineered Automation
Drawer D. 152 Cregar Rd.
High Bridge, NJ 08829

Everett/Charles
700 East Harrison Ave.
Pomona, CA 91767
Phone: 714-625-5571

Excellon Automation
23915 Garnier St.
Torrance, Ca 90509
Phone: 213-325-8000

Fuji America
805 Bonnie Lane
Elk Grove Village, IL 60007
Phone: 312-437-8844

Icon Corporation
156 Sixth St.
Cambridge, MA 02142
Phone: 617-878-5400

Ismeca USA Inc.
1857 Charter Lane, Suite A
Lancaster, PA 17601
Phone: 717-397-8855

Manix (Henry Mann, Inc.)
Mann Road
Huntingdon Valley, PA 19006
Phone: 215-355-7200

MCT Circuit Assembly Division
P.O. Box 64013
St. Paul MN 55164
Phone: 612-482-5990

MTI Corporation
55 Industrial Drive
Ivyland, PA 18975

Nitto Kogyo Co., Ltd.
Bosl & Roundy Engineering
4000 Transportation Drive
Fort Wayne, IN 46808

Panasonic
1 Panasonic Way
Secaucus, NJ 07094
Phone: 201-348-5343

Philips
P.O. Box 2087
Milwaukee, WI 53201
Phone: 414-228-7632

Quad Systems Corporation
2 Electronic Drive
Horsham, PA 19044
Phone: 215-657-6202

Siemens Aktiengesellschaft
186 Wood Ave.
Iselin, NJ 08830
Phone: 201-321-3400

Teledyne TAC
10 Forbes Road
Woburn, MA 01801
Phone: 800-TEL-E-TAC

TDK
4709 W. Golf Road, Suite 300
Skokie, IL 60076
Phone: 312-679-8200

Universal Instruments Corp.
Box 825
Binghampton, NY 13902

Zevatech-Switzerland
EPA Co.
5207 Walnut Ave.
Downers Grove, IL 60515

REWORK TOOLS

Manix Division of Henry Mann Inc.
Mann Rd.
Huntingdon Valley, PA 19002

Nu-Concept Computer Systems Inc.
Rte. 309 and Advance Lane
Colmer, PA 18915
Phone: 215-822-8400/275-3200

Pamran Co. Inc.
1101 Cedar Creek
Racine, WI 53402
Phone: 414-639-9076

ROBOTS

Anorad Corporation
110 Oser Ave.
Hauppauge, NY 11788
Phone: 516-231-1990

API (Automated Process Inc.)
P.O. Box 23217
Milwaukee, WI 53223-0217
Phone: 414-354-4370

Automation Systems
1900 Politt Drive
Fair Lawn, NJ 07410
Phone: 201-797-8200

Automation Unlimited, Inc.
10 Roessler Road
Woburn, MA 01801
Phone: 617-933-7288

Automatix
1000 Technology Park Drive
Billerica, MA 01821
Phone: 617-667-7900

IBM
Advanced Manufacturing Sytems
1000 NW 51st St.
Boca Raton, FL 33432
Phone: 305-998-2000

Intelledex, Inc.
33840 Eastgate Circle
Corvallis, OR 97333
Phone: 404-475-6100

Manca, Inc.
P.O. Box 738
Westwood, NJ 07675
Phone: 201-666-4100

Microbot, Inc.
453-H Ravendale Drive
Mountain View, CA 94043
Phone: 201-968-8888

Panasonic
1 Panasonic Way
Secaucus, NJ 07094
Phone: 201-348-5343

Rhino Robots, Inc.
2505 S. Neil
Champaign, IL 61820
Phone: 217-352-8485

United States Robots
650 Park Ave.
King of Prussia, PA 19406
Phone: 313-588-1255

SCREEN PRINTERS

Alfra America Inc.
655 West Wise Rd.
Schaumburg, IL 60193
Phone: 312-893-9407

Aremco Products, Inc.
P.O. Box 429
Ossining, NY 10562
Phone: 914-762-0685

Autoroll Dennison
1345 West Mason St.
Green Bay, WI 54303
Phone: 414-494-5166

BMC (Buckbee-Mears Company)
130 W. Garden Blvd.
Gardena, CA 90248
Phone: 213-770-4340

Cugher,Inc.
1975 Annapolis Lane
Minneapolis, MN 55441
Phone: 612-559-0200

deHaart, Inc.
12 Wilmington Rd.
Burlington, MA 01803
Phone: 617-272-0794

ETP
(Engineered Technical Products)
3380 US Hiway 22, North Branch
Somerville, NJ 08876
Phone: 201-725-0330

Fineling
1121-17 Lincoln Ave., Holbrook
Long Island, NY 11741
Phone: 517-563-2390

SCREEN PRINTERS

Forslund Crystal Mark Inc.
613 Justin Ave.
Glendale, CA 91201
Phone: 213-240-7520

Martec Intl. Electronics Corp.
3285 Scott Blvd.
Santa Clara, CA 95051
Phone: 408-727-8447

MPM Corp.
71 West St.
Medfield, MA 02052
Phone: 617-359-7928

Precision Screen Machines, Inc.
44 Utter Ave.
Hawthorne, NJ 07506-2199
Phone: 201-427-5100

AMI
P.O. Box 5049
North Branch, NJ 08876
Phone: 201-722-7100

C.W.Price Co., Inc.
Box 366
Hampton, NJ 08827
Phone: 201-735-9797-9798

Screen Printing Sytems, Inc.
(Reinke brand)
527 Dunn Circle
Sparks, NV 89431
Phone: 702-359-6000

Svecia Silkscreen Maskiner AB
Svecia USA, Inc.
9557 Candida St.
Miramar, CA 92126
Phone: 714-582-6270

Universal Instruments Corp.
Box 825
Binghampton, NY 13902
Phone: 607-772-7522

Utz Engineering
101 Industrial EAST
Clifton, NJ 07012
Phone: 201-778-4560

Weltid Division
Wells Electronics, Inc.
1701 South Main
South Bend, IN 46613
Phone: 219-287-5941

SOLDER REFLOW

Argus International
P.O. Box 38
Hopewell, NJ 08525-0038
Phone: 609-466-1677

Corpane Industries, Inc.
Bluegrass Industrial Park
250 Production Court
Louisville, KY 40299
Phone: 502-491-4433

HTC
Pond Lane
Concord, MA 01742
Phone: 617-369-1110

Intex
40 Tower Rd.
Brookfield, CT 06805

Manix, Division of Henry Mann Inc.
Mann Road
Huntingdon Valley, PA 19006

MCT Circuit Assembly Division
P.O. Box 64013
St. Paul, MN 55164
Phone: 612-482-5990

Multicore
Cantiague Road
Westbury, NY 11590
Phone: 516-334-7997

RTC (Radiant Technology Corp.)
13856 Bettencourt St.
Cerritos, CA 90701
Phone: 213-404-3526

Sikama International
437 Via Roma
Santa Barbara, CA 93110
Phone: 805-683-2626

Vitronocis Corporation
4 Mulliken Way
Newburyport, MA 01950
Phone: 617-465-7026

SOCKETS

AMP Incorporated
Box 3608
Harrisburg, PA 17105
Phone: 717-564-0100

Burndy
Richards Ave.
Norwalk, CT 06856
Phone: 203-838-4444

3M Textool Products
1410 W. Pioneer Dr.
Irving, TX 75601
Phone: 214-259-2676

Wells Electronics
1710 S. Main
South Bend, IN 46613
Phone: 219-287-5941

SOLDER PASTE

Alphametals Inc.
600 Route 440
Jersey City, NJ 07304
Phone: 201-434-6778

Cermalloy Corp.
Union Hall Industrial Park
Conshohocken, PA 19578
Phone: 215-825-6050

ESL
P.O. Box 596
Pennsauken, NJ 08110
Phone: 609-663-7777

Fry Metals Inc.
6th Ave. & 41st St.
Altoona, PA 16602
Phone: 814-946-1611

Gardiner Solder Co.
Div. Of Gardiner Metals
4820 S. Campbell Ave.
Chicago, IL 60632
Phone: 312-847-0100

Indium Corp. of America
1676 Lincoln Ave.
Utica, NY 13502
Phone: 315-979-1630

Kester Solder
4201 Wrightwood Ave.
Chicago, IL 60639
Phone: 312-235-1600

Multicore
Cantiague Rock Rd.
Westbury, NY 11590
Phone: 516-334-7997

SOLVENTS FOR REFLOW

Fluorinert

3M International
3M Center, 220-3W-01
St. Paul, MN 55144

Flutec

Manchem Incorporated
105 College Road East
Princeton Forrestal Center
Princeton, NJ 08540
Phone: 609-734-4966

TEST HANDLERS

Delta Design Inc.
5772 Kearny Villa Rd.
San Diego, CA 92123

Manix Div. of Henry Mann Inc.
Huntingdon Valley, PA 19006
Phone: 215-355-7200

MCT, Inc.
599 Cardigan Road
P.O. Box 43013
St. Paul, MN 55164
Phone: 612-482-5100

Sym-Tech
3912 Calle Fortunada
San Diego, Ca 92123
Phone: 714-569-6800

TEST SOCKETS

AMP, Inc.
Harrisburg, PA 17105
Phone: 717-564-0100

Augat, Inc.
Interconnection Components Div.
33 Perry Ave.
Attleboro, MA 02703

Surface Mount Devices, Inc.
P.O. Box 6818
Stamford, Ct 06903

Texas Instruments Incorporated
Mellurgical Components Div.
Attleboro, MA 02703

SERVICES

Accutronix of ARZ Inc.
17650 N. 25th Ave.
Phoenix, AZ 85023

Advance Engineering Services
2074 B. Walsh Ave.
Santa Clara, CA 95050

A.J. Electronix
20945 Plummer St.
Chatsworth, CA 91311

Allen-Bradley Electronics Corp.
1201 S. Second St.
Milwaukee, WI 53204

Altron, Inc.
6700 Industry Ave. N.W.
Anoka, MN 55303

Andrews Glass Company
3740 Northwest Blvd.
Vineland, NJ 08360

Apogee Engineering
429 Reynolds Circle
San Jose, CA 95112

Arrays Technology
1297 Parkmoore Ave.
San Jose, CA 95126

ASMD, Inc.
11180-G Roseloe St.
San Diego, CA 92121

Avco Electronics
4805 Bradford Dr.
Huntsville, AL 35805

AWI
558 Oakmead Parkway
Sunnyvale, CA 94096

Bally Electronics
10750 W. Grand Ave.
Franklin Park, IL 60131

Boeing Electronics Company
P.O. Box 3707, MS 9A-10
Seattle, WA 98124

C.E. Elgin Electronics
802 Walnut St.
Waterford, PA 16441

Centralab
5855 N. Glen Park Rd.
Milwaukee, WI 53209

Century Circuits
9880 Chartwell Dr.
Dallas, TX 75243

Century Circuits & Elec. Inc.
155 Eaton St.
St. Paul, MN 55107

Comptronx
1800 Gunter Ave.
Guntersville, AL 35976

Computer Assembly Systems
1245 California Ave.
Brockville, Ontario, Canada

Computrol, Inc.
P.O. Box 4442
Boise, ID 83714

CVI, Ltd.
Flowerfield, Bldg. 7 Suite 7
St. James, NY 11007

Electronic Associates
185 Monmouth Parkway
West Long Branch, NJ 07764

Electronics Inc.
802 West Brazos Park Dr.
Clute, Texas 77531

Elemex SA
Box 10015
El Paso, TX 79991

EMC/Operations
P.O. Box 242
Hunt Valley, MD 21030

Empire Electronics, Inc.
629 E. Elmwood
Troy, MI 48083

Flextronics
35325 Fair Crest St.
Newar, CA 94560

G.T.I.
404 Amour St.
Davidson, NC 28036

H.H. Electronics Corp.
105 Norton St.,P.O. Box 271
Newark, NJ 14513

IMI
P.O. Box 604, 23 Olney Ave.
Cherry Hill, NJ 14513

Indy Electronics
P.O. Box 2301
Manteca, CA 95336

Insouth Microsystems
P.O. Box 1209
Auburn, AL 36831-0601

Interconics
3602 West Watters Ave.
Tampa, FL 33682

Jabil Circuits Company
32275 Mally Road
Madison Heights, MI 48071

Jabil Circuits Company
10880 Roosevelt Blvd.
St. Petersburg, FL 33702

Kepro Circuit Systems, Inc.
630 Axminister Dr.
Fenton, MO 63026

Kimball International
1038 East 15th St.
Jasper, IN 47546

King Hi-Tech Inc.
15345 Bonanza Rd.
Victorville, CA 92392

Krueger Technology
2219 S. 48th St. Suite J
Tempe, AZ 85282

Lake Center Industries
P.O. Box 649
Winona, MN 55987

Manufacturing Resources Electronics
70 Industrial Ave.
Kenosha, WI 58140

MCM Industries
401 Dearborn St.
Great Falls, SC 29055

Manutronics
9115 26th St.
Kenosha, WI 58140

MCM Industries
4451 La Palma
Anaheim, CA 92807

Methode Electronics, Inc.
1 Industrial Park
Willingboro, NJ 08046

M.H.I.
1170 Terra Bella
Mt. View, CA 94043

Minco Products,Inc.
7300 Commerce Lane
Minneapolis, MN 55432

Micro Industries
691 Greencrest Dr.
Westerville, OH 43081

Montalbano Engineering Co.
1665 Jarvis Ave.
Elk Grove Village, IL 60007

Novatech, Inc.
3935 Varsity Dr.
Ann Arbor, MI 48104

NuGraphix
59 N. Santa Cruz Ave.
Los Gatos, CA 95030

Omega Circuits & Tng. Corp.
8 Terminal Rd.
New Brunswick, NJ 08901

On Line Designs,Inc.
8259 N. Military Trail #10
Palm Beach Gardens, FL 33410

Optifab, Inc.
1550 W. Van Buren
Phoenix, AZ 85077

Orchard Electronics
9280 Davidson Highway
Concord, NC 28025

Owl Electronics Laboratories, Inc.
233 Boston Post Rd.
Old Saybrook, CT 06475

PC Artwork Design, Inc.
154 McKay Rd.
Huntington Station, NY 11746

SCI Systems
400 South Memorial Pkwy.
Huntsville, AL 35902

Selectron
2001 Forturne Dr.
San Jose, CA 95131

Simmonds Precision
Panton Rd.
Vergennes, VT 05491

S.M.T. CAD
1471 171st Ave.
Hayward, CA 94541

Solectron Corp.
2001 Fortune Dr.
San Jose, CA 95131

SPI
2230 W. First St.
Loveland, CO 80537

Steller Systems
1944 Harrison Ave.
Cincinnati, OH 45214

Sur Mount Corp.
718 Market St.
Brooksville, OH 45309

Surface Mount Tech.,Inc.
P.O. Box 35346
Edina, MN 55435

Tandon
49 Strathearn Pl.
Simi Valley, CA 93065

Tekcom, Inc.
9837 W. 69th St.
Eden Prairie, MN 55344

Triad Engineering Corp.
One North Ave.
Northwest Industrial Park
Burlington, MA 01803

Trinity Tool & Die
342 West Ridge Pike
Limeric, PA 19468

Universal Instru. Corp.-MEA
498 Conklin Ave.
P.O. Box 825
Binghamton, NY 13902

Valtronic, Inc.
2200 South Main St.
Lombard, IL 60148

Vandor Manufacturing
P.O. Box 960
Plumsteadville, PA 18999

Vestal Electronics Devices, Inc.
213 Tracy Creek Road
Vestal, NY 13850

Vicab Electronic Services
44 Old State Rd. Unit 14
New Milford, Ct. 06776

White Technology, Inc.
4246 E. Wood St.
Phoenix, AZ 85040

XETEL
202 Depot St.
Elgin, Texas 78621

ZETRON
2894 Aiello Dr.
San Jose, CA 95111

Seminars/Consultants

AWI
408-720-8860

D. Brown Associates
215-343-0123

SMD Technology Center
414-228-7632

S.M.T. CON
408-354-0700

Worldwide Convention Mgmt.
312-362-8711

Engineerd Circuit Research
408-293-1181

J.T. Stimey Inc.
215-244-1157

Surface Mount Design(CAD/CAE)

CADNETIX
5757 Central Ave.
Boulder, CO 80301

Micropak Systems Inc.
2410 Armstrong St.
Livermore, CA 94550

Northern Colorado Design Service
1204 Gregory Road
Fort Collins, CO 80524

Royal Digital Systems
Automate Div.
3600 W. Bayshore
Palo Alto, CA 94303

SMT CAD
1471 171st Ave.
Hayward, CA 94541

SMT Plus
2001 Gateway Pl. Suite 740
San Jose, CA 95110-1010

Surface Mount Directories

A.W.I.
408-720-8860

D. Brown Associates
215-343-0123

I.C. Master
408-249-6800

Nu Graphix
408-395-5859

S.M.T.A.
408-354-0700

appendix II

SURFACE MOUNT DEVICE BIBLIOGRAPHY

Circuits Manufacturing

Anonymous. 1985. *Shopper's Guide to Surface Mount Assembly*. Circuits Manufacturing, Vol. 25, No. 5, (May): 84
Brisky, Michael. 1985. *Planning Your Surface Mount Manufacturing Line*. Circuits Manufacturing, Vol. 25, No. 5, (May): 77
Ross-Kozel, Dr. Barbara L. 1984. *Controlling Pad Bridging in SMD Attachment*. Circuits Manufacturing, Vol. 24, No. 12, (December): 68
Tuck, John. 1984. *Cleaning Surface-Mounted Assemblies*. Circuits Manufacturing, Vol. 24, No. 1, (January): 40

EDN

Ormond, Tom. 1985. *Surface-Mount Technology*. EDN, Vol. 30, No. 9, (April 18): 105

Electronic Design

Beedle, Mitch. 1985. *Surface-Mounted Packaging*. Electronic Design, Vol. 33, No. 1, (January 10): 232
Biancomano, Vincent. 1983. *Surface Mounting Profits From Material Gains*. Electronic Design, Vol. 31, No. 14, (July 7): 85
Hutchins, Charles. 1983. *VLSI Prompts Designers to Ponder Packaging*. Electronic Design, Vol. 31, No. 26, (December 22): 167

Electronic Engineering Times

Margolin, Bob. 1985. *Surface-Mount Component Technology*. Electronic Engineering Times.

Electronic Packaging and Production

Brown, Don, et al. 1984. *Surface Mountable Components—An Update*. Electronic Packaging and Production, Vol. 24, No. 1, (January): 76
Brown, Don. 1985. *U.S. Surface Mounting Makes Its Move*. Electronic Packaging and Production, Vol. 25, No. 2, (February): 208
Buck, Thomas J. 1985. *A Discrete-Wired Solution For High Speed Surface-Mount Packaging*. Electronic Packaging and Production, Vol. 25, No. 6, (June): 136
Dixon, Tom. 1984. *SMT, Shielding Needs Drive Changes in Chemicals and Materials*. Electronic Packaging and Production, Vol. 24, No. 11, (November): 62
Dixon, Tom. 1985. *Unresolved Issues Hinder Full Realization of SMT's Benefits*. Electronic Packaging and Production, Vol. 25, No. 1, (January): 40

Dixon, Tom. 1985. *SMT Calls for a Re-examination of Soldering Techniques*. Electronic Packaging and Production, Vol. 25, No. 2, (February): 96
Ginsberg, Gerald L. 1984. *Surface Mounting Impacts Printed Wiring Connectors*. Electronic Packaging and Production, Vol. 24, No. 11, (November): 96
Hinch, Stephen W., et al. 1984. *Setting Up Production of Surface Mount Assemblies*. Electronic Packaging and Production, Vol. 24, No. 1, (January): 66
Jones, P.R. 1985. *Solder Joint Integrity is Key to Temperature Cycling Resistance*. Electronic Packaging and Production, Vol. 25, No. 1, (January): 84
Kastner, Mark. 1985. *Surface Mount Offers Choice of Design Solutions*. Electronic Packaging and Production, Vol. 25, No. 1, (January): 96
Leibowitz, Joseph. 1985. *Graphite Layers in SMT Boards Control Thermal Mismatch*. Electronic Packaging and Production, Vol. 25, No. 6, (June): 98
Marcoux, Phil. 1984. *Surface Mount Assemblies Shrink Circuitry*. Electronic Packaging and Production, Vol. 24, No. 1, (January): 82
Markstein, Howard W. 1983. *Surface-Mount Substrates: The Dey in Going Leadless*. Electronic Packaging and Production, Vol. 23, No. 6, (June): 50
Markstein, Howard W. 1984. *Pick-and-Place Swings Toward Surface Mount*. Electronic Packaging and Production, Vol. 24, No. 1, (January): 56
Markstein, Howard W. 1985. *Low TCE Metals and Fibers Prove Viable for SMT Substrates*. Electronic Packaging and Production, Vol. 25, No. 1, (January): 52
Markstein, Howard W. 1985. *Automation Thrives as Companies Strive for SMT Capability*. Electronic Packaging and Production, Vol. 25, No. 5, (May): 68
Pound, Ronald. 1985. *Pick-and-Place Faces a Variety of SMC's*. Electronic Packaging and Production;, Vol. 25, No. 1, (January): 74
Reyonolds, Robert A. 1985. *SMDs Invade Military and Commercial Equipment*. Packaging and Production, Vol. 25, No. 2, (February): 130

Electri*Onics

Dance, Francis J. 1983. *Part 1: Can PWBS Meet the Surface Mount Challenge?* Electri*onics (May).
Lancaster, Michael C. 1984. *Surface Mount Technology, Plastic Chip Carrier Trends*. Electri*onics, Vol. 30, No. 13, (December)

Electronics

Bierman, Howard. 1985. *Packaging Changes Make Automatic Testing Tougher, More Costly*. Electronics, Vol. 58, No. 28, (July 15): 48
Lyman, Jerry. 1984. *Surface Mounting is Changing the PC-Board Landscape*. Electronics, Vol. 57, No. 3, (February 9): 113

Evaluation Engineering

Anonymous. 1985. *Handlers for ATE—Dips Still Reign As SMT Struggles With Standards*. Evaluation Engineering, Vol. 24, No. 5, (May): 50
Dixon, Tom, Sr. Ed. 1983. *Components Address Surface Mount Technology Needs*. Evaluation Engineering, (November): 36
Fitts, Michael. 1985. *Surface Mounted Devices: A PC Board Testing Challenge*. Evaluation Engineering, Vol. 23, No. 1, (January): 40

Electronics Week

Lyman, Jerry. 1985. *Components for SMA Arrive. Electronics Week, Vol. 58, No. 14, (April 8): 49*

IEEE Transactions on Components, Hybrids and Manufacturing Technology

Knausenberger, W.H., et al. 1983. *High Pin-out IC Packaging and the Density Advantagae of Surface Mounting*. IEEE Transactions on Components, Hybrids, And Manufacturing Technology, Vol. CHMT-6, No. 3, (September): 298
Waller, D.L., et al. 1983. *Analysis of Surface Mount Thermal and Thermal Stress Performance*. IEEE Transactions on Components, Hybrids, and Manufacturing Technology, Vol. CHMT-6, No. 3, (September): 257

Journal of Vacuum Science and Technology

Montante, J. M. 1985. **Summary Abstract: An Overview of Surface Mount Technology for Microciruit Assembly.** Journal Vacuum Science and Technology, Vol. 3, No. 3, (May/June): 779

Index